WiMAX/MobileFi

WiMAX/MobileFi

Advanced Research and Technology

Edited by Yang Xiao

CRC Press
Taylor & Francis Group
Boca Raton London New York

CRC Press is an imprint of the
Taylor & Francis Group, an **informa** business

AN AUERBACH BOOK

CRC Press
Taylor & Francis Group
6000 Broken Sound Parkway NW, Suite 300
Boca Raton, FL 33487-2742

First issued in paperback 2019

ISBN-13: 978-1-4200-4351-8 (hbk)
ISBN-13: 978-0-367-38802-7 (pbk)

Library of Congress Cataloging-in-Publication Data

WiMAX/MobileFi : advanced research and technology / Yang Xiao, [editor].
   p. cm.
  Includes bibliographical references and index.
  ISBN 978-1-4200-4351-8 (alk. paper)
  1. Wireless communication systems. 2. Mobile communication systems. 3. IEEE 802.16 (Standard) I. Xiao, Yang.

TK5103.2.W55 2008
621.384--dc22   2007025386

Visit the Taylor & Francis Web site at
http://www.taylorandfrancis.com

and the CRC Press Web site at
http://www.crcpress.com

Contents

Contents

Preface

IEEE 802.16 working group was set up in 1999 to develop a new standard for broadband wireless access (BWA) and published the first IEEE 802.16 standard in October 2001. An industrial association, namely worldwide interoperability for microwave access (WiMAX) forum was formed to promote the 802.16 standards by defining the interoperability specifications between 802.16 products from different vendors. Thus, the IEEE 802.16 networks are also often referred to as WiMAX networks. In October 2004, the new standard 802.16-2004 was published, which is actually an amalgamation of 802.16 and 802.16a. Recently, the 802.16e standard was also ratified in December 2005 by allowing the upgrade from fixed BWA systems to mobile service provision up to vehicular speeds.

WiMAX provides a worldwide revolution in BWA including both fixed and mobile handsets. This book addresses many research-oriented issues in WiMAX such as connection admission control, integration with WiFi networks, quality of service (QoS) support, security issues, mobility support, terminal energy consumption, frequency reuse, configuration issues, QoS architecture, handoff management, architecture for wireless IP networks, scheduling algorithms, mesh networks, load-balancing, radio resource management, systems deployment, performance modeling and analysis, and simulations. The goal of this edited book is to provide an excellent reference for students, faculty, researchers, and people in the industry related to these fields.

This edited book contains chapters written by experts on a wide range of topics that are associated with novel methods, techniques and applications of WiMAX systems. The book contains 17 chapters from prominent researchers working in these areas around the world. Although the covered topics may not be an exhaustive representation of all issues in WiMAX, they do represent a rich and useful sample of the strategies and contents.

This book has been made possible by the great efforts and contributions of many people. First of all, we would like to thank all the contributors for putting together excellent chapters that are very comprehensive and informative. Second, we would like to thank the staff members for putting this book together. Finally, I would like to dedicate this book to my family (Cary, Kevin, and Yan).

Yang Xiao

About the Editor

Yang Xiao is currently with the Department of Computer Science at The University of Alabama. He worked at Micro Linear as a MAC (Medium Access Control) architect involved with IEEE 802.11 standard enhancement work before he joined the Department of Computer Science at The University of Memphis in 2002. Dr. Xiao is the director of W^4-Net Lab, and was with CEIA (Center for Information Assurance) at The University of Memphis. He is an IEEE Senior member and is a member of the American Telemedicine Association. He was a voting member of IEEE 802.11 Working Group from 2001 to 2004. He currently serves as Editor-in-Chief for *International Journal of Security and Networks (IJSN)*, *International Journal of Sensor Networks (IJSNet)*, and *International Journal of Telemedicine and Applications (IJTA)*. He serves as an associate editor or on editorial boards for the following refereed journals: *IEEE Transactions on Vehicular Technology*, *International Journal of Communication Systems*, *Wireless Communications and Mobile Computing (WCMC)*, *EURASIP Journal on Wireless Communications and Networking*, *Security and Communication Networks*, *International Journal of Wireless and Mobile Computing*, *Research Letters in Communications*, and *Recent Patents on Engineering*. Dr. Xiao serves as a (lead) guest editor for the *Journal of Wireless Communications and Mobile Computing*, special issue on "Wireless Monitoring and Control" and in 2008, as a (lead) guest editor for *EURASIP Journal on Wireless Communications and Networking*, special issue on "Wireless Telemedicine and

Applications" in 2007, a guest editor for *IEEE Network*, special issue on "Advances on Broadband Access Networks" in 2007, a guest editor for *IEEE Wireless Communications*, special issue on "Radio Resource Management and Protocol Engineering in Future Broadband and Wireless Networks" in 2006, a (lead) guest editor for *International Journal of Security in Networks (IJSN)*, special issue on "Security Issues in Sensor Networks in 2005, a (lead) guest editor for *EURASIP Journal on Wireless Communications and Networking*, special issue on "Wireless Network Security" in 2005, a (sole) guest editor for *Computer Communications journal*, special issue on "Energy-Efficient Scheduling and MAC for Sensor Networks, WPANs, WLANs, and WMANs" in 2005, a (lead) guest editor for *Journal of Wireless Communications and Mobile Computing*, special issue on "Mobility, Paging and Quality of Service Management for Future Wireless Networks" in 2004, a (lead) guest editor for *International Journal of Wireless and Mobile Computing*, special issue on "Medium Access Control for WLANs, WPANs, Ad Hoc Networks, and Sensor Networks" in 2004, and an associate guest editor for *International Journal of High Performance Computing and Networking*, special issue on "Parallel and Distributed Computing, Applications and Technologies" in 2003. He served as editor/ co-editor for eleven edited books: *Mobile Telemedicine: A Computing and Networking Perspective*, *WiMAX/MobileFi: Advanced Research and Technology*, *Security in Distributed and Networking Systems*, *Security in Distributed, Grid, and Pervasive Computing*, *Security in Sensor Networks*, *Wireless Network Security*, *Adaptation Techniques in Wireless Multimedia Networks*, *Wireless LANs and Bluetooth*, *Security and Routing in Wireless Networks*, *Ad Hoc and Sensor Networks*, and *Design and Analysis of Wireless Networks*. He serves as a referee/reviewer for many funding agencies, as well as a panelist for the U.S. National Science Foundation (NSF) and a member of the Canada Foundation for Innovation (CFI)'s Telecommunications expert committee. He serves on the Technical Program Committee (TPC) for more than 90 conferences such as INFOCOM, ICDCS, ICC, GLOBECOM, WCNC, etc. His research areas are security, telemedicine, sensor networks and wireless networks. He has published more than 200 papers in major journals (more than 50 in various IEEE journals/magazines) and refereed conference proceedings related to these research areas. Dr. Xiao's research has been supported by the U.S. National Science Foundation (NSF). E-mail: yangxiao@ieee.org.

Contributors

Oreste Andrisano
WiLab
University of Bologna
Bologna, Italy

Alessandro Bazzi
WiLab
University of Bologna
Bologna, Italy

Hsi-Lu Chao
Department of Computer
   Science
National Chiao
   Tung University
Hsinchu, Taiwan, ROC

Shafaq B. Chaudhry
School of Electrical Engineering
   and Computer Science
University of Central Florida
Orlando, Florida

Imrich Chlamtac
CREATE-NET
Trento, Italy

Claudio Cicconetti
Dipartimento di Ingegneria
   dell'Informazione
University of Pisa
Pisa, Italy

Todor Cooklev
School of Engineering
San Francisco State University
San Francisco, California

Suman Das
Bell Labs, Alcatel-Lucent
Murray Hill, New Jersey

Petar Djukic
Edward S. Rogers Sr.
   Department of Electrical and
   Computer Engineering
University of Toronto
Toronto, Ontario, Canada

S.-E. Elayoubi
Research and Development Division
France Telecom
Issy-les-Moulineaux, France

Stanislav Filin
JSC Kodofon
Voronezh, Russia

Bernard Fong
Department of Electronic and
    Information Engineering
Hong Kong Polytechnic University
Kowloon, Hong Kong

Mo-Han Fong
Nortel
Ottawa, Ontario, Canada

B. Fourestié
Research and Development
    Division
France Telecom
Issy-les-Moulineaux, France

Alexandre Garmonov
JSC Kodofon
Voronezh, Russia

E. Gregori
Italian National Research
    Council
Pisa, Italy

Ratan K. Guha
School of Electrical Engineering
    and Computer Science
University of Central Florida
Orlando, Florida

O. Ben Haddada
Research and Development
    Division
France Telecom
Issy-les-Moulineaux,
    France

Matthias Hollick
Technische Universität
Darmstadt, Multimedia
    Communications Lab
Darmstadt, Germany

Ekram Hossain
TRLabs
and
University of Manitoba
Winnipeg, Manitoba, Canada

Rose Qingyang Hu
Nortel
Richardson, Texas

Thierry E. Klein
Bell Labs, Alcatel-Lucent
Murray Hill, New Jersey

Mikhail Kondakov
JSC Kodofon
Voronezh, Russia

Giacomo Leonardi
WiLab
University of Bologna
Bologna, Italy

Kin K. Leung
Imperial College
London, United Kingdom

Yi-Bing Lin
Department of Computer Science
National Chiao Tung University
Hsinchu, Taiwan, ROC

David G. Michelson
Department of Electrical and
    Computer Engineering
University of British Columbia
Vancouver, British Columbia, Canada

D. Miorandi
CREATE-NET
Trento, Italy

Parag S. Mogre
Technische Universität
Darmstadt, Multimedia
    Communications Lab
Darmstadt, Germany

Sergey Moiseev
JSC Kodofon
Voronezh, Russia

Sayandev Mukherjee
Marvell Semiconductor
Santa Clara, California

Dusit Niyato
TRLabs
and
University of Manitoba
Winnipeg, Manitoba, Canada

David Paranchych
Nortel
Richardson, Texas

Gianni Pasolini
WiLab
University of Bologna
Bologna, Italy

Francesco De Pellegrini
CREATE-NET
Trento, Italy

Subbu Ponnuswamy
School of Engineering
San Francisco State University
San Francisco, California

Yi Qian
Department of Electrical and
    Computer Engineering
University of Puerto Rico at
    Mayagüez
Mayagüez, Puerto Rico

R. Riggio
CREATE-NET
Trento, Italy

George E. Rittenhouse
Alcatel-Lucent
Whippany, New Jersey

Louis G. Samuel
Bell Labs, Alcatel-Lucent
Swindon, United Kingdom

Nicola Scalabrino
Italian National Research Council
Pisa, Italy

Christian Schwingenschlögl
Department of Information and
    Communications
Siemens AG, Corporate Technology
Munich, Germany

Ralf Steinmetz
Technische Universität
Darmstadt, Multimedia
    Communications Lab
Darmstadt, Germany

Yu-Chee Tseng
Department of Computer
    Science
National Chiao Tung
    University
Hsinchu, Taiwan, ROC

Shahrokh Valaee
Edward S. Rogers Sr.
    Department of Electrical and
    Computer Engineering
University of Toronto
Toronto, Ontario, Canada

Krishna Sumanth Velidi
Department of Computer Science
The University of Alabama
Tuscaloosa, Alabama

Harish Viswanathan
Bell Labs, Alcatel-Lucent
Murray Hill, New Jersey

Geng Wu
Nortel
Richardson, Texas

Yang Xiao
Department of Computer Science
The University of Alabama
Tuscaloosa, Alabama

Wen-Hsin Yang
Department of Computer Science
National Chiao Tung University
Hsinchu, Taiwan, ROC

Yonghong Zhang
Department of Electrical and
    Computer Engineering
University of British Columbia
Vancouver, British Columbia, Canada

Haitao Zheng
University of California, Santa Barbara
Santa Barbara, California

Chapter 1

Connection Admission Control in OFDMA-Based WiMAX Networks: Performance Modeling and Analysis

Dusit Niyato and Ekram Hossain

Orthogonal frequency division multiple access (OFDMA)-based WiMAX technology has emerged as a promising technology for broadband access in a wireless metropolitan area network (WMAN) environment. In this chapter, we address the problem of queueing theoretic performance modeling and analysis of OFDMA-based broadband wireless networks. We first review the related works on queueing analysis of wireless transmission systems. To provide a background for the presented queueing model, an introduction to stochastic process and Discrete-Time Markov Chain (DTMC) is presented. We consider a single-cell WiMAX environment in which the base station allocates subchannels to the subscriber stations in its coverage area. The subchannels allocated to a subscriber station are shared by multiple connections at that subscriber station. To ensure the quality-of-service (QoS) performances, two connection admission control (CAC) schemes, namely, threshold-based and queue-aware CAC schemes, are considered at a subscriber station. A queueing analytical framework for these admission control schemes is presented considering OFDMA-based transmission at the physical layer. Then, based on the queueing model, both the connection-level and the packet-level performances are studied.

1.1 Introduction

OFDMA is a promising wireless access technology for the next generation broadband packet networks. With OFDMA, which is based on orthogonal frequency division multiplexing (OFDM), the wireless access performance can be substantially improved by transmitting data via multiple parallel channels, and also it is robust to intersymbol interference and frequency-selective fading. OFDMA has been adopted as the physical layer transmission technology for IEEE 802.16/WiMAX-based broadband wireless networks. Although the IEEE 802.16/WiMAX standard defines the physical layer specifications and the Medium Access Control (MAC) signaling mechanisms, the radio resource management methods such as those for connection admission control (CAC) and dynamic bandwidth adaptation are left open. However, to guarantee QoS performances (e.g., call blocking rate, packet loss, and delay), efficient admission control is necessary in a WiMAX network at both the subscriber and the base stations.

The admission control problem was studied extensively for wired networks (e.g., for ATM networks [1]) and also for traditional cellular wireless systems. The classical approach for CAC in a mobile wireless network is to use the guard channel scheme [2] in which a portion of wireless resources (e.g., channel bandwidth) is reserved for handoff traffic. A more general CAC scheme, namely, the fractional guard scheme, was proposed [3] in which a handoff call/connection is accepted with a certain probability. To analyze various connection admission control algorithms, analytical models based on continuous-time Markov chain, were proposed [4]. However, most of these models dealt only with call/connection-level performances (e.g., new call blocking and handoff call dropping probabilities) for the traditional voice-oriented cellular networks. In addition to the connection-level performances, packet-level (i.e., in-connection) performances also need to be considered for data-oriented packet-switched wireless networks such as WiMAX networks.

In this chapter, we first present the related works in the area of queueing theoretic performance analysis of wireless transmission systems. An introduction to stochastic processes and DTMC are presented to provide a background on queueing analysis. Then we present two connection admission control schemes for a multi-channel and multi-user OFDMA network. The first scheme is threshold-based, in which the concept of guard channel is used to limit the number of admitted connections to a certain threshold. The second scheme is based on the information on queue status and it also inherits the concept of fractional guard channel in which an arriving connection is admitted with certain connection acceptance probability. Specifically, the connection acceptance probability is determined based on the queue status (i.e., the number of packets in the queue). A queueing analytical model is developed based on a two-dimensional DTMC which captures the system dynamics in terms of the number of connections and queue status. Based on this model, various performance measures such as connection blocking probability, average number of ongoing connections, average queue length, probability of packet dropping due to lack of buffer space, queue throughput, and average queueing delay are obtained. The numerical results reveal the comparative performance characteristics of the threshold-based and the queue-aware CAC algorithms in an OFDMA-based WiMAX network. Simulation results are also presented to validate the analytical model.

1.2 Related Work on Queueing Analysis for Wireless Transmission Systems

In general, in a wireless transmission system, arriving packets are buffered into radio link level queue before transmission to the target mobile (in downlink) or to the base station (in uplink). This buffering process causes delay in wireless transmission. Again, the transmission rate of a wireless channel varies due to variations in channel quality and channel errors. Therefore, a queueing analysis which can capture the channel and the radio link level buffer dynamics becomes a useful mathematical tool to investigate the QoS performances of a wireless transmission system. By using a queueing model, the performance results can be obtained more efficiently when compared with simulations. In addition, a queueing analytical model can be used not only to analyze the system's behavior under different parameter settings, but also to optimize the system performances in which several standard techniques in optimization (e.g., Markov decision process) can be applied.

Different queueing models for wireless transmission systems were proposed in the literature. Queueing analyses for a polling system and a system with cyclic service queues with Bernoulli scheduler were presented [5,6]. These works considered only single rate transmission at the physical layer in which the modulation and coding scheme is fixed. On the other hand, multirate transmission based on adaptive modulation and coding (AMC) has been proposed in most of the current wireless standards to archive higher system capacity. Queueing analysis for radio link level scheduling with adaptive modulation in time division multiple access (TDMA) system was proposed [7]. Also, a queueing model was used for optimizing the radio link level system parameters [8].

Multimedia services will be common in next generation wireless systems. A queueing model developed specifically for video source (e.g., MPEG traffic) was presented [9]. In addition, the authors demonstrated how the video source coding parameters can be optimally adjusted to achieve the best performance. Again, in a multiservice traffic scenario, prioritization of real-time traffic over best-effort traffic is required. An analytical model for priority queueing was presented [10].

Combatting transmission errors due to interference and noise is one of the challenging issues in wireless transmission systems. Automatic Repeat reQuest (ARQ) is one of these methods to recover erroneous transmissions. When a transmission fails (i.e., the receiver cannot decode the transmitted information correctly), the receiver requests the transmitter to retransmit. Different variants of ARQ can be used (e.g., probabilistic retransmission, finite and infinite retransmission). Queueing models with ARQ mechanism were proposed [11–20]. For example, queueing models for go-back-N [16] and selective repeat ARQ [17] were presented. Since with AMC multiple packets can be transmitted in one time slot, if an error occurs, N packets up to the last transmission will be retransmitted for go-back-N ARQ and only the erroneous packets will be retransmitted in case of selective repeat ARQ. Due to the time and space-dependent wireless channel errors and the burstiness of the errors, some connections could experience inferior performance compared with the others. Therefore, a compensation mechanism was introduced to maintain fairness (by allowing more transmissions in the current time frame to compensate errors in the previous time frame) and guarantee the target QoS performance. Also, a queueing model for this compensation mechanism was proposed [21].

In a multi-user wireless system, packet scheduling is required to allocate the available transmission resources to the ongoing users in a fair and efficient manner. Various packet scheduling policies were proposed in the literature. The most common policy is fair scheduling in which the ongoing users receive services based on their preassigned weights. A queueing model for this scheduling policy in a wireless transmission environment was presented [22]. On the other hand, opportunistic scheduling was developed specifically for wireless transmission systems. This scheduling policy takes advantage of multi-user diversity to improve the throughput of the entire system. In particular, the user who has the best channel quality in the current time slot will be selected to transmit. A queueing model of this scheduling policy was proposed [18,23]. In addition, queueing analyses for different resource-sharing schemes (i.e., max-min fairness, proportional fairness, and balanced fairness) were presented [24]. Also, the stability condition for each of the schemes was studied.

While most of the queueing models in the literature considered only the variation of wireless channel on system performance, a few works considered the impacts of resource allocation and admission control on queueing performance. For example, impacts of resource allocation and admission were considered in the queueing model for a TDMA-based cellular wireless system [25] using adaptive modulation and coding, and a code division multiple access (CDMA)-based system [26] with rate adaptation. These investigations showed that resource reservation for handoff connections (i.e., through guard channel) as well as the transmission rate adaptation can impact the queueing performance of the mobile users significantly.

Queueing analysis can be also used for the development of admission control mechanisms. This approach was presented in the literature [27–29]. In particular, given the traffic parameters and the estimated wireless channel quality, for a given medium access control mechanism, information on queueing delay and packet loss can be obtained and used by the admission controller to decide whether a new connection can be accepted or not. The decision on acceptance or rejection of a new connection is based on whether the QoS performances of both the ongoing connections and the new connection can be maintained at the target level or not. A queueing model was presented which could be used for admission control in Bluetooth-based wireless personal area networks (PANs) [30]. A queueing model for IEEE 802.11-based wireless local area networks (WLANs) was proposed [31,32]. In these works, the MAC protocol was assumed to be carrier sense multiple access/collision avoidance (CSMA/CA).

In addition to the traditional wireless systems which rely on the Single-Input Single-Output (SISO) transmission, Multiple-Input Multiple-Output (MIMO) system has been developed to provide better performance in terms of error or transmission rate. A queueing model for MIMO system was proposed [33,34] in which the advantages of spatial diversity in terms of smaller queueing delay and packet loss were demonstrated.

Multihop communication will be a significant component in the next generation wireless systems. In a multihop network, the transmission range of a transmitter (e.g., in the base station) can be extended by relaying the traffic through multiple intermediate nodes. This multihop transmission is a common feature in wireless *ad hoc*, mesh, and sensor networks. Queueing models for this multihop wireless communication were proposed in the literature. For example, end-to-end performances of a multihop wireless network in terms of latency, reliability, and throughput were studied through a queueing model [35] considering adaptive modulation and coding at the physical layer. A tandem queueing model for multihop transmission was presented in [36]. For sensor networks, energy conservation is one of the challenging issues. Since the amount of energy available at a sensor node is limited (e.g., due to battery size or energy-harvesting technology such as a solar cell), an energy saving mechanism is required and it can impact the wireless transmission performance significantly. A queueing model for sensor networks with energy conservation feature through sleep and wake-up mechanism was presented [37]. A vacation queueing model was used to investigate the inter-relationship between the transmission performance and the energy consumption.

1.3 Preliminaries on Queueing Analysis

Queueing theory is a mathematical tool used to analyze the arrival process, buffering or storing process, and departure process of a queue. Based on queueing theory, several performance measures including probability that a queue is full or empty, average number of items/customers in queue, waiting time, throughput,

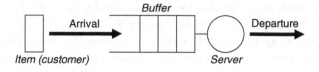

Figure 1.1   General model of a queueing system.

and rejection (i.e., loss) rate can be obtained analytically. Queueing theory has many applications in business, commerce, industry, and engineering. Especially, in engineering disciplines, queueing theory is used in to analyze the performances of computer and communication systems.

The general components of a queueing system are shown in Figure 1.1 [38]. In this queueing model, the arrival and the departure processes can be random in which arrival of customers and service completion are described by some probability distributions. Also, the number of items in the queue becomes random due to these random arrival and departure processes. Therefore, a queueing system can be well studied in the context of a stochastic process.

1.3.1   Stochastic Process

A stochastic process is defined as a set of random variables $\mathscr{X}(t)$ defined on a common probability space. In general, t is considered as time where $t \in \mathbb{T}$ and $\mathbb{T}$ is the set of points in time. A stochastic process can be classified according to the set $\mathbb{T}$ which could be discrete or continuous. $\mathscr{X}(t)$ and $\mathscr{X}_i$, in general, denote the value of a random variable in continuous and discrete time domains, respectively. For example, if the stochastic process is defined in discrete time, the value of the random variable is determined at the specific time step as follows: $\mathscr{X}_0$, $\mathscr{X}_1$, $\mathscr{X}_2$, ...

In the context of a queueing system, the state of the system is defined as the number of items in the queue which is a random variable $\mathscr{X}_i$. Let us consider the number of items over the time period $[0,t]$. It is clear that for $t_i > t_j > 0$, there is a relationship between $\mathscr{X}(t_i)$ and $\mathscr{X}(t_j)$. In particular, $\mathscr{X}(t_i)$ is equal to $\mathscr{X}(t_j)$ plus the number of items arriving in interval $[t_j,t_i]$ minus those departing in the same period.

A stochastic process can be characterized based on the relationships among the corresponding random variables. This can be classified as follows:

1.3.1.1   Stationary Process

A stochastic process is stationary if the complete joint distribution function of the random variables is invariant of time.

$$P[\mathscr{X}(t_1) = x_1, \dots, \mathscr{X}(t_n) = x_n] = P[\mathscr{X}(t_1 + \tau) = x_1, \dots, \mathscr{X}(t_n + \tau) = x_n] \quad (1.1)$$

where τ denotes the amount of shift in time.

1.3.1.2 Markov Process

A Markov process is a stochastic process with a discrete state space. It has the memoryless property; that is, the future state of the system depends only on the present state, not on the past states. The transition from the current state to a next future state is determined by a transition probability which is a function of the current state only. Mathematically, for a Markov process,

$$P[\mathscr{X}(t_{n+1}) = x_{n+1} \mid \mathscr{X}(t_n) = x_n, \mathscr{X}(t_{n-1}) = x_{n-1}, ..., \mathscr{X}(t_1) = x_1]$$
$$= P[\mathscr{X}(t_{n+1}) = x_{n+1} \mid \mathscr{X}(t_n) = x_n]. \tag{1.2}$$

If a Markov process is defined in discrete time, it is also called a Markov chain. Note that, this Markov process is very important for the queueing analysis as the state of the queue can be modeled as a random variable with Markov property.

1.3.1.3 Birth–Death Process

A birth–death process is a special case of Markov process with the property that the transitions are allowed only among adjacent (i.e., neighboring) states. Therefore,

$$P[\mathscr{X}(t_{n+1}) = x_n + i \mid \mathscr{X}(t_n) = x_n] \begin{cases} \geq 0, & i \in \{-1, 0, 1\} \\ = 0, & \text{otherwise.} \end{cases} \tag{1.3}$$

The birth–death process has an important role in queueing analysis as in some of the queueing systems, the change in number of items in the queue is either one or zero (i.e., single arrival and single departure).

1.3.2 Discrete Time Markov Chain

We describe the basics of a DTMC which is the basis for the queueing analysis presented in the next section of this chapter.

Let $P[\mathscr{X}_n = j \mid \mathscr{X}_{n-1} = i]$ denote the transition probability from state i at time step $n - 1$ to state j at time step n. This transition probability is time homogeneous if

$$p_{i,j} = P[\mathscr{X}_n = j \mid \mathscr{X}_{n-1} = i] = P[\mathscr{X}_{n+\tau} = j \mid \mathscr{X}_{n+\tau-1} = i] \tag{1.4}$$

for $n = 1, 2, ...,$ and $\tau \geq 0$. Note that, $i, j \in \mathbb{S}$, where $\mathbb{S}$ denotes the state space of the system which can be either finite or infinite. Based on this transition probability, we can establish the probability transition matrix for the Markov chain as follows:

$$\mathbf{P} = \begin{bmatrix} p_{0,0} & p_{0,1} & \cdots & p_{0,j} & \cdots \\ p_{1,0} & p_{1,1} & \cdots & p_{1,j} & \cdots \\ \vdots & \vdots & \cdots & \vdots & \cdots \\ p_{i,0} & p_{i,1} & \cdots & p_{i,j} & \cdots \\ \vdots & \vdots & \cdots & \vdots & \cdots \end{bmatrix}. \tag{1.5}$$

If $p_{i,j} > 0$, the system can change from state i to state j directly, and the total transition probability from state i to any state must be one, that is, $\Sigma_{j \in S} \, p_{i,j} = 1$.

In the state space of the Markov chain, we can classify different types of states depending on the transition path as follows [39].

- *Absorbing state*: when the system moves to an absorbing state, it will be in this state forever.
- *Transient state*: if the system starts at a transient state, there is a non-zero probability that the system will never reach this transient state again.
- *Recurrent state*: if the system starts at a recurrent state, then the system will eventually return to this state.
- *Periodic state*: a state is periodic if there exists an n-step transition from state i to state i. Here n is defined as follows: $n = kd$, where $d > 1$ is fixed (known as the period of this periodic state) and k can assume any integer value.

One of the properties of Markov chain is irreducibility. A Markov chain is irreducible if the system can transit from and to all states. Otherwise, the Markov chain is reducible.

The main objective of establishing a Markov chain and the corresponding probability transition matrix $\mathbf{P}$ is to obtain the probability for the system to be in a particular state. From the state probabilities, various performance measures related to queueing system (e.g., average number of items in queue, waiting time, probabilities that the queue is full and empty, and so on) can be obtained. There are two types of state probabilities, that is, transient and steady-state probabilities. Note that, this transient state and steady state refer to the state of the system, not a state in the state space of the Markov chain.

1.3.2.1 Transient State Behavior

A queueing system moves to transient state momentarily when there is a change in the system parameters. In transient state, the system output will fluctuate. However, if the system is stable, as time passes, the steady state will be reached after this transient period is over.

One of the methods to obtain the transient state probability is by using Chapman-Kolmogorov equation. Let $\pi^{(n)} = [\pi_0^{(n)}, \pi_1^{(n)}, \ldots, \pi_i^{(n)}, \ldots]$ denote the state probability vector at time step n and $\pi_i^{(n)}$ is transient state probability of state i. Then

$$\pi_i^{(1)} = \sum_{j=0}^{\infty} \pi_j^{(0)} p_{j,i} \tag{1.6}$$

or in matrix form, we have

$$\pi^{(1)} = \pi^{(0)}\mathbf{P}. \tag{1.7}$$

Because the system has Markov property, the future state depends only on the current state. Equation 1.7 can be generalized as follows.

$$\pi^{(n)} = \pi^{(n-1)}\mathbf{P} \tag{1.8}$$
$$= \pi^{(0)}\mathbf{P}^n. \tag{1.9}$$

Based on this state probability at time step n, system behavior at transient state can be obtained. For example, if the system parameter changes at time n, the performance measures obtained from state probability at time step $n + 1$, $n + 2$, $n + 3$, ... are at the transient state performances of the system. However, at $n + \tau$, where $\tau \rightarrow \infty$, the system will reach steady state again.

1.3.2.2 Steady-State Behavior

If state i is aperiodic and state j is recurrent and the condition

$$\lim_{n\to\infty} p_{i,j}^{(n)} = \pi_j \tag{1.10}$$

holds, then the probability that the system is in state j is determined by π_j. This probability is called steady-state probability and in matrix form it is denoted by $\pi = [\pi_0, \pi_1, \ldots, \pi_i, \ldots]$. For a system in steady state,

$$\pi = \pi\mathbf{P} \tag{1.11}$$

where π is stationary in time. From the law of probability,

$$\sum_{i\in S} \pi_i = 1. \tag{1.12}$$

Equations 1.11 and 1.12 provide a set of $s + 1$ linear equations for s unknown variables, where s is the total number of states in state space $\mathbb{S}$. The steady-state probabilities (and hence the performance measures when the system is in steady state) can be obtained by solving a set of linear equations using standard methods such as matrix inversion.

1.4 Connection Admission Control for OFDMA-Based WiMAX Networks

1.4.1 System Model

We consider a single cell in a WiMAX network with a base station and multiple subscriber stations (Fig. 1.2). Each subscriber station serves multiple connections. Admission control is used at each subscriber station to limit the number of ongoing connections through that subscriber station. At each subscriber station, traffic from all uplink connections are aggregated into a single queue [40]. The size of this queue is finite (i.e., X packets) in which some packets will be dropped if the queue is full upon their arrivals. The OFDMA transmitter at the subscriber station retrieves the head of line packet(s) and transmits them to the base station. The base station may allocate different number of subchannels to different subscriber stations. For example, a subscriber station with higher priority could be allocated more number of subchannels.

1.4.2 OFDMA Transmission

We consider time division duplex (TDD)–OFDMA-based uplink transmission from a subscriber station to the base station. The base station allocates a set of subchannels $\mathbb{C}$ to a particular subscriber station, and each subscriber station services multiple connections. The frame structure is shown in Figure 1.3 [41].

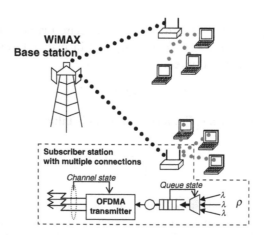

Figure 1.2 System model.

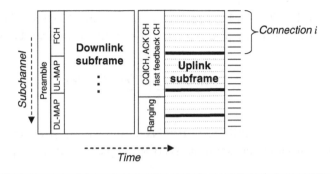

Figure 1.3 Frame structure of WiMAX in time division duplex–orthogonal frequency division multiple access mode.

We consider a Nakagami-m fading model for each subchannel [27]. The channel quality is determined by the instantaneous received Signal-to-Noise Ratio (SNR) γ in each time slot. We assume that the channel is stationary over the transmission frame time. The OFDMA transmitter uses adaptive modulation to achieve different transmission rates at the different transmission modes as shown in Table 1.1.

To determine the mode of transmission (i.e., modulation level and coding rate), an estimated value of SNR at the receiver is used. In this case, the SNR at the receiver is divided into $N+1$ nonoverlapping intervals (i.e., $N=7$ in WiMAX) by thresholds Γ_n ($n \in \{0,1,\ldots,N\}$) where $\Gamma_0 < \Gamma_1 < \ldots < \Gamma_{N+1} = \infty$. The subchannel is said to be in state n (i.e., *rate ID* $= n$ will be used) if $\Gamma_n \leq \gamma < \Gamma_{n+1}$. To avoid possible transmission error, no packet is transmitted when $\gamma < \Gamma_0$. Note that, these thresholds correspond to the required SNR specified in the WiMAX standard,

Table 1.1 Modulation and Coding Schemes in IEEE 802.16

Rate ID	Modulation Level (Coding)	Information Bits/Symbol	Required SNR (dB)
0	BPSK (1/2)	0.5	6.4
1	QPSK (1/2)	1	9.4
2	QPSK (3/4)	1.5	11.2
3	16QAM (1/2)	2	16.4
4	16QAM (3/4)	3	18.2
5	64QAM (2/3)	4	22.7
6	64QAM (3/4)	4.5	24.4

SNR, Signal-to-Noise Ratio.

that is, $\Gamma_0 = 6.4$, $\Gamma_1 = 9.4$, ..., $\Gamma_N = 24.4$ (as shown in Table 1.1). From Nakagami-m distribution, the probability of using rate ID = n [i.e., Pr(n)] can be obtained as follows [8].

$$\Pr(n) = \frac{\Gamma(m, m\Gamma_n/\bar{\gamma}) - \Gamma(m, m\Gamma_{n+1}/\bar{\gamma})}{\Gamma(m)} \qquad (1.13)$$

where $\bar{\gamma}$ is the average SNR, m is the Nakagami fading parameter ($m \geq 0.5$), $\Gamma(m)$ is the Gamma function, and $\Gamma(m, \gamma)$ is the complementary incomplete Gamma function.

Then, we can define row matrix $\mathbf{r}_s$ whose elements r_{k+1} correspond to the probability of transmitting k packets in one frame on one subchannel s ($s \in \mathbb{C}$) as follows:

$$\mathbf{r}_s = [r_0 \, ... \, r_k \, ... \, r_9] \qquad (1.14)$$

where

$$r_{(I_n \times 2)} = \Pr(n) \qquad (1.15)$$

in which I_n is the number of transmitted bits per symbol corresponding to *rate ID* = n, and $r_0 = 1 - \sum_{k=1}^{9} r_k$. We assume that each subchannel is allocated to only one subscriber station.

The matrix for *pmf* of total packet transmission rate can be obtained by convoluting matrices $\mathbf{r}_s$ as follows: $R = \odot_{\forall s \in \mathbb{C}} \mathbf{r}_s$, where $\mathbf{a} \odot \mathbf{b}$ denotes discrete convolution [42] between matrices $\mathbf{a}$ and $\mathbf{b}$, and can be expressed as

$$[\mathbf{a} \odot \mathbf{b}]_{i+1} = \sum_{j=0}^{i} a_j b_{i-j} \qquad (1.16)$$

for $i = 0,1,2,...,u + v - 1$ and $[a]_i$ denotes the element at column i of row matrix $\mathbf{a}$. Note that, matrix $\mathbf{R}$ has size $1 \times R + 1$, where $R = (9 \times |\mathbb{C}|)$ indicates the maximum number of packets that can be transmitted in one frame where $|\mathbb{C}|$ denotes the number of elements in set $\mathbb{C}$.

The total packet transmission rate per frame can be obtained as follows:

$$\phi = \sum_{k=1}^{U} k \times [\mathbf{R}]_{k+1}. \qquad (1.17)$$

1.4.2.1 CAC Policies

The main objective of a CAC mechanism is to limit the number of ongoing connections/flows so that the QoS performances can be guaranteed for all the ongoing connections. The basic components in an admission controller are shown in Figure 1.4. While the performance estimator is used to obtain the current state of the system, resource allocator

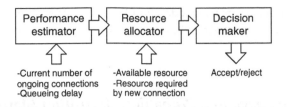

Figure 1.4   Major components in admission control.

uses this state to reallocate available radio resource (e.g., time slot, code and power, and subchannel for TDMA, CDMA, and OFDMA networks, respectively). Then, the admission control decision is made to accept or reject an incoming connection.

To ensure the QoS performances of the ongoing connections, the following two CAC schemes for subscriber stations are proposed.

1.4.2.2   Threshold-Based CAC

In this case, a threshold C is used to limit the number of ongoing connections. When a new connection arrives, the CAC module checks whether the total number of connections including the incoming one is less than or equal to the threshold C. If it is true, then the new connection is accepted, otherwise it is rejected.

1.4.2.3   Queue-Aware CAC

This scheme works based on connection acceptance probability α_x which is determined based on the queue status. Specifically, when a connection arrives, the CAC module accepts the connection with probability α_x, where x ($x \in \{0, 1, ...,X\}$) is the number of packets in the queue in the current time slot. Here, X denotes the size of the queue of the subscriber station under consideration. Note that the value of the parameter α_x can be chosen based on the radio link level performance (e.g., packet delay, packet dropping probability) requirements.

1.4.3   Formulation of the Queueing Model

An analytical model based on DTMC is presented to analyze the system performances at both the connection-level and at the the packet-level for the connection admission schemes described before. To simplify the analytical model, we assume that packet arrival for a connection follows a Poisson process with arrival rate λ which is identical for all connections in the same queue. The connection interarrival time and the duration of a connection are assumed to be exponentially distributed with average $1/\rho$ and $1/\mu$, respectively.

Note that, similar queueing models for OFDM/TDMA system can be found in the literature [43]. However, admission control was not considered in those models.

1.4.4 State Space and Probability Transition Matrix

Assuming that the state of the system is observed at the end of every time slot, the state space of the system for both the CAC schemes is given by

$$\Delta = \{(\mathscr{C}, \mathscr{X}), 0 \le \mathscr{C}, 0 \le \mathscr{X} \le \mathscr{X}\} \tag{1.18}$$

where $\mathscr{C}$ is the number of ongoing connections and $\mathscr{X}$ is the number of packets in the aggregated queue. For both the CAC algorithms, the number of packet arrivals depends on the number of connections. However, for the queue-aware CAC algorithm, the number of packets in the queue affects the acceptance probability for a new connection. The state transition diagram is shown in Figure 1.5. Here, λ and ρ denote rates, not probabilities.

Note that the probability that a Poisson events with average rate λ occur during an interval T can be obtained as follows:

$$f_a(\lambda) = \frac{e^{-\lambda T} (\lambda T)^a}{a!}. \tag{1.19}$$

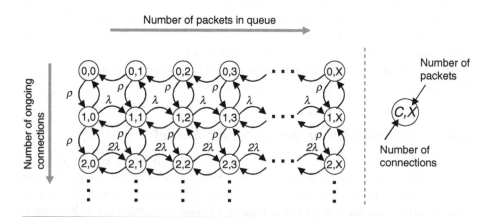

Figure 1.5   State diagram of discrete time Markov chain.

This function is required to determine the probability of both connection and packet arrivals.

1.4.4.1 Threshold-Based CAC Algorithm

In this case, the transition matrix $\mathbf{Q}$ for the number of connections in the system can be expressed as follows:

$$
\mathbf{Q} = \begin{bmatrix}
q_{0,0} & q_{0,1} & & & \\
q_{1,0} & q_{1,1} & q_{1,2} & & \\
\ddots & \ddots & \ddots & & \\
& q_{C-2,C-1} & q_{C-1,C-1} & q_{C-1,C} & \\
& & q_{C-1,C} & q_{C,C}
\end{bmatrix}
\tag{1.20}
$$

where each row indicates the number of ongoing connections (i.e., c). As the length of a frame T is very small compared with connection arrival and departure rates, we assume that the maximum number of arriving and departing connections in a frame is one. Therefore, the elements of this matrix can be obtained as follows:

$$
q_{c,c+1} = f_1(\rho) \times (1 - f_1(c\mu)), \quad c = 0, 1, \ldots, C-1
$$

$$
q_{c,c-1} = (1 - f_1(\rho)) \times f_1(c\mu), \quad c = 1, 2, \ldots, C
\tag{1.21}
$$

$$
q_{c,c} = f_1(\rho) \times f_1(c\mu) + (1 - f_1(\rho)) \times (1 - f_1(c\mu)), \quad c = 0, 1, \ldots, C
$$

where $q_{c,c+1}$, $q_{c,c-1}$, and $q_{c+1,c+1}$ represent the cases that the number of ongoing connections increases by one, decreases by one, and does not change, respectively.

1.4.4.2 Queue-Aware CAC Algorithm

Because the admission of a connection in this case depends on the current number of packets in the queue, the transition matrix can be expressed based on the number of packets (x) in the queue as follows:

$$
\mathbf{Q}_x = \begin{bmatrix}
q_{0,0}^{(x)} & q_{0,1}^{(x)} & & \\
q_{1,0}^{(x)} & q_{1,1}^{(x)} & q_{1,2}^{(x)} & \\
& q_{2,1}^{(x)} & q_{2,2}^{(x)} & q_{2,3}^{(x)} \\
& \ddots & \ddots & \ddots
\end{bmatrix}
\tag{1.22}
$$

where

$$q_{c,c+1}^{(x)} = f_1(\alpha_x \rho) \times (1 - f_1(c\mu)), \quad c = 0, 1, \ldots$$

$$q_{c,c-1}^{(x)} = (1 - f_1(\alpha_x \rho)) \times f_1(c\mu), \quad c = 1, 2, \ldots \quad (1.23)$$

$$q_{c,c}^{(x)} = f_1(\alpha_x \rho) \times f_1(c\mu) + (1 - f_1(\alpha_x \rho)) \times (1 - f_1(c\mu)), \quad c = 0, 1, \ldots$$

in which α_x is the connection acceptance probability when there are x packets in the queue.

1.4.4.3 Transition Matrix for the Queue

The transition matrix **P** of the entire system can be expressed as in Equation 1.24. The rows of matrix **p** represent the number of packets (x) in the queue.

$$\mathbf{P} = \begin{bmatrix} \mathbf{P}_{0,0} & \cdots & \mathbf{P}_{0,A} & & & \\ \vdots & \vdots & \ddots & \ddots & & \\ \mathbf{P}_{R,0} & \cdots & \mathbf{P}_{R,R} & \cdots & \mathbf{P}_{R,R+A} & \\ \ddots & \ddots & \ddots & \ddots & \ddots & \ddots \\ & \mathbf{P}_{x,x-R} & \cdots & \mathbf{P}_{x,x} & \cdots & \mathbf{P}_{x,x+A} \\ & \ddots & \ddots & \ddots & \ddots & \ddots \end{bmatrix} \quad (1.24)$$

Matrices $\mathbf{p}_{x,x'}$ represent the changes in the number of packets in the queue (i.e., the number of packets in the queue changing from x in the current frame to x' in the next frame). To construct matrix $\mathbf{p}_{x,x'}$, we first establish matrices $\mathbf{v}_{x,x'}$, where the diagonal elements of the matrices $\mathbf{v}_{x,x'}$ are given in Equations 1.25 through 1.27 for $r \in \{0, 1, 2, \ldots, D\}$ and $a \in \{0, 1, 2, \ldots (c \times A)\}$, $n = 1, 2, \ldots, D$, and $m = 1, 2, \ldots, (c \times A)$. The nondiagonal elements of $\mathbf{v}_{x,x'}$ are all zero.

$$\left[\mathbf{v}_{x,x-n} \right]_{c+1,c+1} = \sum_{a-r=n} f_a(c\lambda)[\mathbf{R}]_r \quad (1.25)$$

$$\left[\mathbf{v}_{x,x+m} \right]_{c+1,c+1} = \sum_{r-a=m} f_a(c\lambda)[\mathbf{R}]_r \quad (1.26)$$

$$\left[\mathbf{v}_{x,x} \right]_{c+1,c+1} = \sum_{r=a} f_a(c\lambda)[\mathbf{R}]_r. \quad (1.27)$$

Here A is the maximum number of packets that can arrive from one connection in one frame and D is the maximum number of packets that can be transmitted in one frame by all of the allocated subchannels allocated to that particular queue, and it can be obtained from $D = \min(R,x)$. This is due to the fact that the maximum number of transmitted packets depends on the number of packets in the queue and the maximum possible number of transmissions in one frame. Note that, $[\mathbf{v}_{x,x-n}]_{c+1,c+1}$, $[\mathbf{v}_{x,x+m}]_{c+1,c+1}$, and $[\mathbf{v}_{x,x}]_{c+1,c+1}$ represent the probability that the number of packets in the queue increases by m, decreases by m, and does not change, respectively, when there are c ongoing connections. Here, $[v]_{i,j}$ denotes the element at row i and column j of matrix $\mathbf{v}$, and these elements are obtained based on the assumption that the packet arrivals for the ongoing connections are independent of each other.

Finally, we obtain the matrices $\mathbf{p}_{x,x'}$ by combining both the connection-level and the queue-level transitions as follows:

$$\mathbf{P}_{x,x'} = \mathbf{Q}\mathbf{v}_{x,x'} \tag{1.28}$$

$$\mathbf{P}_{x,x'} = \mathbf{Q}_x\mathbf{v}_{x,x'} \tag{1.29}$$

for the cases of threshold-based (Equation 1.28) and queue-aware (Equation 1.29) CAC algorithms, respectively.

1.4.5   QoS Measures

The connection-level and the packet-level performance measures (i.e., connection blocking probability, average number of ongoing connections in the system, and average queue length) are obtained for both the CAC schemes.

For the threshold-based CAC scheme, all of the above performance measures are calculated from the steady state probability vector of the system states π, which is obtained by solving $\pi\mathbf{P} = \pi$ and $\pi\mathbf{1}=1$, where $\mathbf{1}$ is a column matrix of ones. However, for the queue-aware CAC algorithm, the size of the matrix $\mathbf{Q}_x$ needs to be truncated at C_{tr} (i.e., the maximum number of ongoing connections at the subscriber station). Also, the size of the matrix $\mathbf{P}$ needs to be truncated at X (i.e., the maximum number of packets in the queue) for both the schemes.

The steady-state probability, denoted by $\pi(c,x)$ for the state that there are c connections and x ($x = \{0, 1, ..., X\}$) packets in the queue, can be extracted from matrix π as follows:

$$\pi(c,x) = [\pi]_{x\times(C'+1)+c}, \quad c = 0, 1, ..., C' \tag{1.30}$$

where $C' = C$ and $C' = C_{tr}$ for the threshold-based and the queue-aware CAC algorithms, respectively. Using these steady state probabilities, the various performance measures can be obtained. Note that, the subscripts *tb* and *qa* are used to indicate the performance measures for the threshold-based and the queue-aware CAC schemes, respectively.

1.4.5.1 Connection Blocking Probability

It refers to the probability that an arriving connection will be blocked due to the admission control decision. This performance measure indicates the accessibility of the wireless service, and for the threshold-based CAC scheme. It can be obtained as follows:

$$P_{tb}^b = \sum_{x=0}^{X} \pi(C,x).$$

(1.31)

The above probability refers to the probability that the system serves the maximum allowable number of ongoing connections. The blocking probability for the queue-aware CAC is obtained from

$$P_{qa}^b = \sum_{x=0}^{X} \sum_{c=1}^{C_{tr}} ((1-\alpha_x) \times \pi(c,x))$$

(1.32)

in which the blocking probability is the sum of the probabilities of rejection for all possible number of packets in the queue.

1.4.5.2 Average Number of Ongoing Connections

It can be obtained as

$$\overline{C}_{tb} = \sum_{x=0}^{X} \left(c \sum_{c=0}^{C} \pi(c,x) \right)$$

(1.33)

$$\overline{C}_{qa} = \sum_{x=0}^{X} \left(c \sum_{c=0}^{C_{tr}} \pi(c,x) \right).$$

(1.34)

1.4.5.3 Average Queue Length

It is given by

$$\overline{x}_{tb} = x \sum_{x=0}^{X} \sum_{c=0}^{C} \pi(c,x)$$

(1.35)

$$\overline{x}_{qa} = x \sum_{x=0}^{X} \sum_{c=0}^{C_{tr}} \pi(c,x).$$

(1.36)

1.4.5.4 Packet Dropping Probability

This performance measure indicates the probability that an incoming packet will be dropped due to the unavailability of buffer space. It can be derived from the average number of dropped packets per frame. Given that there are x packets in the queue and the number of packets in the queue increases by v, the number of dropped packets is $m - (X - x)$ for $m > X - x$, and zero otherwise. The average number of dropped packets per frame is obtained as follows:

$$\overline{x}_{\text{drop}} = \sum_{s=1}^{C'} \sum_{x=0}^{X} \sum_{m=X-x+1}^{A} \pi(x,s) \left(\sum_{j=1}^{C'} [\mathbf{p}_{x,x+m}]_{s,j} \right) (m - (X - x))$$

(1.37)

where the term $\left(\sum_{j=1}^{C'} [\mathbf{p}_{x,x+m}]_{s,j} \right)$ in Equation 1.37 indicates the total probability that the number of packets in the queue increases by m at every arrival phase. Note that, we consider probability $\mathbf{p}_{x,x+m}$ rather than the probability of packet arrival as we have to consider the packet transmission in the same frame as well. After calculating the average number of dropped packets per frame, we can obtain the probability that an incoming packet is dropped as follows:

$$P_{\text{drop}} = \frac{\overline{x}_{\text{drop}}}{\overline{\lambda}}$$

(1.38)

where $\overline{\lambda}$ is the average number of packet arrivals per frame and it can be obtained from

$$\overline{\lambda} = \lambda \overline{c}.$$

(1.39)

1.4.5.5 Queue Throughput

It measures the number of packets transmitted in one frame and can be obtained from

$$\eta = \lambda(1 - P_{\text{drop}}).$$

(1.40)

1.4.5.6 Average Delay

It is defined as the number of frames that a packet waits in the queue since its arrival before it is transmitted. We use Little's law to obtain average delay as follows:

$$\overline{w} = \frac{\overline{x}}{\eta}.$$

(1.41)

1.4.6 Numerical Results

1.4.6.1 Parameter Setting

We consider one queue (which corresponds to a particular subscriber station) for which five subchannels are allocated and we assume that the average SNR is the same for all of these subchannels. Each subchannel has a bandwidth of 160 kHz. The length of a subframe for downlink transmission is one millisecond, and therefore, the transmission rate in one subchannel with rate ID $= 0$ (i.e., BPSK modulation and coding rate is 1/2) is 80 kbps. We assume that the maximum number of packets arriving in one frame for a connection is limited to 30 (i.e., $A = 30$).

For the threshold-based CAC mechanism, the value of the threshold C is varied according to the evaluation scenarios. For the queue-aware CAC mechanism, the value of the connection acceptance probability is determined as follows:

$$\alpha_x = \begin{cases} 1, & 0 \leq x < B_{th} \\ 0, & B_{th} \leq x \leq X. \end{cases} \tag{1.42}$$

In the performance evaluation, we use $B_{th} = 80$.

For performance comparison, we also evaluate the queueing performance in the absence of any CAC mechanism. For the queue-aware CAC and for the case with no CAC, we truncate the maximum number of ongoing connections at 25 (i.e., $C_{tr} = 25$) so that $\pi(C_{th}, x) < 0.0002$, $\forall x$. The average duration of a connection is set to ten minutes (i.e., $\mu = 10$) for all the evaluation scenarios. The queue size is 150 packets (i.e., $X = 150$). The parameters are set as follows. The connection arrival rate is 0.4 connections per minute. Packet arrival rate per connection is one packet per frame. Average SNR on each subchannel is 5 dB. Note that, we vary some of these parameters depending on the evaluation scenarios whereas the others remain fixed.

To validate the correctness of the analytical model, an event-driven simulator is used to obtain the performance results to be compared with the results obtained from the analytical model. We consider a single transmitter and a single subscriber station (i.e., single queue) in the simulation. Connection inter-arrival time and connection duration are assumed to be exponentially distributed. While connection arrival and departures can occur at any arbitrary point in time, packet transmissions occur on a frame-by-frame basis. That is, packet arrivals and packet transmissions are determined for each frame with fixed length T. While packet arrivals follow a Poisson process, the number of transmitted packets per frame depends on the set of allocated subchannels and the instantaneous SNR for each subchannel. We assume that each subchannel experiences Rayleigh fading.

1.4.6.2 Performance of CAC

We first examine the impact of connection arrival rate on connection-level performances. Variations in average number of ongoing connections and connection

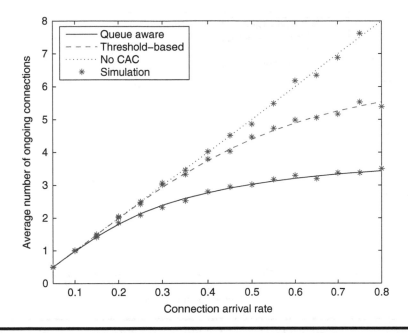

Figure 1.6   Average number of ongoing connections under different connection arrival rates.

blocking probability with connection arrival rate are shown in Figures 1.6 and 1.7, respectively. As expected, when the connection arrival rate increases, the number of ongoing connections and connection blocking probability increase.

The packet-level performances under different connection arrival rates are shown in Figures 1.8 through 1.11 for average number of packets in the queue, packet dropping probability, queue throughput, and average queueing delay, respectively. These performance measures are significantly impacted by the connection arrival rate. Because both the CAC schemes limit the number of ongoing connections, packet-level performances can be maintained at the target level. In this case, both the CAC schemes result in better packet-level performances compared with those without any CAC scheme.

Variations in packet dropping probability, queue throughput, and average queueing delay with channel quality are shown in Figures 1.12 and 1.13, respectively. As expected, the packet-level performances become better when channel quality becomes better. Also, we observe that while the connection-level performances for the threshold-based CAC scheme and those without any CAC scheme are not impacted by the channel quality, connection blocking probability decreases significantly for the queue-aware CAC mechanism when the channel quality becomes better (Fig. 1.14). Based on these observations, we can conclude

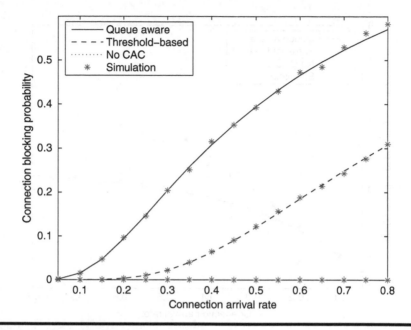

Figure 1.7 Connection blocking probability under different connection arrival rates.

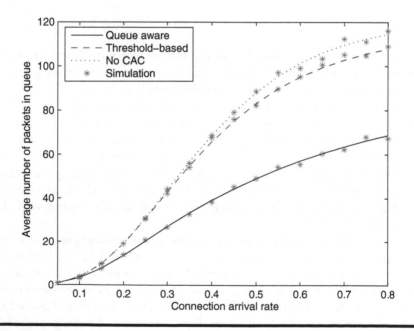

Figure 1.8 Average number of packets in queue under different connection arrival rates.

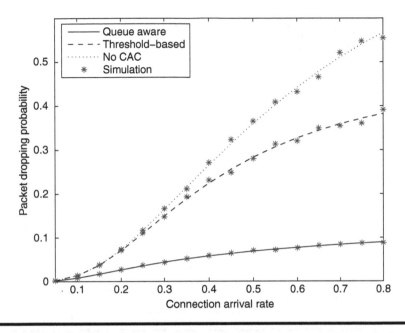

Figure 1.9 **Packet dropping probability under different connection arrival rates.**

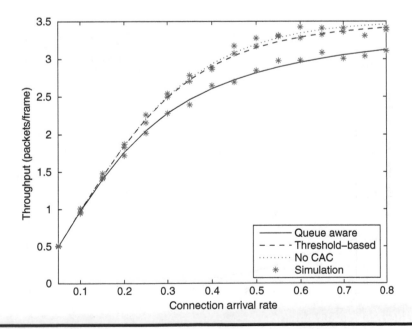

Figure 1.10 **Queueing throughput under different connection arrival rates.**

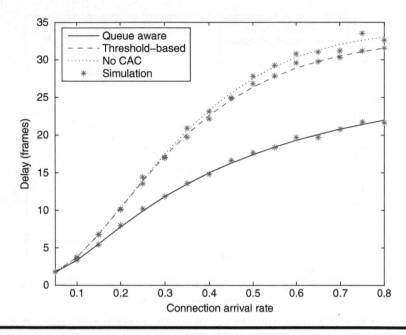

Figure 1.11   Average delay under different connection arrival rates.

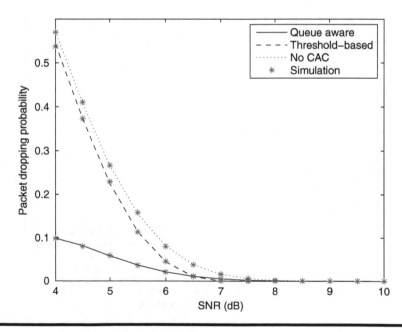

Figure 1.12   Packet dropping probability under different channel qualities.

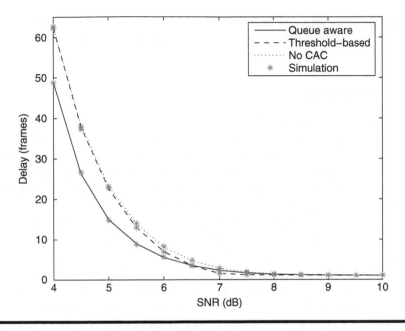

Figure 1.13   Average delay under different channel qualities.

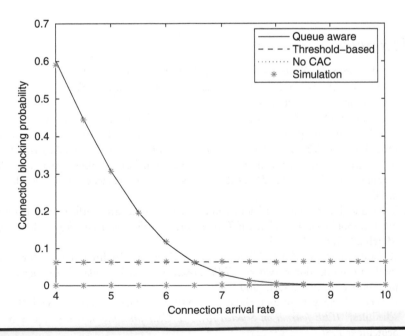

Figure 1.14   Connection blocking probability under different channel qualities.

that the queue-aware CAC can adapt the admission control decision based on the queue status which is desirable for a system with high traffic fluctuations.

1.5 Summary

In this chapter, we have addressed the problem of queueing theoretic performance modeling and analysis of OFDMA transmission under admission control. We have considered a WiMAX system model in which a base station serves multiple subscriber stations and each of the subscriber stations is allocated with a certain number of subchannels by the base station. There are multiple ongoing connections at each subscriber station. We have first reviewed the related works on queueing analysis for wireless transmission systems. Then, the preliminaries on stochastic processes and DTMC have been discussed as the background for the developed queueing model to analyze the system model under consideration.

For CAC, we have considered two schemes, namely, queue-aware scheme and threshold-based scheme. While the threshold-based CAC scheme simply fixes the number of ongoing connections, the queue-aware CAC scheme considers the number of packets in the queue for the admission control decision of a new connection. The connection-level and packet-level performances of these CAC schemes have been studied based on the queueing model. Also, we have presented simulation results to validate the analysis.

References

1. K. Shiomoto, N. Yamanaka, and T. Takahashi, "Overview of measurement-based connection admission control methods in ATM networks," *IEEE Communications Surveys*, pp. 2–13, First Quarter 1999.
2. D. Hong and S. S. Rappaport, "Traffic model and performance analysis for cellular mobile radio telephone systems with prioritized and nonprioritized handoff procedures," *IEEE Transactions on Vehicular Technology*, pp. 77–92, August 1986.
3. R. Ramjee, R. Nagarajan, and D. Towsley, "On optimal call admission control in cellular networks," in *Proc. IEEE INFOCOM'96*, vol. 1, San Francisco, CA, March 1996, pp. 43–50.
4. Y. Fang and Y. Zhang, "Call admission control schemes and performance analysis in wireless mobile networks," *IEEE Transactions on Vehicular Technology*, vol. 51, no. 2, March 2002, pp. 371–382.
5. R. Fantacci and L. Zoppi, "Performance evaluation of polling systems for wireless local communication networks," *IEEE Transactions on Vehicular Technology*, vol. 49, pp. 2148–2157, November 2000.
6. L. Servi, "Average delay approximation of M/G/1 cyclic service queues with Bernoulli schedules," *IEEE Journal on Selected Areas in Communications*, vol. 4, pp. 813–822, September 1986.

7. L. B. Le, E. Hossain, and A. S. Alfa, "Queuing analysis for radio link level scheduling in a multi-rate TDMA wireless network," in *Proc. IEEE GLOBECOM'04*, vol. 6, pp. 4061–4065, November–December 2004.

8. Q. Liu, S. Zhou, and G. B. Giannakis, "Queuing with adaptive modulation and coding over wireless links: cross-layer analysis and design," *IEEE Transactions on Wireless Communications*, vol. 4, no. 2, pp. 1142–1153, May 2005.

9. L. Galluccio, F. Licandro, G. Morabito, and G. Schembra, "An analytical framework for the design of intelligent algorithms for adaptive-rate MPEG video encoding in next-generation time-varying wireless networks," *IEEE Journal on Selected Areas in Communications*, vol. 23, no. 2, pp. 369–384, February 2005.

10. Y. Wu, Z. Li, Z. Chen, Y. Wu, L. Wang, and T. Lu, "Analysis of real-time communication system with queuing priority," in *Proc. IEEE ICCNMC'05*, vol. 2, pp. 863–866, September 2005.

11. R. Fantacci, "Queuing analysis of the selective repeat automatic repeat request protocol wireless packet networks," *IEEE Transactions on Vehicular Technology*, vol. 45, no. 2, pp. 258–264, May 1996.

12. M. Zorzi and R. R. Rao, "On the use of renewal theory in the analysis of ARQ protocols," *IEEE Transactions on Communications*, vol. 44, no. 9, pp. 1077–1081, September 1996.

13. J. G. Kim and M. M. Krunz, "Delay analysis of selective repeat ARQ for a Markovian source over a wireless channel," *IEEE Transactions on Vehicular Technology*, vol. 49, no. 5, pp. 1968–1981, September 2000.

14. Q. Liu, S. Zhou, and G. B. Giannakis, "Cross-layer combining of adaptive modulation and coding with truncated ARQ over wireless links," *IEEE Transactions on Wireless Communications*, vol. 3, no. 5, pp. 1746–1755, September 2004.

15. J.-B. Seo, S.-Q. Lee, N.-H. Park, H.-W. Lee, and C.-H. Cho, "A queueing model of an adaptive type-I hybrid-ARQ with a generalized Markovian source and channel," in *Proc. IEEE CCECE'05*, May 2005.

16. L. B. Le and E. Hossain, "Queueing analysis of go-back-N ARQ protocol in multi-rate wireless networks with feedback delay," in *Proc. IEEE GLOBECOM'05*, vol. 6, November–December 2005.

17. L. B. Le and E. Hossain, "Delay statistics for selective repeat ARQ protocol in multi-rate wireless networks with non-instantaneous feedback," in *Proc. IEEE GLOBECOM'05*, vol. 6, November–December 2005.

18. T. Issariyakul and E. Hossain, "Channel-quality-based opportunistic scheduling with ARQ in multi-rate wireless networks: modeling and analysis," *IEEE Transactions on Wireless Communications*, vol. 5, no. 4, pp. 796–806, April 2006.

19. L. Badia, M. Rossi, and M. Zorzi, "SR ARQ packet delay statistics on markov channels in the presence of variable arrival rate," *IEEE Transactions on Wireless Communications*, vol. 5, no. 7, pp. 1639–1644, July 2006.

20. M. Rossi, L. Badia, and M. Zorzi, "SR ARQ delay statistics on N-state Markov channels with non-instantaneous feedback," *IEEE Transactions on Wireless Communications*, vol. 5, no. 6, pp. 1526–1536, June 2006.

21. L. Zhang and T. T. Lee, "Performance analysis of wireless fair queuing algorithms with compensation mechanism," in *Proc. IEEE ICC'04*, vol. 7, pp. 4202–4206, June 2004.

22. D. Niyato and E. Hossain, "Analysis of fair scheduling and connection admission control in differentiated services wireless networks," in *Proc. IEEE ICC'05*, vol. 5, pp. 3137–3141, May 2005.

23. L. B. Le, E. Hossain, and A. S. Alfa, "Queueing analysis and admission control for multi-rate wireless networks with opportunistic scheduling and ARQ-based error control," in *Proc. IEEE ICC'05*, vol. 5, pp. 3329–3333, May 2005.

24. T. Bonald, L. Massoulie, A. Proutiere, and J. Virtamo, "A queueing analysis of max-min fairness, proportional fairness and balanced fairness," *Queueing Systems*, vol. 53, no. 1–2, June 2006.

25. D. Niyato and E. Hossain, "Call-level and packet-level performance analysis of call admission control and adaptive channel allocation in cellular wireless networks," in *Proc. IEEE GLOBECOM'05*, vol. 5, November–December 2005.

26. D. Niyato and E. Hossain, "Call-level and packet-level quality of service and user utility in rate-adaptive cellular CDMA networks: a queuing analysis," *IEEE Transactions on Mobile Computing*, vol. 5, no. 12, pp. 1749–1763, December 2006.

27. D. Niyato and E. Hossain, "Delay-based admission control using fuzzy logic for OFDMA broadband wireless networks," in *Proc. IEEE ICC'06*, June 2006.

28. D. Niyato and E. Hossain, "Queue-aware uplink bandwidth allocation and rate control for polling service in IEEE 802.16 broadband wireless networks," *IEEE Transactions on Mobile Computing*, vol. 5, no. 6, pp. 668–679, June 2006.

29. D. Niyato and E. Hossain, "A queuing-theoretic and optimization-based model for radio resource management in IEEE 802.16 broadband wireless networks," *IEEE Transactions on Computers*, vol. 55, no. 11, pp. 1473–1488, November 2006.

30. J. Misic, K. L. Chan, and V. B. Misic, "Admission control in Bluetooth piconets," *IEEE Transactions on Vehicular Technology*, vol. 53, no. 3, pp. 890–911, May 2004.

31. O. Tickoo and B. Sikdar, "Queueing analysis and delay mitigation in IEEE 802.11 random access MAC based wireless networks," in *Proc. IEEE INFOCOM'04*, vol. 2, pp. 1404–1413, March 2004.

32. C. G. Park, H. S. Jung, and D. H. Han, "Queueing analysis of IEEE 802.11 MAC protocol in wireless LAN," in *Proc. IEEE ICN/ICONS/MCL'06*, pp. 139, April 2006.

33. M. Airy, S. Shakkottai, and R. W. Heath, Jr., "Limiting queueing models for scheduling in multi-user MIMO systems," in *Proc. IASTED*, pp. 418–423, November 2003.

34. I. S. Ahmed and M. Ibnkahla, "Performance analysis of multiuser diversity in MIMO channels," in *Proc. IEEE CCECE'04*, vol. 2, pp. 1199–1202, May 2004.

35. M. J. Neely, "Exact queueing analysis of discrete time tandems with arbitrary arrival processes," in *Proc. IEEE ICC'04*, vol. 4, pp. 2221–2225, June 2004.

36. T. Issariyakul, E. Hossain, and A. S. Alfa, "End-to-end batch transmission in a multihop and multirate wireless network: latency, reliability, and throughput analysis," *IEEE Transactions on Mobile Computing*, vol. 5, no. 9, pp. 1143–1155, September 2006.

37. A. Fallahi, E. Hossain, and A. S. Alfa, "QoS and energy trade off in distributed energy-limited mesh/relay networks: a queuing analysis," *IEEE Transactions on Parallel and Distributed Systems*, vol. 17, no. 6, pp. 576–592, June 2006.

38. L. Kleinrock, *Queueing Systems Volume 1: Theory*, Wiley-Interscience, 1975.
39. R. Nelson, *Probability, Stochastic Processes and Queueing Theory*, Springer-Verlag, 1995.
40. D. Niyato and E. Hossain, "Connection admission control algorithms for OFDMA wireless networks," in *Proc. IEEE GLOBECOM'05*, St. Louis, MO, USA, 28 November–2 December 2005.
41. T. Kwon, H. Lee, S. Choi, J. Kim, D.-H. Cho, S. Cho, S. Yun, W.-H. Park, and K. Kim, "Design and implementation of a simulator based on a cross-layer protocol between MAC and PHY layers in a WiBro compatible IEEE 802.16e OFDMA system," *IEEE Communications Magazine*, vol. 43, no. 12, pp. 136–146, December 2005.
42. C. D. Meyer, *Matrix Analysis and Applied Linear Algebra*, Siam Philadelphia, 2000.
43. D. Niyato and E. Hossain, "Queueing analysis of OFDM/TDMA systems," in *Proc. IEEE GLOBECOM'05*, St. Louis, MO, USA, 28 November–2 December 2005.

Chapter 2

Mobile WiMAX Networks and Their Integration with WiFi Networks

Wen-Hsin Yang and Yu-Chee Tseng

Among many wireless networking technologies, the WiMAX technology is one promising solution to support broadband wireless access. The first goal of this chapter is to review the specifications of WiMAX. In addition, before the deployment of WiMAX networks achieves popularity, a hybrid environment of WiFi and WiMAX networks seems to be inevitable. The second goal of this chapter is to study the vertical handover issue in such a hybrid environment. We present an energy- and bandwidth-aware vertical handover scheme, which is able to take into consideration of both Mobile Stations' (MS) energy consumption and applications' bandwidth requirements. By providing multicasting and buffering mechanisms, the proposed vertical handover scheme has potential to provide ubiquitous broadband access for real-time streaming services.

2.1 Introduction

There have been growing demands for broadband wireless applications, such as audio and video streaming, near real-time downloading services, broadband Internet access services, and Voice-over-Internet Protocol (VoIP) services. On the other hand, mobile devices, such as smart phones, personal digital assistants (PDAs), and laptops, are all quite ready to embrace such services. While all these developments are almost ready to support broadband services, the wireless access bandwidth has become a bottleneck. The WiMAX technology [1,2] is one of the potential solutions to respond to this demand.

Although there are some legitimate concerns over the different usages of WiMAX networks (e.g., licensed or license-exempt concerns), from the technical aspect, three usage types, fixed, nomadic, and mobile users, are the target customers of WiMAX networks. The first goal of this chapter is to review the standards of WiMAX technologies to support these usage types. Furthermore, from a practical point of view, before the deployment of WiMAX networks achieves its popularity, a hybrid environment of WiMAX and WiFi networks is more likely to coexist for a while. Therefore, one of the primary concerns is to design a seamless handover process in this hybrid network environment. The second goal of this chapter is to dedicate to this issue.

The rest of this chapter is organized as follows. Section 2.2 provides an overview of the mobile WiMAX specification, including the IEEE 802.16 standard and the 802.16e-2005 specification. In Section 2.3, we discuss the mobility management-related schemes in the mobile WiMAX specification. In Section 2.4, we propose an energy- and bandwidth-aware handover scheme for the hybrid of WiFi and WiMAX networks. Concluding remarks and future directions are presented in Section 2.5.

2.2 Overview of the Mobile WiMAX Specification

2.2.1 IEEE 802.16 Standard

The IEEE 802.16 standard [3], which includes Medium Access Control (MAC) and physical (PHY) layer specifications, aims at supporting Internet services over wireless metropolitan area networks (WMAN). It is also an alternative to traditional wired networks, such as Digital Subscriber Line (DSL) and cable-modem. There are two modes defined in WiMAX networks: point-to-multiple-points (PMP) mode and mesh mode.

In the PHY layer, the IEEE 802.16 standard adopts the orthogonal frequency division multiplexing (OFDM), which is a multicarrier modulation scheme. The IEEE 802.16 standard has two OFDM-based modes: OFDM and orthogonal frequency division multiplexing access (OFDMA). Both of these technologies allow subcarriers to be adaptively modulated (e.g., QPSK, 16-QAM, and 64-QAM), depending on transmission distance and noise. Moreover, OFDMA has scalability to provide efficient use of bandwidth.

The MAC layer of IEEE 802.16 standard was originally designed for the PMP mode. On the later amendments of the IEEE 802.16a and the IEEE 802.16d, the mesh mode was included. The IEEE 802.16a adopts OFDM to provide greater spectral efficiency and to mitigate interference. IEEE 802.16b covers most of the

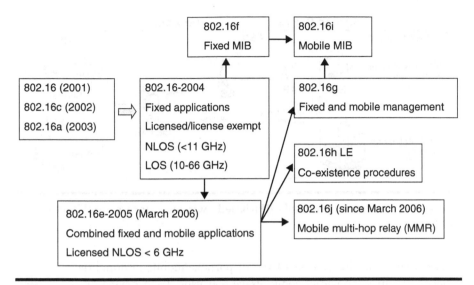

Figure 2.1   Evolution of the IEEE 802.16 standard.

quality of service (QoS) aspects. The IEEE 802.16e introduces scalable OFDMA into the standard, and supports mobile communications. With handover mechanisms, WiMAX is thus able to support mobile communications at vehicular speeds. We summarize the history of the evolution of the IEEE 802.16 standard in Figure 2.1.

The IEEE 802.16 working groups on broadband access standards developed the IEEE 802.16 WirelessMAN standard for WMANs. On the other hand, the WiMAX forum was formed in June of 2001 to ensure interoperability among 802.16 products from different vendors. These groups and their activities may help popularizing WiMAX networks and systems by bringing vendors together and improving the specifications.

According to the IEEE 802.16 specification, the non-line-of-sight (NLOS) transmission range is 4 miles. With the combination of soft-switch technologies, a WiMAX network can work as a wireless "last mile" and make a viable alternative to the Public switched telephone network (PSTN) for VoIP services. In addition, a WiMAX network can work as a point-to-point backhaul trunk with a transmission capability of 72 Mbps at a transmission distance over 30 miles. With its technological advantages of throughput, power, transmission range, and versatility, WiMAX might be a strong competitor of other technologies, such as WiFi and 3G. Therefore, from both economical and technical points of view, WiMAX could be an appealing choice for broadband wireless services.

2.2.2   WiMAX Network Architecture

WiMAX has an IP-based wireless access architecture, which contains three parts: user terminal devices, access service network (ASN), and core service network (CSN).

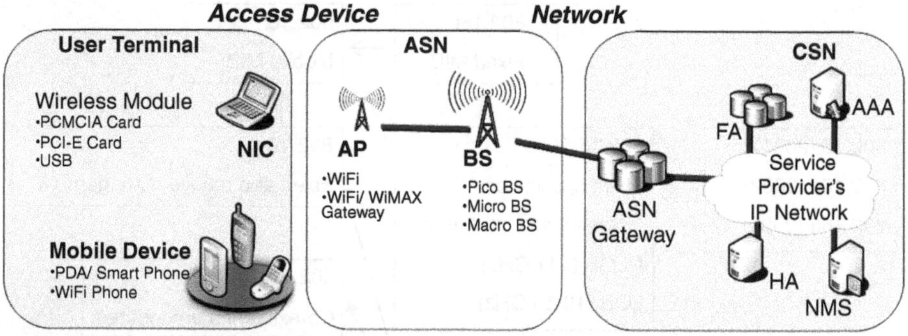

Figure 2.2 Internet Protocol (IP)-based wireless access architecture of WiMAX.

A user terminal device can be a fixed or portable/mobile terminal device, which supports the fixed/nomadic/mobile usage scenarios. Each device can establish a connection link to a WiMAX Base Station (BS), and perform authentication and registration through an access gateway in the CSN. The system architecture is illustrated in Figure 2.2.

A mobile WiMAX network has a similar architecture as a cellular network, where PMP links are between each BS and multiple Subscriber Stations (SSs). Each BS provides frequency and timing reference to SSs for synchronization purpose. The detailed MAC layer protocols and message sequences will be described later.

2.2.3 IEEE 802.16e-2005 Specification

The IEEE 802.16e-2005 specification defines the mobile WiMAX network protocols and the related message sequences. The specification consists of the PHY and the MAC layers, as shown in Figure 2.3. The MAC layer is composed of a security sublayer, a MAC common part sublayer, and a convergence sublayer.

Data transmission between a BS and a SS at the PHY layer relies on the resource allocation of data burst through the OFDMA scheme. A BS can transmit to multiple SSs concurrently in the downlink (DL) direction in separate subchannels or separate symbols; similarly, multiple SSs can transmit to the same BS concurrently in the uplink (UL) direction in separate subchannels or separate symbols. Each channel width is from 1.25 to 20 MHz, which spreads to different frequency bands. Therefore, the PHY layer supports orthogonal subchannels for multiple accesses. This design can reduce interference and improve capacity. Moreover, it has the advantages of flexible subchannelization and bandwidth allocation.

The security sublayer is to ensure the privacy of subscribers across the WiMAX network by encrypting connections between a SS and a BS. In addition, a BS can protect against unauthorized access to data transport services by enforcing encryption

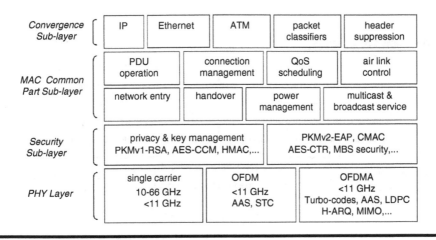

Figure 2.3   IEEE 802.16e-2005 protocol layers.

of the associated service flows across the network. Privacy employs an authenticated client/server key management protocol in which a BS, the server, controls distribution of keying material to a SS, the client. Additionally, the basic privacy mechanisms are strengthened by adding digital-certificate-based SS authentication to its key management protocol.

Over the security sublayer, there are the MAC common part sublayer and the convergence sublayer. The MAC common part sublayer utilizes a shared medium mechanism to efficiently handle the resource of communication links. On top of the MAC common part sublayer is the convergence sublayer, which includes MAC service access points (APs). The MAC layer functionalities in IEEE 802.16e-2005 specification are illustrated in Figure 2.4.

In summary, the IEEE 802.16e-2005 specification offers improvements over the technology specified by the original fixed WiMAX standard. These significant improvements can cost-effectively deliver broadband services to end-users, offering increased performance in NLOS environments for mobility and fixed indoor applications. These improvements can be categorized as follows [4].

- *Mobility*: The support for mobility is the major feature of mobile WiMAX, which introduces new MAC for handover and allows a SS to maintain a connection when moving from one BS to another. Mobile WiMAX is designed to support mobility applications up to 160 km/h.
- *High availability*: High connection availability in NLOS environments can be supported in mobile WiMAX by using advanced antenna, channel coding, sub-channelization, and dynamic modulation technologies to increase link budget.
- *NLOS performance*: New technologies have been introduced in mobile WiMAX. These include support for intelligent antenna technology, such as

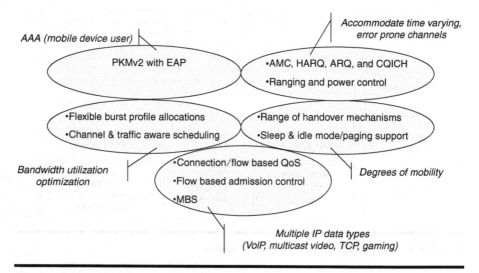

Figure 2.4 Medium Access Control (MAC) layer functionalities in the IEEE 802.16e-2005 specification.

Multiple-Input Multiple-Output (MIMO) and adaptive antenna system (AAS), high-performance coding, such as turbo coding (TC), and a Hybrid Automatic Repeat reQuest (HARQ) mechanism for increasing NLOS performance.

■ *Security*: Based on the security features of the fixed WiMAX standard, the mobile WiMAX specification introduces a number of enhancements. For example, the AES as well as 3DES are now a mandatory feature. New high-performance coding schemes, such as TC and low-density parity check (LDPC), are included. These features enhance the security of the mobile WiMAX air interface.

■ *QoS*: Both the connection and service-type-based QoS are designed to meet the requirements of mobile broadband services. These two QoS mechanisms manage both UL and DL directions and support two-way traffic, such as VoIP. The mobile WiMAX QoS has the features of service multiplexing, low data latency, and varying granularity to support real-time broadband multimedia applications.

2.3   WiMAX Mobility Management

The mobility feature in IEEE 802.16e-2005 specification refers to a MS changing its point of attachment when remaining connected to the network. Such a process of handling the change of point of attachment of a MS is called handover. Next, we describe how to support handover in mobile WiMAX.

2.3.1 Network Entry and Initialization

A SS needs to successfully complete the network entry process with a desired BS to join the network. The network entry process is composed of four stages. The first stage is capability negotiation. After successful completion of initial ranging, the SS will request the BS to describe its available modulation capability, coding schemes, and duplexing methods. During this stage, the SS shall acquire a DL channel. Once the SS finds a DL channel and synchronizes with the BS at the PHY level, the MAC layer will look for DCD (downlink channel descriptor) and UCD (uplink channel descriptor) to get modulation and other parameters. The SS remains in synchronization with the BS as long as it continues to receive the DL-medium access protocol (MAP) and DCD messages. Finally, the SS will receive a set of transmission parameters from UCD as its UL channel. If no UL channel can be found after a suitable timeout period, the SS shall continue scanning to find another DL channel. Once the UL parameters are obtained, the SS shall perform the ranging process.

The second stage is authentication. In this stage, the BS authenticates and authorizes the SS. Then the BS performs key exchange with the SS, such that the provided keys can enable the ciphering of transmission data. The third stage is registration. To register with the network, the SS and the BS will exchange registration request/response messages. The last stage is to establish IP connectivity. The SS gets its IP address and other parameters to establish IP connectivity. After this step, operational parameters can be transferred and connections can be set up.

2.3.2 Ranging Process

Ranging is the process of acquiring the correct timing offset and power adjustments such that the SS's transmission is aligned to the BS's timing. Two types are supported for ranging process in the IEEE 802.16e-2005 specification. One is initial ranging, and the other is periodic ranging.

Initial ranging is performed during network initialization and registration/reregistration to allocate CDMA codes in UL ranging opportunities. Then the SS is allowed to join the network to acquire correct TX parameters (timing offset and TX power level). On the other hand, periodic ranging is performed when transmission is on-going on a periodic basis. It uses regular UL burst to allow SS to adjust TX parameters so that the SS can maintain UL communications with the BS.

2.3.3 Handover Process

According to the scope of node movement, mobility can be divided into micro-mobility and macro-mobility. On the link layer, most access networks provide mobility by having an access router keep track of the specific AP to which a MS is attached. The localized mobility between pico-cells (probably heterogeneous cells) in the same subnet and the mobility between subnets in one domain is called micro-mobility, whereas the mobility between domains in wide-area wireless networks is called macro-mobility. The mobility solutions like Mobile IP are classified as macro-mobility. But Mobile IP is not suitable for micro-mobility due to its signaling overhead, handover latency, and transient packet loss.

2.3.3.1 Hard Handover and Soft Handover

Hard handover is mandatory to be supported in mobile WiMAX networks. Hence, break-before-make operations may happen during the handover process. In other words, link disconnection may occur and throughput may degrade. Therefore, various levels of optimization are demanded to reduce association and connection establishment with the target BS. These optimization methods are not clearly defined in the IEEE 802.16e specification, so they should be supported on specific WiMAX systems and products.

On the contrary, soft handover is optional in mobile WiMAX networks. Two schemes, macro-diversity handover (MDHO) and fast Base Station switching (FBSS) are supported. In case of MDHO, MS receives from multiple BSs simultaneously during handover, and chooses one as its target BS. As for FBSS, the MS receives from/transmits to one of several BSs (determined on a frame-by-frame basis) during handover, such that the MS can omit the decision process of selecting the target BS to shorten the latency of handover.

2.3.3.2 MAC Layer Handover Procedure

The handover procedure in IEEE 802.16e-2005 is divided into MAC- and PHY-layer handover. Looking at the MAC-layer handover procedure, it is divided into the network topology acquisition phase and the handover process phase according to its performing sequence.

In the network topology acquisition phase, as illustrated in Figure 2.5, three functions are performed, namely network topology advertisement, MS scanning for neighboring BSs, and association procedure. After receiving a neighbor advertisement message broadcast from the serving BS, the MS gets all the neighboring BSs of its current serving BS. The MS can then perform synchronization with each neighboring BS, and then continue to the handover process phase.

During the handover procedure, the process includes handover decision, handover initiation, and ranging procedures, followed by authorization and registration procedures. These procedures include cell reselection, handover decision and handover initiation, synchronization with new DL, acquisition of UL parameters, ranging, MS reauthorization, reregistration, and termination with the serving BS. These are shown in Figures 2.6 and 2.7.

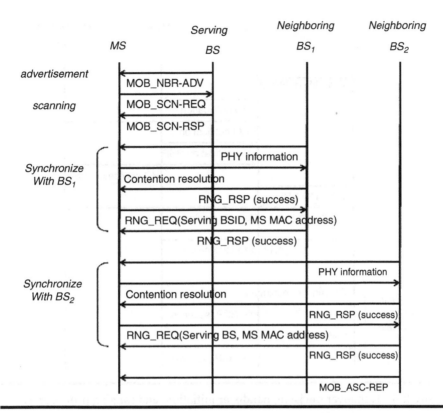

Figure 2.5   Network topology acquisition phase for handover.

When the MS migrates from its serving BS to its target BS, the following process is executed. First, the MS conducts cell reselection based on the information obtained from the network topology acquisition stage. The handover decision and the handover initiation can be originated by both MS and BS using the *MOB_MSHO-REQ/ MOB_BSHO-REQ* message. When the target BS is decided, the MS sends a *MOB_HO-IND* message to the serving BS and the actual handover process begins as illustrated in Figure 2.6.

In the ranging process, the MS can synchronize to the DL of the target BS and obtain DL and UL parameters using the *DCD/UCD* message. Then *RNG_REG/ RNG_RSP* messages are exchanged to complete the initial ranging process. It may be done in a contention-based or non-contention-based manner.

If the *RNG_REG* contains the serving BSID, the target BS can obtain the MS information from the serving BS through the backbone network. If the MS is already associated with the target BS at the previous stage, some steps may be omitted. Therefore, the neighboring BS scanning and association should be done right after the handover initiation by utilizing preobtained information before the channel condition changes.

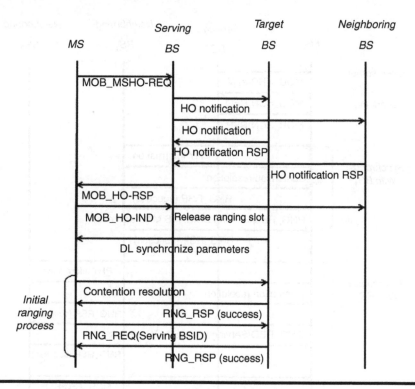

Figure 2.6 Handover decision, handover initiation, and ranging procedures.

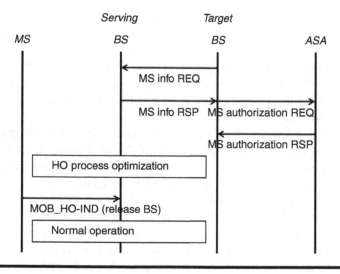

Figure 2.7 Authorization and registration procedure.

If all physical parameter adjustments are done successfully, the network re-entry process is initiated. Figure 2.7 shows this procedure. It includes MS authorization and new BS registration. The target BS requests MS authorization information via its backbone network. The new BS registration is performed by *REG_REQ* and *REG_RSP* messages. This includes capabilities negotiation, MS authorization, key exchange, and registration. After successful registration with the target BS, the MS can send a *MOB_HO-IND* message to the serving BS to indicate that handover is completed.

2.3.4 Mobile Multi-Hop Relay Network

The Mobile Multi-Hop Relay (MMR), which is a new initiative of IEEE 802.16, was kicked off by the IEEE 802.16j subtask group in March 2006. The purpose of MMR is to propose and develop relay mode for fixed/mobile WiMAX terminals. It enhances the normal frame structure at the PHY layer and adds new protocols for relay networking.

In the *status quo* of the MMR initiative, two architectures are proposed. One is a tree of point-to-multipoint (PMP) network between BS and SS, whereas the other is the hybrid PMP-mesh architecture with a mesh network between BS and SS. It is proposed to develop the concept of Relay Station (RS) with the following capabilities. First, it can enhance the network throughput and extend the network coverage with simple implement, at low cost, and with ease of deployment. Second, RS should be compatible with 802.16e (OFDMA) specification and capable of supporting end-to-end QoS. Last, mobile RS should be self-organized, not just with amplify-and-forward or decode-and-forward capabilities. That is, a RS has the intelligence to schedule transmission for cooperative communications. However, a RS has a shorter radio range as well as a lower transmission power, and needs to handle less traffic, which is different from BS in MMR.

The pros and cons of the PMP-tree relay architecture and the PMP-mesh relay architecture are summarized in Table 2.1. In summary, the PMP-mesh architecture

Table 2.1   Comparison of PMP-Tree and PMP-Mesh Architectures

	PMP-Tree	PMP-Mesh
Pros	Better efficiency and control when the depth of tree is small No hidden-node and exposed-node problems	Robust, no single point of failure More flexible in allocation of resources
Cons	Single point of failure	More control messages required, less efficient More costly

PMP, point-to-multipoint.

is more flexible than the PMP-tree architecture. The former is more suitable for robust multi-hop communications, and can easily adapt to network topology change. Therefore, it enables peer-to-peer communication between RSs.

Although the PMP-mesh architecture is advocated as a flexible and efficient approach for MMR WiMAX networks, some open problems still have not been solved yet. To name some examples, the reliability of establishing communication links between BS and MS, the performance of scheduling UL and DL bandwidths on relay links, and the practical usage scenarios and their technical requirements still need to be addressed.

2.4   Techniques for Integrating Mobile WiMAX and WiFi Networks

2.4.1   Service Scenarios

There are several mobile usage scenarios of WiFi- and WiMAX-integrated networks. The potential application targets include public safety market, energy/utility market, intelligent transportation market, network provisioning market, entertainment, and military [4]. These usage scenarios and their service requirements can be divided into five categories as shown in Figure 2.8. One important design issue is to establish connection links smoothly for mobile devices to satisfy their application requirements when they move around in such a heterogeneous environment.

CLASS DESCRIPTION	REAL TIME?	APPLICATION	BANDWIDTH
Information technology	No	Instant messaging	<250 byte messages
		Web browsing	>500 kbps
		e-mail (attachment)	>500 kbps
Media content download (store and forward)	No	Bulk data, movie download	>1 Mbps
		Peer-to-peer	>500 kbps
Interactive gaming	Yes	Interactive gaming	5 – 128 kbps
VoIP, video conference	Yes	VoIP	4 – 64 kbps
		Video phone	32 – 384 kbps
Streaming media	Yes	Music/speech	5 – 128 kbps
		Video clips	20 – 384 kbps
		Movie streaming	> 2 Mbps

Figure 2.8   Categorized services of WiFi- and WiMAX-integrated networks.

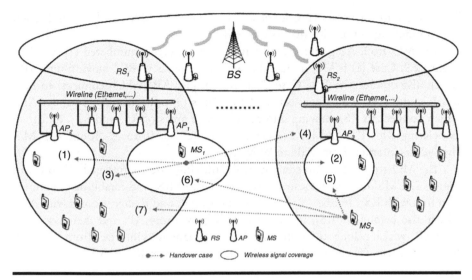

Figure 2.9   Architecture of the WiFi- and WiMAX-integrated network.

2.4.2   Network Architecture

The architecture of a WiFi- and WiMAX-integrated network is shown in Figure 2.9. Mobile devices are all dual-mode devices with a WiFi and a WiMAX interfaces. Each WiMAX BS is connected to multiple WiMAX RSs. Specifically, there are four kinds of devices: WiMAX BS, WiMAX RS, WiFi AP, and WiFi/WiMAX dual-mode MS. The operational functions of these devices in the WiFi- and WiMAX-integrated network are described as follows:

1. There may be several RSs in the coverage area of one WiMAX BS. Each RS can establish a communication link with its BS and can serve several MSs in its signal coverage. In addition, a RS can connect to multiple APs through wirelines (such as Ethernet, Fiber, etc.), that is, this RS serves as the access network of these APs.
2. A dual-mode MS may connect to an AP or establish a communication link with a RS. To save energy, a MS typically prefers to turn on its 802.11b interface rather than its 802.16e interface, unless when limited by bandwidth constraint. Therefore, a MS normally works on WiFi mode, and switches to WiMAX mode when it is unable to find any AP in its neighborhood.

2.4.3   Mobility Patterns

The mobility patterns of a MS in a WiFi- and WiMAX-integrated network can be categorized into two groups according to the MS's presence before and after moving.

The first group is a MS moving out of an AP (MO-from-AP), and the second group is a MS moving out of a RS (MO-from-RS).

The MO-from-AP group contains four moving cases, numbered (1) to (4) in Figure 2.9. Case (1) is MS_1 under AP_1, which connects to RS_1, moving into AP_2, which also connects to RS_1. Case (2) is MS_1 under AP_1, which connects to RS_1, moving into AP_3, which connects to RS_2. Cases (3) and (4) are MS_1 under AP_1, which connect to RS_1, moving into the coverage areas of RS_1 and RS_2, and then associating with RS_1 and RS_2, respectively. Note that in cases (3) and (4), MS_1 changes its interface from WiFi to WiMAX.

The MO-from-RS group contains three moving cases, numbered (5) to (7) in Figure 2.9. MS_2 originally locates in the coverage of RS_2 and establishes a connection link with RS_2. In case (5), MS_2 moves into AP_3 and then associates with AP_3, which also belongs to RS_2. In case (6), MS_2 moves into AP_1 and then associates with AP_1, which belongs to a different RS_1. MS_2 changes its interface from WiMAX to WiFi in cases (5) and (6). In case (7), MS_2 moves from RS_2 to RS_1.

2.4.4   Technical Challenges for Handover

Technical challenges of handover include connectivity, location management, routing, seamless mobility, mobility context management, paging, network composition, and migration [5]. Next, we concentrate on the connectivity requirements. The radio link condition can be one of the essential events to trigger handover. So triggering process contains collecting and identifying various signal strength-related events. A MS in homogeneous networks can initiate a handover when the carrier-to-interference ratio or the received signal strength (RSS) falls below a specified threshold. This is known as horizontal handover. However, when a MS can move between WiFi and WiMAX networks, vertical handover happens.

Vertical handover has the following characteristics [6,7]. First, when a MS moves from a WiMAX network to a WiFi network, it may not know that it has entered a WiFi network. Some decision algorithms based on subjective criteria, such as energy cost, would be necessary for the handover decision. Second, in vertical handover, because the signal qualities in WiFi and WiMAX networks are heterogeneous, these signals cannot be compared directly. Finally, the metrics for making a handover decision may involve not only the PHY layer parameters, but also the network conditions from upper layers, for example, available bandwidth and delay, and user preference.

Based on the aforementioned observations, we list some design requirements for handover between WiFi and WiMAX networks:

■ Reducing unnecessary handovers to avoid overloading the network with signaling traffic.
■ Maximizing the network utilization and minimizing users' energy consumption.

- Avoiding moving into a congested network.
- Smooth and fast handover.
- Providing applications with the required degrees of QoS.

2.4.5 Energy-Aware Vertical Handover

Next, we propose a vertical handover scheme, which is able to efficiently use MS's energy while considering its bandwidth requirements. Our scheme is both energy- and bandwidth-aware. Let us consider the MO-from-AP mobility group. When a MS moves out of its serving AP, it has to choose another target access station. Two possible target points of attachments can be chosen: AP and RS. The handover process consists of the following phases:

- *Triggering phase*: When the MS finds that the value of RSS from its serving AP is lower than some thresholds, it will enter the next scanning neighboring AP phase.
- *Scanning neighboring AP phase*: Because the MS prefers to use the WiFi channel over the WiMAX channel for the sake of saving power, the MS will scan neighboring APs first. If an AP is found, it will go to the handover decision phase. Otherwise, the next scanning neighboring RS phase will be executed.
- *Scanning neighboring RS phase*: When the MS finds that its energy remains is above a threshold, it will turn on its WiMAX interface to scan neighboring RSs. If a RS is found, it will go to the handover decision phase. Otherwise, it will get back to the triggering phase.
- *Handover decision phase*: A decision algorithm will be applied to select a target point of attachment for handover.
- *Performing handover phase*: The MS initiates handover to the target access station, which contains re-establishment of connection links, data streams synchronization, and cross networks authentication between WiFi and WiMAX networks.

The handover process for the MO-from-RS mobility group is similar to the MO-from-AP group, except that in the triggering and scanning phases, the MS will decide to check neighboring APs only if its remaining energy is above a threshold.

Next, we will explain the details of triggering process and handover decision. Depending on the MS's current network, the triggering process splits into two branches as illustrated in Figure 2.10. The following parameters are used in the process:

- $Power_{MS}$: The remaining energy of a MS.
- $P\text{-}MO_{AP}$: The probability that a MS may move out of its current serving AP.
- T_{AP}: The probability threshold to trigger handover in the MO-from-AP mobility.

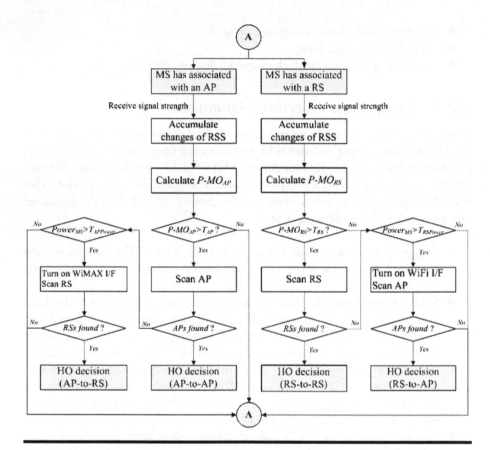

Figure 2.10 Proposed triggering process for energy-aware vertical handover.

- T_{APPower}: The power threshold for a MS to turn on its WiMAX interface in the MO-from-AP mobility case.
- $P\text{-}MO_{\mathrm{RS}}$: The probability that a MS may move out of its current serving RS.
- T_{RS}: The probability threshold to trigger handover in the MO-from-RS mobility.
- T_{RSPower}: The power threshold of a MS to turn on its WiFi interface in the MO-from-RS mobility case.

Consider the MO-from-AP mobility scenario (refer to the left branch in Fig. 2.10). An MS will periodically measure the RSS to estimate its current $P\text{-}MO_{\mathrm{AP}}$. The MS then compares the values of $P\text{-}MO_{\mathrm{AP}}$ and T_{AP}. If $P\text{-}MO_{\mathrm{AP}} > T_{\mathrm{AP}}$, the MS will scan neighboring APs. If any AP is found, the triggering process is terminated and a handover decision will be made. Otherwise, the MS will compare its $Power_{\mathrm{MS}}$ against T_{APPower}. The MS will turn on its WiMAX interface to search for neighboring RSs if $Power_{\mathrm{MS}} > T_{\mathrm{APPower}}$.

Next, consider the MO-from-RS mobility scenario (refer to the right branch in Fig. 2.10). Similarly, a MS will periodically measure the RSS to estimate its $P\text{-}MO_{RS}$. If it finds that $P\text{-}MO_{RS} > T_{RS}$, the MS will scan neighboring RSs. If any RS is found, the triggering process is terminated and a handover decision will be made. If no RS is found or $P\text{-}MO_{RS} \leqq T_{RS}$, the MS will check whether $Power_{MS}$ is above the threshold $T_{RSPower}$ or not. If so, the MS will turn on its WiFi interface to search for neighboring APs.

After this triggering, the handover decision algorithm will select a target point of attachment for the MS. There are four cases of handover, namely AP-to-RS, AP-to-AP, RS-to-RS, and RS-to-AP. The decision will be based on the scanning results. We suggest to use two cost functions, f and g, to estimate the qualities of the scanned AP and RS, respectively. The bandwidth BW_{AP_i} of AP_i can be estimated by its current contention probability (P_C), and its current workload (α): $BW_{AP_i} = f\{P_C(AP_i), \alpha(AP_i)\}$. The BW_{RS_i} of RS_i can be estimated by the density of deployed APs inside the coverage area of RS_i (D_{AP}), and the current workload of RS_i (α): $BW_{RS_i} = g\{D_{AP}(RS_i), \alpha(RS_i)\}$. Then the AP or RS with the highest BW value will be selected as the target point of attachment of the MS. One possible way to design functions f and g is to give some weights to the parameters in them. However, this still deserves further investigation.

2.5   Conclusion and Future Work

We have reviewed important aspects of the current development of WiMAX. We have also pointed out an important direction of WiFi and WiMAX integration. In the converged WiFi- and WiMAX-integrated networks, designing an efficient vertical handover scheme is critical for the overall system performance. We have proposed a bandwidth- and energy-aware vertical handover scheme, which takes into account both MSs' remaining energy and their bandwidth requirements. The proposed scheme has potential to provide ubiquitous broadband access for real-time streaming services. For future work, one direction is to apply adaptive thresholds for MS's remaining energy to balance energy and bandwidth considerations. The other direction is to consider the density of APs inside the coverage area of RS as a parameter to decide how frequently a MS should scan neighboring APs.

References

1. G. Nair, J. Chou, T. Madejski, K. Perycz, D. Putzolu, and J. Sydir, IEEE 802.16 Medium Access Control and Service Provisioning, *Intel Technology Journal*, Vol. 8, Issue 3 (2004), pp. 213–228.
2. E. Agis, H. Mitchel, S. Ovadia, S. Aissi, S. Bakshi, P. Iyer, M. Kibria, C. Rogers, and J. Tsai, Global, Interoperable Broadband Wireless Networks: Extending WiMAX Technology to Mobility, *Intel Technology Journal*, Vol. 8, Issue 3 (2004), pp. 173–187.

3. LAN/MAN Standards Committee, *IEEE Std 802.16e-2005 and IEEE Std 802.16-2004/Cor1-2005*, IEEE Computer Society and IEEE Microwave Theory and Techniques Society.

4. WiMAX Forum, *Mobile WiMAX Usage Scenarios*, WiMAX Forum (April 2006).

5. S. Hussain, Z. Hamid, and N. S. Khattak, Mobility Management Challenges and Issues in 4G Heterogeneous Networks, *Proceedings of the First International Conference on Integrated Internet Ad Hoc and Sensor Networks* (2006).

6. J. Nie, X. He, Z. Zhou, and C. L. Zhao, Benefit-Driven Handoffs between WMAN and WLAN, *IEEE Military Communications Conference* (2005).

7. A. H. Zahran, B. Liang, and A. Saleh, Signal Threshold Adaptation for Vertical Handoff in Heterogeneous Wireless Networks, *Mobile Network Application*, Vol. 11 (2006), pp. 625–640.

Chapter 3

QoS Support in IEEE 802.16-Based Broadband Wireless Networks

Hsi-Lu Chao and Yi-Bing Lin

This chapter focuses on quality of service (QoS) support in IEEE 802.16-based systems. We first introduce the WiMAX architecture, and the components of Medium Access Control (MAC) and physical (PHY) layers. Then we describe some protocols proposed to fulfill QoS support, including goodput estimation, link adaptation, connection admission control (CAC), and scheduling. In addition, we show the performances of the introduced protocols through simulation results.

3.1 Introduction

The rapid growth in demanding high-speed and high-quality multimedia communications creates opportunities and challenges for next-generation wireless network designs. Multimedia communications entail diverse QoS requirements for applications including, for example, voice and real-time video. Therefore, providing QoS-guaranteed services is essential for future wireless networks. IEEE 802.16 is a next-generation broadband wireless access (BWA) network standard, which defines PHY and MAC layers to support multiple services with point-to-multi-point (PMP) BWA [1,2]. IEEE 802.16 is broadly applicable to systems operating between 2 and 66 GHz and is capable of supporting large capacity, high-data rate, and advanced multimedia services. However, some QoS-related issues such as bandwidth allocation (BA), CAC, and scheduling are still open to IEEE 802. 16-based wireless systems.

In this chapter, we introduce general concepts of PHY and MAC layers, and then explore protocols designed to achieve QoS guarantee.

3.2 IEEE 802.16 Architecture

The IEEE 802.16 architecture includes the following advantages:

1. *Flexibility*: The MAC defined in IEEE 802.16 is capable of working with multiple PHY technologies.
2. *Modularity*: Both IEEE 802.16 MAC and PHY have a set of mandatory and optional features for fixed and mobile configurations. The optional features are negotiable between Base Stations (BSs) and Subscriber Stations (SSs).
3. *Versatility*: The IEEE 802.16 MAC supports various payload types by defining several convergence sublayers (CSs) and a standard interface between the core MAC and CSs.
4. *Subscriber-level PHY adaptation*: The IEEE 802.16 PHY allows a variety of parameters setups for an SS to adaptively select a suitable service level.
5. *QoS capability*: The IEEE 802.16 defines multiple QoS service types. For each service type, several parameters (such as minimum reserved traffic rate and maximum sustained traffic rate) are used to define the behavior of a specific flow.

3.2.1   Reference Model

The IEEE 802.16 standard defines both the MAC and PHY layers through the data and the management planes. The data plane defines how information is encapsulated/decapsulated in the MAC layer and modulated/demodulated by the PHY layer. A set of control functions supports various configurations and operations. The management plane is responsible for classification, security, QoS, and connection setup. Figure 3.1 shows the IEEE 802.16 reference model. There are three sublayers in IEEE 802.16 MAC layer: service-specific CS (Fig. 3.1a), the MAC common part sublayer (MAC CPS; Fig. 3.1b), and the security sublayer (Fig. 3.1c).

A service-specific CS performs functions that are specific to the higher layer protocols it supports. The IEEE 802.16 currently supports a set of CSs to interface with IP, Ethernet, and asynchronous transfer mode (ATM) protocol layers.

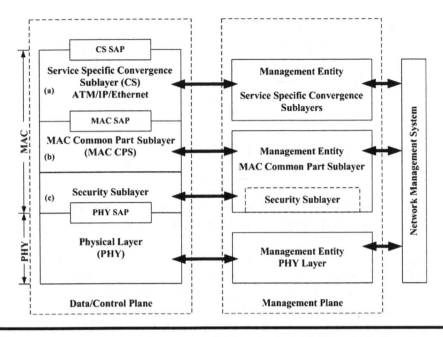

Figure 3.1 IEEE 802.16 reference model for physical (PHY) and Medium Access Control (MAC) layers.

The MAC CPS provides the medium access, connection management, and QoS functions that are independent of specific CSs. In particular, the MAC CPS receives MAC service data units (MSDUs) from the CS and transforms them into MAC protocol data units (MPDUs).

The security sublayer provides privacy and protection through encryption, decryption, and authentication. Details will be elaborated in Section 3.3.

The MAC CPS interfaces with several PHYs through the PHY service access point (SAP). The MAC CPS may receive MSDUs from multiple MAC CSs. Each MAC CPS instance is expected to support only one specific PHY implementation.

The management plane consists of four management entities corresponding to the CS, CPS, security sublayer, and the PHY. The IEEE 802.16 does not specify the details of the management plane. However, specific interfaces and messages may be discussed and then standardized to support management functions in IEEE 802.16 amendments. For example, IEEE 802.16g [3] amendment emphasizes on the procedures and services of the management plane.

3.2.2 Frame Structure

IEEE 802.16 supports both time division duplex (TDD) and frequency division duplex (FDD). As shown in Figure 3.2, a TDD frame is divided into a downlink

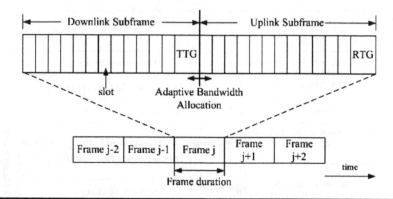

Figure 3.2 IEEE 802.16 time division duplex (TDD) frame structure.

(DL) subframe and an uplink (UL) subframe. The numbers of time slots in UL and DL subframes are adjustable on a per-frame basis according to system parameter setups (such as QoS demands of DL and UL traffic). The BS broadcasts downlink map (DL-MAP) and uplink map (UL-MAP) messages at the beginning of each frame to inform the SSs of DL and UL allocation. There is a receive-transmit transition gap (RTG) between the UL subframe and the subsequent DL subframe, and the duration of RTG depends on the selected PHY implementation. A BS will switch from the receiving mode to the transmitting mode, and a SS will switch from the transmitting mode to the receiving mode within this gap. Similarly, a transmit-receive transition gap (TTG) is required between a DL subframe and the subsequent UL subframe.

In an FDD system, the UL and DL channels are located in separate frequency bands. To simplify the BA algorithm, frame duration is fixed for both UL and DL transmissions. The system can simultaneously support full-duplex and half-duplex SSs. A full-duplex SS can transmit and receive data at the same time (Fig. 3.3a). Contrarily, a half-duplex SS can either transmit or receive data but not both (Fig. 3.3b). Therefore, the bandwidth controller shall not allocate UL bandwidth for a half-duplex SS when it is expected to receive data on the DL channel. A BS uses broadcast time period to send all SSs the DL-MAP and UL-MAP (Fig. 3.3c).

3.3 MAC Layer

The IEEE 802.16 MAC layer is similar to those in the Data Over Cable Service Interface Specification (DOCSIS) standards [3,4]. Operations defined in the IEEE 802.16 MAC include network entry and initiation, PHY maintenance, QoS service flows, security, Automatic Repeat reQuest (ARQ), and Hybrid Automatic Repeat reQuest (HARQ).

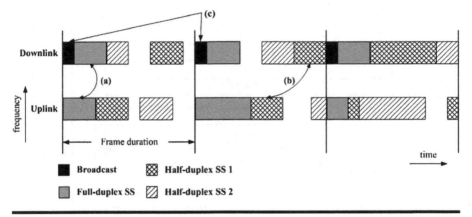

Figure 3.3 Example of bandwidth allocation in a frequency division duplex (FDD) system.

3.3.1 Security

To provide privacy, authentication, and protection, the IEEE 802.16 security sublayer defines two protocols:

1. An encryption protocol is utilized for encrypting MAC PDU payload. All MAC management messages are sent in plain text.
2. A private key management (PKM) protocol is for secure distribution of key establishment between a BS and an SS.

In IEEE 802.16, the PKM protocol is a two-tier mechanism. The SS and the BS establish a shared secret [i.e., authorization key (AK)] through the first-tier mechanism; the shared secret is then used to secure subsequent PKM exchange in the second-tier mechanism. The security tasks are achieved through PKM request (PKM-REQ) and PKM response (PKM-RSP) message exchange. PKM-REQ messages are sent from an SS to the BS; PKM-RSP is sent from the BS to the SS.

3.3.2 Network Entry and Initiation

When an SS is powered on, it starts the network entry and initiation process. Through this process, the SS obtains all required addresses and parameters used to communicate with the network. The process consists of six phases described as follows:

1. *DL channel scanning and synchronization.* As shown in Figure 3.4a, an SS first scans the possible channels of the operating DL frequency band until a valid DL signal is found. A BS shall periodically transmit downlink channel

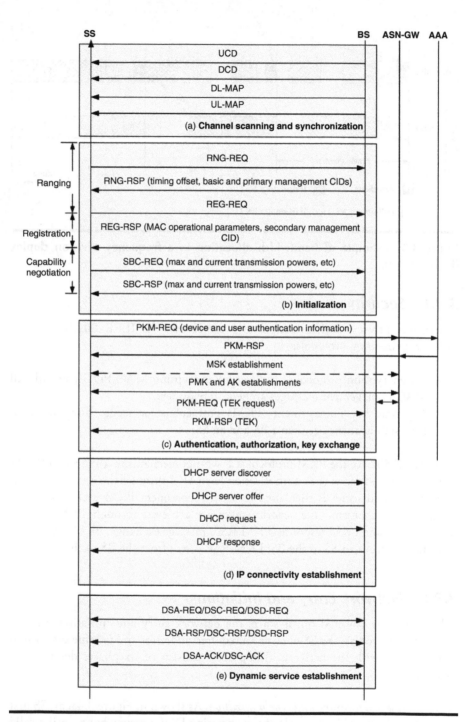

Figure 3.4   Network entry and initiation process.

descriptor (DCD) and uplink channel descriptor (UCD) messages to define the characteristics of a DL channel and an UL channel, respectively. From the received DCD and UCD messages, the SS determines the DL and the UL transmission parameters.

An SS maintains synchronization at both the PHY and the MAC layers. To maintain PHY synchronization, the SS keeps receiving DCD or UCD messages. Similarly, to achieve MAC synchronization, the SS must continue receiving the DL-MAP and DCD messages. If the DL-MAPs or DCD messages are not received for a predefined time interval, the SS should conduct the synchronization procedure again.

2. *Initial ranging and registration.* The ranging process acquires the correct timing offset and power adjustment for an SS to align transmissions. The BS allocates an initial ranging interval which consists of several time slots, and an SS will send a ranging request (RNG-REQ) to the BS at a randomly selected slot. If this RNG-REQ is correctly received, the BS replies a ranging response (RNG-RSP) to specify the timing offset and power for the SS (Fig. 3.4b). After exchanging RNG-REQ and RNG-RSP messages, two management connections identified by unique connection identifications (CIDs) are established between the BS and the SS. One is the basic management connection used to exchange short and time-urgent MAC management messages; the other is the primary management connection used to exchange longer and more delay-tolerant MAC management messages.

It is possible that two SSs send their RNG-REQs at the same slot, and the BS can decode none of the messages. In such situation, the BS will not respond, and the SSs need to resend the RNG-REQ after waiting for a randomly selected number of initial ranging slots.

An SS also needs to perform registration by sending a registration request (REG-REQ) message. The BS will authorize the registration with a response (REG-RSP) message (Fig. 3.4b). In this process, the SS and the BS can further negotiate operational MAC parameters and an optional secondary management CID. Besides the primary management connection, a secondary management connection is used to transfer standard-based protocol messages by the BS and the SS. Examples of standard-based protocols are dynamic host configuration protocol (DHCP) and trivial file transfer protocol (TFTP).

3. *Capability negotiation.* After adjusting timing and power, the SS and the BS need to negotiate optional parameters including maximum transmit power, modulation schemes, forward error correction (FEC) codes, and BA schemes. An SS informs the BS of its basic capabilities by sending an SS basic capability request (SBC-REQ). After successful decoding of that message, the BS replies with an SS basic capability response (SBC-RSP) to the SS (Fig. 3.4b).

4. *Authorization, security association (SA) establishment, and key exchange.* An SS starts the device authorization process by sending an authentication information message, which contains the SS manufacturer's X.509 [4] certificate,

to the BS (Fig. 3.4c). After passing the authentication, the SS and the authentication, authorization, and accounting (AAA) server independently create a same master session key (MSK) to proceed to the authorization process, and this MSK is further sent from the AAA server to the authenticator [i.e., the access service network gateway (ASN-GW)]. Upon the receipt of MSK, the ASN-GW generates a pair-wise master key (PMK). Together with both MSK and PMK and other information, the SS and the ASN-GW independently generate an authentication key (AK). The ASN-GW then sends the AK to the BS. Based on the generated AK, the SS further sends a digitally signed key request message to the BS to acquire a traffic encryption key (TEK) for each SA. After receiving the key reply message sent from the BS, the SS can start the next steps.

5. *IP connectivity establishment.* As shown in Figure 3.4d, through the DHCP mechanism, an SS obtains an IP address and other parameters needed to establish IP connectivity. The SS first broadcasts a DHCP request. The BS then offers the SS a list of DHCP servers. After choosing a DHCP server, the SS sends a DHCP request message to that server. The IP connectivity is established when the SS receives the corresponding DHCP response message.

6. *Dynamic service establishment.* Before data delivery, dynamic service establishment is exercised to activate service flows. There are three types of dynamic service establishment: dynamic service addition (DSA), dynamic service change (DSC), and dynamic service deletion (DSD). Both DSA and DSC are three-way handshakes (Fig. 3.4e). Either an SS or a BS can initiate a DSA (or a DSC) to create (or change parameters) a new (or an existing) service flow. An SS or a BS deletes an existing service flow by sending a DSD-REQ message, and a DSD-RSP message is generated in response to a received DSD-REQ message.

3.3.3   Scheduling Services

Scheduling services represent data handling mechanisms supported in IEEE 802.16 MAC. To support QoS, IEEE 802.16 MAC utilizes two types of scheduling: outbound transmission scheduling and UL request/grant scheduling.

1. Outbound transmission scheduling is utilized by an SS or a BS to select a data for UL or DL transmission in a particular frame. The SS and the BS can choose a same or different transmission scheduling mechanisms.

2. UL request/grant scheduling is performed at a BS to provide each subordinate SS with bandwidth for UL transmissions or opportunities to send bandwidth-request (BW-REQ) messages. A service flow's service class and QoS parameters are carried in the BW-REQ. Based on the received message, the BS provides polls or grants at the appropriate times.

In IEEE 802.16, each connection is associated with one of five service classes:

1. Unsolicited Grant Service (UGS) supports constant bit-rate (CBR) or CBR-like flows such as VoIP. These applications require constant BA.
2. Real-Time Polling Service (rtPS) supports real-time variable bit-rate (VBR)-like flows such as Moving Picture Experts Group (MPEG) video, or teleconference. These applications have specific bandwidth requirements as well as a deadline (maximum delay). Late packets that miss the deadlines will be discarded.
3. Non-Real-Time Polling Service (nrtPS) supports non-real-time flows which require better QoS than the Best Effort (BE) service. The nrtPS applications are time-insensitive and require minimum BA. An example is bandwidth-insensitive file transfer.
4. Best Effort (BE) service supports BE traffic such as Hyper Text Transfer Protocol (HTTP). There is no QoS guarantee. An application in this service flow is allocated the available bandwidth only after the bandwidth requests from the previous three service flows have been satisfied.
5. Enhanced Real-Time Polling Service (ertPS) is defined in IEEE 802.16e to support VoIP with silence suppression or real-time services with variable bit-rates. The difference between UGS and ertPS is that in ertPS a BS provides a mechanism for an SS or a Mobile Station (MS) to dynamically change its UL allocation to improve the overall radio resource utilization.

For UGS, the amount of allocated bandwidth is fixed, and BW-REQ is not required; for rtPS class, a BS provides unicast request opportunities for an SS to send its BW-REQ at a predefined interval. In other words, the BS periodically polls the SS to allocate the UL bandwidth demand. In rtPS, an SS can transmit a BW-REQ message for the nrtPS class through polling mechanism. In addition, the SS is allowed to use contention request opportunities to send a BW-REQ message. BW-REQ messages of BE class can only be transmitted through contention request opportunities. For ertPS, BW-REQ is used for an MS to modify its UL allocation demand. The current packet queue size is represented as the bandwidth demand in a BW-REQ message. An SS transmits a BW-REQ message for each connection and the BS only replies one bandwidth-grant message for each class of the SS.

The IEEE 802.16 defines the signaling mechanism for information exchange between BS and SS (such as connection setup, BW-REQ, and UL-MAP), and the UL scheduling for UGS service flows. However, IEEE 802.16 does not define the UL scheduling for rtPS, nrtPS, and BE service flows, neither the admission control nor the traffic policing process.

Figure 3.5 shows the existing IEEE 802.16 QoS architecture, where the blocks with dashed line show the undefined parts in IEEE 802.16. An application is treated as a traffic flow (Fig. 3.5a) in this architecture. As the MAC protocol defined in IEEE 802.16 is connection oriented, each traffic flow needs to establish the

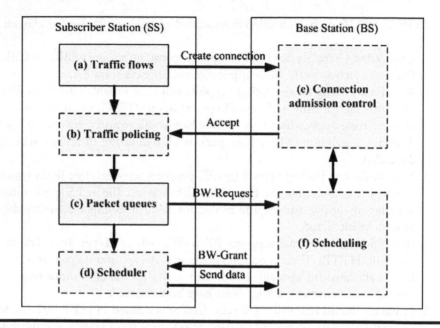

Figure 3.5 IEEE 802.16 quality-of-service (QoS) architecture.

connection with its BS as well as the associated service flow (UGS, ertPS, rtPS, nrtPS, or BE), and specifies its QoS demand through parameters such as minimum reserved rate, maximum sustained rate, delay, and jitter. The BS performs CAC (Fig. 3.5e) mechanism to ensure that the QoS of the new connection is guaranteed. If the new connection can be accepted, the BS assigns it a unique CID. After obtaining a valid CID, the connection starts to generate packets and these packets are processed through traffic policing (Fig. 3.5b) to fit the agreed QoS profile. All packets are then classified based on their CIDs and forwarded to the appropriate packet queues (Fig. 3.5c). At the SS, the scheduler (Fig. 3.5d) will retrieve the packets from the queues and transmit them to the network in the time slots specified by the UL-MAP sent from the BS to the SS. The UL-MAP is determined by the BS's scheduler (Fig. 3.5f) based on the BW-REQ messages that report the current queue size of each connection in the SS.

3.4 PHY Layer

There are four PHY specifications for IEEE 802.16 BWA systems, including WirelessMAN-SC PHY, WirelessMAN-SCa PHY, WirelessMAN-OFDM PHY (OFDM-PHY), and WirelessMAN-OFDMA PHY [1,2]. In this chapter, we focus on WirelessMAN OFDM PHY that is based on orthogonal frequency division multiplexing (OFDM) modulation and designed for non-line-of-sight (NLOS) operation in frequency bands below 11 GHz.

Table 3.1   Seven PHY Modes of IEEE 802.16 OFDM-PHY

Mode	Data Rate (Mbps)	Modulation	Bytes Per Symbol	Overall Code Rate	RS Code	CC Code Rate
1	3.9544	BPSK	12	1/2	(12,12,0)	1/2
2	7.9096	QPSK	24	1/2	(32,24,4)	2/3
3	11.8648	QPSK	36	3/4	(40,36,2)	5/6
4	15.82	16-QAM	48	1/2	(64,48,8)	2/3
5	23.7304	16-QAM	72	3/4	(80,72,4)	5/6
6	31.6408	64-QAM	96	2/3	(108,96,6)	3/4
7	35.596	64-QAM	108	3/4	(120,108,6)	5/6

Abbreviations: PHY, physical; OFDM, orthogonal frequency division multiplexing; RS, Reed-Solomon; CC, convolutional code.

To support link adaptation, the IEEE 802.16 OFDM-PHY provides seven mandatory PHY modes with various modulation schemes and coding rates. Details of these PHY modes are listed in Table 3.1.

In OFDM-PHY, the forward error control (FEC) consists of a Reed-Solomon (RS) outer code and a rate-compatible convolutional inner code. The block diagram of FEC operations is shown in Figure 3.6, where input data bits are first coded in sequence by the RS encoder and convolutional-code encoder (CC encoder). After interleaving, these encoded data frames are transmitted on the radio channel with an error probability P_e. At the receiving side, a reverse process (including deinterleaving and decoding) is executed to obtain the original data bits. As the deinterleaving process only changes the order of received data, the error probability is intact. When passing through the CC-decoder and the

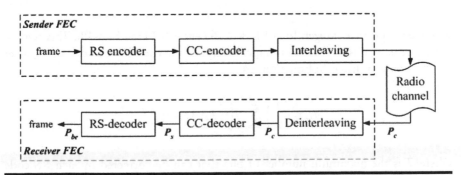

Figure 3.6   Block diagram of forward error correction (FEC) operations.

RS-decoder, some errors may be corrected, which results in lower error rates, indicated as P_s and P_{be}, respectively. These two measures can be used to estimate the wireless link's goodput.

The goodput is the expected bandwidth excluding all overheads (such as MAC/PHY overheads, acknowledgments, and retransmissions). The goodput of IEEE 802.16 OFDM-PHY can be investigated through the bit error probability in the additive white Gaussian noise (AWGN) channel model. Based on this bit error rate, the corresponding bit error probability P_{cb} after CC-decode processing can be approximated as [5–7]

$$P_{cb} \approx \left(\frac{1}{a}\right) \sum_{k=d_f}^{d_f+N} \left(c_k \times P_k\right),$$

(3.1)

where a is the number of bits input into the CC-encoder, d_f is the free distance of the convolutional code in a specific PHY mode, c_k is the number of error bits that occur in all incorrect paths in the trellis that differ from the correct path in exactly k positions, and P_k is the probability that an incorrect path at distance k from the correct path is chosen by the Viterbi decoder. From [5], P_k is expressed as

$$P_k = \begin{cases} \displaystyle\sum_{i=(k+1)/2}^{k} \binom{k}{i} P_c^i \left(1-P_c\right)^{k-i}, & \text{if } k \text{ is odd} \\ \\ \displaystyle\frac{1}{2}\binom{k}{k/2} P_c^{k/2} \left(1-P_c\right)^{k/2} + \sum_{i=k/2+1}^{k} \binom{k}{i} P_c^i \left(1-P_c\right)^{k-i}, & \text{if } k \text{ is even.} \end{cases}$$

(3.2)

An upper bound of the RS symbol error rate before RS-decoding for symbol in Galois Field, $F(2^b)$ is given as

$$P_S \le b P_{cb},$$

(3.3)

where b is a positive integer. In IEEE 802.16 OFDM–PHY, $b = 8$ [8]. This symbol error rate is used to calculate an upper bound of the overall bit error rate P_{be} after RS-decoding, as listed in Equation 3.4.

$$P_{be} \le \left(\frac{1}{n}\right) \sum_{i=T+1}^{n} i\binom{n}{i} P_s^i \left(1-P_s\right)^{n-i},$$

(3.4)

where T is the maximum number of bits that can be corrected in each RS code block, and n is the number of bits in each block.

We can use Equation 3.4 to estimate packet error probability P_{pe} as

$$P_{pe} = 1 - (1-P_{be})B,\tag{3.5}$$

where B is the packet size.

For PHY mode m, the estimated goodput G_m is expressed in Equation 3.6 by using Equation 3.5:

$$G_m = \left(\left\lfloor \left(S_{m,\text{frame\_UL}} - S_{m,\text{control\_UL}}\right) \times l_m / B \right\rfloor + \right.$$

$$\left. \left\lfloor \left(S_{m,\text{frame\_DL}} - S_{m,\text{control\_DL}}\right) \times l_m / B \right\rfloor \right) \times \left(1 - P_{pe}\right) \times \frac{B_L}{T_{\text{frame}}}.\tag{3.6}$$

In Equation 3.6, $S_{m,\text{frame\_UL}}$ and $S_{m,\text{frame\_DL}}$ represent the numbers of symbols of an UL and a DL subframes at PHY mode m, respectively. Similarly, $S_{m,\text{control\_UL}}$ and $S_{m,\text{control\_DL}}$ are the numbers of symbols of UL and DL control overheads at PHY mode m. T_{frame} is the frame duration and B_L is the payload size within a packet, and l_m is the number of bytes per symbol of PHY mode m. The simulated error performances for all PHY modes are shown in Figure 3.7. The x- and y-axes are the measured Signal-to-Noise Ratios (SNR) (denoted as E_b/N_0) and the packet error rate (PER), respectively. As expected, for the same PER value, high-data-rate PHY

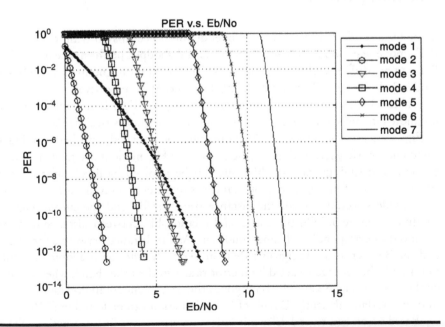

Figure 3.7 Error performance in the additive white Gaussian noise (AWGN) channel.

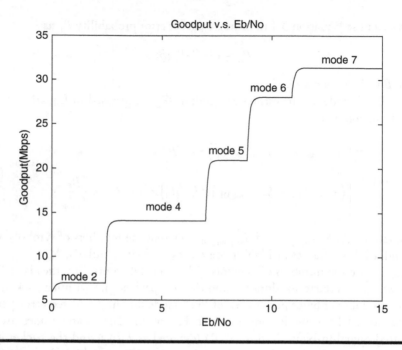

Figure 3.8 Goodput performance versus E_b/N_0 in the additive white Gaussian noise (AWGN) channel.

mode (e.g., mode 7) needs better channel quality as compared with low-data-rate PHY mode (e.g., mode 1). Figure 3.8 shows the goodput performance of OFDM-PHY upon different channel conditions. The maximum goodputs in Mbps of modes 2, 4, 5, 6, and 7 are 6.9, 14, 20.9, 27.9, and 31.3, respectively.

Table 3.1 indicates that some modulations (e.g., 64-QAM) are capable of sending more bits per symbol and thus have higher throughputs (see the mode 7 curve in Fig. 3.7). However, these modulations must be operated with better SNRs to overcome interferences and fading. Selection of PHY mode impacts on a system's goodput and QoS guarantee. This issue can be resolved by link adaptation that selects one out of multiple available transmission rates at a given time. Factors for PHY mode selection include the distance from the BS and the current channel condition. Figure 3.9 illustrates a general relationship between distance from the BS and the corresponding channel condition. As the distance between a specific SS and the BS increases, the signals fade more seriously, and lower adopted modulation should be selected to avoid high error rate. On the other hand, when the link quality between an SS and the BS is good, the SS can choose high-data-rate mode to transmit data efficiently. Details of link adaptation can be found in [9,10].

Based on an end-to-end delay minimization criterion, a link adaptation scheme was proposed in [8]. This scheme uses Transport Control Protocol (TCP) delay as the link quality indicator to determine the SNR operation ranges for all typical

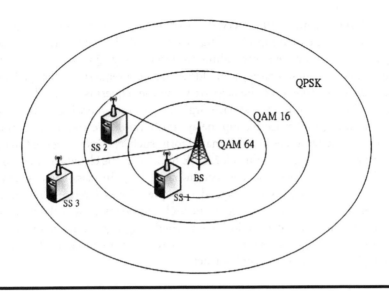

Figure 3.9 General relationship between distance and modulation.

PHY transmission modes. By setting an acceptable end-to-end delay on the AWGN channel model, the study in [11] obtained the required SNR range for each PHY mode. Therefore, based on current measured SNR value, the BS can suggest a suitable PHY mode to satisfy the QoS demand of the SS.

In [9], an SS informs the BS of a connection QoS demand by specifying PER. The BS then calculates the corresponding SNR operation range for each PHY mode. In this approach, the BS should keep measuring SNR value of the radio channel of the SS, and dynamically suggest appropriate PHY modes for the SS.

3.5 CAC and Scheduling

Based on an admission control policy in the BS, the CAC scheme ensures that the QoS of existing connections are not degraded significantly while new connection's QoS is satisfied.

In [10], a simple admission control was proposed by using the minimum reserved rate as the control criterion. The BS collects all DSA/DSC/DSD requests and updates the estimated available bandwidth (C_a) based on bandwidth change. Suppose that there are I classes of service and the ith class of service has J_i connections. The available bandwidth can be expressed as

$$C_a = C_{\text{total}} - \sum_{i=0}^{I-1} \sum_{j=0}^{J_i-1} r_{\min}(i, j), \tag{3.7}$$

where $r_{min}(i,j)$ is the minimum reserved rate of the *j*th connection in the *i*th service class, and C_{total} is the total capacity of the wireless link. For connections whose r_{min} are equal to zero, they are always admitted without any QoS guarantee. When a new service flow request arrives or an old service flow requests to change its QoS, the recalculated C_a should not be negative when the request is accepted.

Scheduling determines the service order of packets according to the selected metrics including packet timestamp, delay, and cumulative credit. It allows data flows to share the resources in a designated form such as fair or weighted share.

Existing scheduling algorithms for IEEE 802.16 can be classified into two categories: class-based [12,13] and connection-based [14]. In class-based scheduling algorithms, each service class is assigned a unique priority level, and resources are first allocated to the flows belonging to the highest priority class, followed by those with lower priorities. For connection-based scheduling algorithms, resource allocation is based on connection's satisfaction, where the most unsatisfied connection has the highest priority to transmit packets.

3.5.1 Class-Based Scheduling

To support various types of service flows (UGS, rtPS, nrtPS, and BE), an UL scheduling was introduced in [12,13]. This scheduling protocol combines Strict priority service discipline, earliest deadline first (EDF), and weight fair queue (WFQ), as illustrated in Figure 3.10. Strict priority discipline ensures that the available bandwidth is always allocated to the data flows from the highest priority to the lowest priority, that is, UGS, rtPS, nrtPS, and then BE. A major drawback of strict priority is that higher priority connections may cause bandwidth starvation for the lower priority connections. To resolve this issue, one can implement traffic policing to force every connection to follow the agreed traffic behavior.

The BS allocates bandwidth to UGS flows at periodic intervals and this allocation is sufficient to guarantee their QoS requirements. As the BS offers fixed-size grants on a periodic basis [12,13], does not design a scheduling algorithm for UGS class (Fig. 3.11a). The intraclass scheduling protocol of rtPS service class is EDF, where all queued packets are sorted based on their deadlines in the increasing order; that is, the one with the earliest deadline will be scheduled first (Fig. 3.11b). For nrtPS service class, the intraclass scheduling adopts WFQ, where resource share is based on a preset weight assigned to each connection (Fig. 3.11c). Finally, the remaining bandwidth is equally distributed to all BE connections (Fig. 3.11d).

3.5.2 Connection-Based Scheduling

A two-phase scheduling mechanism for IEEE 802.16, named two-tier slot allocation (2TSA), is introduced in [14]. The first-tier and second-tier scheduling are category-based and satisfaction degree-based, respectively. By comparing QoS

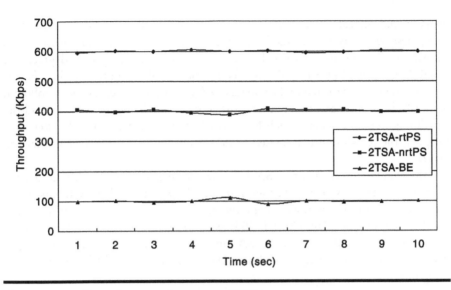

Figure 3.10 Throughput performance of 2TSA.

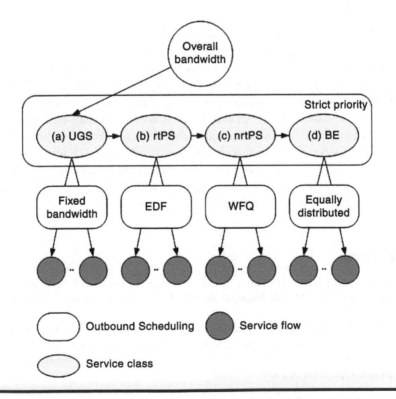

Figure 3.11 Hierarchical structure of class-based scheduling algorithm.

demand and allocated bandwidth, the BS classifies all connections into three categories: over-satisfied, satisfied, and under-satisfied. In this mechanism, each connection is associated with a dynamically calculated satisfaction degree. The scheduling priority in the first-tier adopts the concept of strict priority. Specifically, the BS first allocates the bandwidth to service flows of unsatisfied category, followed by satisfied category, and finally over-satisfied category. To perform the second-tier scheduling, the SS calculates each flow's satisfaction degree. Among the flows of a category, the scheduling is exercised based on the increasing order of satisfaction degree.

The simulation results of 2TSA are shown in Figure 3.10. In this simulation experiment, each service class has seven service flows. The available UL bandwidth is 8 Mbps. The minimum reserved rate and maximum sustained rate of rtPS, nrtPS, and BE are (500 kbps, 700 kbps), (300 kbps, 500 kbps), and (0, 200 kbps), respectively. The figure indicates that 2TSA guarantees each service's minimum QoS demand and appropriately distributes the available bandwidth to the three categories.

3.6 Conclusions

The IEEE 802.16 is a viable technology for wireless local and metropolitan area networks with high transmission rate and flexible QoS support. In particular, it is a promising alternative for last mile access for those areas where installation of a cable-based infrastructure is economically infeasible. However, the IEEE 802.16 specifications only define fundamental components such as frame structures, authorization and authentication, and dynamic service creation change and deletion. Cross-layer protocol designs are not elaborated to improve system throughput and to realize QoS support. In this chapter, we introduced CAC, scheduling goodput estimation, and link adaptation protocol to illustrate the concept of PHY/MAC cross-layer design. The core network design for IEEE 802.16 is out of the scope of this chapter, and the reader is referred to [15] for the details.

References

1. IEEE 802.16-REVd/D5-2004, "IEEE Standard for Local and Metropolitan Area Networks—Part 16: Air Interface for Fixed Broadband Wireless Access Systems," May 13, 2004.
2. IEEE 802.16e-2005, "IEEE Standard for Local and Metropolitan Area Networks—Part 16: Air Interface for Fixed and Mobile Broadband Wireless Access Systems—Amendment 2: Physical and Medium Access Control Layers for Combined Fixed and Mobile Operation in Licensed Bands and Corrigendum 1," February 28, 2006.
3. http://www.ieee802.org/16/netman/16g.html.
4. http://www.ietf.org/rfc/rfc3280.txt.

5. D. Qiao, S. Choi, and K. G. Shin, "Goodput Analysis and Link Adaptation for IEEE 802.11a Wireless LANs," *IEEE Transactions on Mobile Computing*, vol. 1, no. 4, October–December 2002, pp. 278–292.

6. D. Haccoun, "High-Rate Punctured Convolutional Codes for Viterbi and Sequential Decoding," *IEEE Transactions on Communications*, vol. 37, no. 11, November 1989, pp. 1113–1125.

7. P. Frenger, P. Orten, and T. Ottosson, "Convolutional Codes with Optimum Distance Spectrum," *IEEE Communications Letters*, vol. 3, no. 11, pp. 317–319, November 1999, pp. 317–319.

8. G. M. Chiasson and M. B. Pursley, "Concatenated Coding for Frequency-Hop Packet Radio," *IEEE Military Communications Conference*, November 1991, pp. 1235–1239.

9. Y.-T. Yu and H.-L. Chao, "Goodput Analysis and Link Adaptation for IEEE 802.16 Broadband Wireless Access Systems," accepted by the *IEEE International Symposium on Personal, Indoor and Mobile Radio Communications,* 2007.

10. J. Chen, W. Jiao, and H. Wang, "A Service Management Strategy for IEEE 802.16 Broadband Wireless Access System in TDD Mode," *IEEE International Conference on Communications*, 2006, pp. 3422–3426.

11. S. Ramachandran, C. W. Bostian, and S. F. Midkiff, "A Link Adaptation Algorithm for IEEE 802.16," *IEEE International Conference on Wireless Communications and Networking* 2005, pp. 1466–1471.

12. D.-H. Cho, J.-H. Song, M.-S. Kim, and K.-J. Han, "Performance Analysis of the IEEE 802.16 Wireless Metropolitan Area Network," *International Conference on Distributed Frameworks for Multimedia Applications*, 2005, pp. 130–136.

13. J. Chen, W. Jiao, and H. Wang, "A Service Flow Management Strategy for IEEE 802.16 Broadband Wireless Access Systems in TDD Mode," *IEEE International Conference on Communications*, 2005, pp. 3422–3426.

14. L.-F. Chan, H.-L. Chao, and Z.-T. Chou, "Two-Tier Scheduling Algorithm for Uplink Transmissions in IEEE 802.16 Broadband Wireless Access Systems," *IEEE International Conference on Wireless Communications, Networking and Mobile Computing*, 2006.

15. Y.-B. Lin and A.-C. Pang, *Wireless and Mobile All-IP Networks*, Wiley, 2005.

Chapter 4

Security in Fixed and Mobile IEEE 802.16 Networks

Subbu Ponnuswamy, Todor Cooklev, Yang Xiao, and Krishna Sumanth Velidi

IEEE 802.16 standard specifies air interfaces for fixed and mobile point to multipoint broadband wireless systems, including Medium Access Control (MAC) and physical (PHY) layers. The security sublayer of IEEE 802.16 provides subscribers with encryption and authentication services across broadband wireless networks. In addition, a digital-certificate-based Subscriber Station (SS) device authentication model provides operators with strong protection from theft of service. The security sublayer employs an authenticated client/server key management protocol in which a Base Station (BS), the server, controls distribution of keying material to clients, SSs. In this chapter, we provide an overview of the IEEE 802.16 security sublayer and detailed operation of Private Key Management (PKM) protocol versions 1 and 2.

4.1 Introduction

Security threats that can arise in any wireless network include the following:

1. An unauthorized user can get access to a network by masquerading as a legitimate user.
2. An eavesdropper can listen to transmissions and decode the information. This information may be later replayed in its original or possibly modified form.
3. An attacker can flood the network with spurious data, resulting in network overload and denial-of-service to authorized users.
4. An unauthorized user can be the man-in-the-middle, convincing part of the network that it is a client (e.g., SS), and convincing the other part of the network that it is a server (e.g., BS).

The security mechanisms in wireless networks should be designed to provide protection against these security threats. Encryption algorithms are at the foundation of many of these security mechanisms in wireless networks.

4.1.1 Encryption Methods

A message to be encrypted is called plain text and the resulting encrypted message is called cipher text. The objective of encryption is to be practically impossible for an unauthorized party to understand the contents of the cipher text. Two main categories of cryptographic systems are secret key and public key systems.

In secret key cryptographic systems, the sender uses a secret key for the encryption. The intended receiver recovers the plain text using the same key. It is the sharing of a secret (also known as shared key algorithm) that makes the communication secure. If an eavesdropper is able to reconstruct the key, this method will fail to provide security.

The Data Encryption Standard (DES) was adopted in 1977 by the National Bureau of Standards, now National Institute of Standards and Technology (NIST). DES is a shared key cryptographic algorithm. In the encryption process, DES first divides the data into blocks of 64 bits. Each block is separately encrypted into a block of 64-bit cipher text. The DES key is 56 bits. The encryption algorithm has 19 steps. Each step has a 64-bit input and produces a 64-bit output. The first step is a permutation independent of the key. The last step is a permutation that is the inverse of the initial permutation. The stage immediately before the last simply swaps the 32 bits on the left with the 32 bits on the right. Each of the remaining 16 iterations performs the same processing, but uses a different key. The key at every step is generated from the key at the previous step by applying a permutation, circular rotation, and another permutation. Of the 56 bits, only 48 bits are used for the key at all iterations. The following operations are performed during every iteration. First, the 64-bit input is divided into two equal portions, L_{i-1} and R_{i-1}. The output L_i is equal to R_{i-1}. The right part of the output R_i is the result of a bitwise exclusive-or (XOR) of the left part of the input and a function of the key at the given iteration K_i and the right part of the input, that is, $R_i = L_{i-1} \oplus f(R_{i-1}, K_i)$. This is Electronic Codebook (ECB) mode of DES. The ECB mode may not be secure when the structure of a message is known to an attacker, as is typical in communication standards. Improvements to ECB are possible. One way to improve the algorithm is at every step to XOR the current plain text block with the preceding cipher text block. The first plain text block is XORed with an initialization vector (IV). This is the Cipher Block-Chaining (CBC) mode. Another problem with DES is that the 56-bit key is not sufficiently long for protection against brute-force attacks. An improvement called triple DES (3-DES) mitigates this problem. 3-DES uses two keys and therefore the total key length becomes 112 bits. According to 3-DES cipher text C is produced from plain text P by

$$C = E_{K1}(D_{K2}(E_{K1}(P))).$$

The corresponding decryption algorithm is

$$P = D_{K1}(E_{K2}(D_{K1}(C))).$$

If the two keys are identical triple DES reduces to single DES.

In 1997 NIST announced a public contest to select the successor of DES. In 2001, the Rijandel proposal by Belgian scientists Rijmen and Daemen was selected as the Advanced Encryption Standard (AES). The Rijndael algorithm encrypts 128-bit blocks of data with keys of 128, 192, and 256 bits in length and can be implemented in a very efficient manner. AES can be used in many different modes, including ECB, CBC, Cipher Feedback (CFB), Output Feedback (OFB), and Counter (CTR).

In public key cryptography, keys are not shared between a sender and a receiver. It relies on two different keys, a public key and a private key. Encryption is performed using the public key and decryption is performed by using the private key. Privacy is enforced as only the holder of the specific private key can decrypt an encrypted message. The Ron Rivest, Adi Shamir, and Leonard Adleman (RSA) algorithm is one of the most popular public key algorithms. It is named after its inventors Rivest, Shamir, and Adleman. It is based on the fact that while it is simple to find the product of two numbers, factorization is not simple. The public and private keys are generated as follows. First, two large prime numbers p and q are chosen, where $pq = n$. Then, a number e is found, which is relatively prime to $(p-1)$ $(q-1)$, and a number d is found so that $de = 1 \bmod (p-1)(q-1)$. The public key is $\{e, n\}$, and the private key is $\{d, n\}$. This algorithm guarantees that for every number P smaller than n, $P^{de} = P \bmod n$. The encryption operation is defined as $C = P^e \bmod n$ and the decryption operation is defined as $C^d = P \bmod n$.

It should be noted that shared key algorithms are computationally simpler compared with the public key algorithms. However, key management is often complex with shared key algorithms. On the other hand, public key algorithms are computationally intensive. Public key algorithms simplify key management by requiring each user to have a secret private key and a public key that can be freely distributed. In practice, public key algorithms require certificates to verify that a given public key corresponds to a certain user.

4.1.2 Overview of IEEE 802.16

The IEEE 802.16 standard [1–3] defines air interfaces for fixed and mobile broadband wireless systems, including the MAC and PHY layers. The MAC layer is further subdivided into three sublayers, namely convergence (CS) sublayer, common part sublayer (CPS), and security sublayer as shown in Figure 4.1. The CS sublayer receives packets from various higher-layer protocols, performs higher-layer protocol-specific functions such as packet/frame classification and header suppression, and encapsulates these packets into common IEEE 802.16 MAC Service Data Unit (MAC SDU) format. The MAC SDUs are then delivered to the CPS sublayer. Currently, two such CS sublayers are defined, namely the asynchronous transfer mode (ATM) CS and packet CS. Additional CS sublayer can be easily added to the IEEE 802.16 MAC. For example the 802.16g amendment, currently under development, is adding a generic packet convergence sublayer (GPCS) that is independent of the higher-layer protocols, by defining a generic interface and allowing protocol-dependent functions to be performed outside the MAC CS.

The CPS layer performs functions such as connection establishment, connection maintenance, media access, and quality of service (QoS) enforcement [4]. The security sublayer provides authentication, privacy, and confidentiality services by

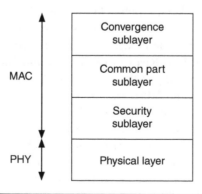

Figure 4.1   IEEE 802.16 sublayers.

authenticating SS or BS and encrypting connections between SS and BS across the broadband wireless network.

4.1.3 Acronyms and Definitions

In this subsection, we briefly discuss some of the commonly used acronyms and definitions in this chapter.

- *Downlink (DL) and uplink (UL)*: It should be noted that the IEEE 802.16 defines a framed PHY, where fixed sized frames that repeat at regular intervals are used. If frequency division duplexing (FDD) is used, a PHY frame may only carry information from the BS to the SS or vice versa. If time division duplexing (TDD) is used, a PHY frame may carry information from BS to SS and vice versa. The term DL refers to the direction from the BS to the SS in fixed and mobile 802.16 wireless networks. Similarly, the term UL refers to direction from the SS to the BS. When TDD is used, a PHY frame may be subdivided into DL and UL subframes, indicating the DL and UL portions, respectively.
- *DL-MAP and UL-MAP*: IEEE 802.16 is a centrally managed system, where the DL and UL transmissions are allocated and scheduled by the BS. The DL-MAP describes the DL allocations in a PHY frame so that the receiving SSs can properly decode the DL portion of the frames. Similarly, the UL-MAP describes UL allocations so that the SSs can transmit at their allocated time.
- *Finite state machine (FSM)*: The term FSM is used in the traditional sense to denote various state machines used to represent and implement various parts of the system. An FSM consists of a finite number of states, state transitions, and specific actions (during entry, exit, input, and transition) at each state.
- *REQ/RSP messages:* The IEEE 802.16 MAC defines a set of MAC management messages to support various MAC and PHY management functions that require exchanging messages between BS and SS. Each MAC management message is referred to by an abbreviated name. Some of the management messages are used for request–response protocols and the suffixes REQ and RSP are used to refer request and response messages, respectively. For example, the ranging request message is referred to as RNG-REQ.

4.2 Security Sublayer Overview

The architecture of the security sublayer is shown in Figure 4.2. The security model of IEEE 802.16 is built around two major components, an encapsulation method and a key management protocol.

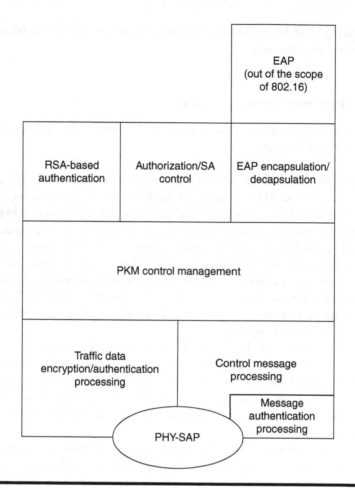

Figure 4.2 Security sublayer architecture.

The encapsulation method defines the encryption and authentication methods. These methods are defined in the form of cryptographic suites, where a suite is a pairing of an encryption algorithm and an authentication algorithm. The rules for applying these algorithms to the MAC Protocol Data Unit (PDU) are also defined by the encapsulation method.

In the IEEE 802.16 authenticated client/server key management model, the logical BS acts as a server for the keying material. The PKM protocol defines methods for secure distribution of keys from the server (BS) to all clients (SSs) and synchronization of keys between the BS and SS. The BS also uses the PKM protocols to enforce conditional access to the network services. Currently, two versions of the PKM protocol, PKMv1 and PKMv2 are defined in the IEEE 802.16 standard.

4.2.1 IEEE 802.16 Encryption Overview

Encryption is only applied to the MAC PDU payload and the generic MAC header is not encrypted. It should be noted that in the IEEE 802.16 MAC, the MAC sub-headers are considered part of the MAC PDU payload for encryption/decryption purposes. The generic MAC header carries information specific to the encapsulation method being used, in the encryption control and encryption key sequence fields. The MAC management messages are not encrypted.

The encryption capabilities of the SSs are communicated during the registration and connection setup process and the BS determines whether to allow a specific encryption for the connection based on the supported cryptographic suites and other provisioned parameters. The one-bit encryption control field of the generic MAC header indicates if the MAC PDU that follows is encrypted or not.

The encryption modes supported in IEEE 802.16 are summarized in Table 4.1. The AES CTR and AES CBC modes were added as part of the IEEE 802.16e amendment. The AES CTR mode is used for encrypting Multicast Broadcast Services (MBS).

The supported cryptographic suites are shown in Table 4.2. A cryptographic suite defines the data encryption method, data authentication method, and the method for encrypting Traffic Encryption Keys (TEK).

4.2.2 PKM Protocol Overview

The PKM protocol is used for multiple purposes including, initial key distribution, key refresh, key synchronization, authentication, and reauthentication. The authentication protocol establishes a share secret between the BS and the SS, known

Table 4.1 Supported Encryption Modes in IEEE 802.16

Encryption Mode	Reference
DES in CBC mode	DES algorithm [FIPS 46-3, FIPS 74, FIPS81]
AES in CCM mode	AES algorithm [NIST Special Publication 800-38C, FIPS-197]
AES in CTR mode	AES algorithm [NIST Special Publication 800-38A, FIPS 197, RFC 3686]
AES in CBC mode	AES algorithm [NIST Special Publication 800-38A, FIPS 197, RFC 3686]

Abbreviations: DES, Data Encryption Standard; AES, Advanced Encryption Standard; CBC, Cipher Block-Chaining; CCM, CTR and CBC modes; CTR, Counter; NIST, National Institute of Standards and Technology.

Table 4.2 **Supported Cryptographic Suites in 802.16**

Data Encryption/Key Length (Bits)	Data Authentication	TEK Encryption/Key Length (Bits)
None	None	3-DES/128
DES CBC/56	None	3-DES/128
None	None	RSA/1024
DES CBC/56	None	RSA/1024
AES CCM/128	AES CCM/128	AES ECB/128
AES CCM/128	AES CCM/128	AES key wrap/128
AES CBC/128	None	AES ECB/128
AES CTR/128	None	AES ECB/128
AES CTR/128	None	AES key wrap/128

Abbreviations: DES, Data Encryption Standard; AES, Advanced Encryption Standard; CBC, Cipher Block-Chaining; CTR, Counter; ECB, Electronic Codebook.

as the Authorization Key (AK). Once this AK is established, other encryption and authentication keys are derived from this AK to secure subsequent exchange of PKM messages. This two-tiered process avoids the computationally expensive process of deriving the AK for every key refresh.

For the initial distribution and subsequent refreshing of keying material, the PKM protocol follows the typical client/server model. The PKM client (SS) requests keying material and the PKM server (BS) responds with the keys, if the client is authorized to obtain the requested keys. The PKM-REQ and PKM-RSP MAC management messages are used to exchange PKM messages between the client and the server.

The PKMv1 protocol defined in the IEEE 802.16-2004 1 supports RSA-based authentication of the SS, key distribution, key refresh, key synchronization, and reauthentication. The PKMv2 protocol, introduced as part of the IEEE 802.16e-2005 amendment 1, supports additional features such as a new key hierarchy, Extensible Authentication Protocol (EAP)-based authentication, mutual authentication, new encryption methods, AES key wraps for secure distribution of keys, and security for MBS.

4.2.3 Authentication Overview

The PKM protocol supports both unilateral and mutual authentication. In the unilateral authentication model, the BS may authenticate the SS, but not vice versa. In mutual authentication, both BS and SS authenticate each other. The PKM protocol supports RSA authentication (PKM RSA) and EAP-based authentication (PKM EAP) of the SS.

When a SS authenticates with a BS, it presents its credentials, which will be a X.509 digital certificate from the manufacturer for PKM-RSA authentication or an operator-specified credential for PKM EAP authentication. The PKMv1 protocol only supports the PKM-RSA authentication using X.509 digital certificates. The IEEE 802.16e amendment added the PKMv2 protocol, which supports the optional PKM-EAP authentication, as many mobile operators have already invested in many EAP authentication infrastructures. Therefore, the PKM-RSA is mandatory if PKMv1 is used and is optional otherwise. The PKM-EAP is an optional feature in PKMv2.

The X.509 digital certificate used in the PKM-RSA authentication protocol consists of the SS's public key and SS's 48-bit IEEE MAC address. SSs using the PKM-RSA authentication also require that they have factory-installed RSA private/public key pairs or provide an algorithm to generate such key pairs dynamically. If the RSA private/public key pairs are factory-installed, the X.509 digital certificate also has to be factory-installed. If an algorithm is provided to dynamically generate RSA private/public key pairs, the SS has to support a mechanism for installing the X.509 digital certificates followed by the key generation.

The PKM EAP authentication protocol uses the EAP as defined in RFC 3748 [5]. The specific type of credential used in this method of authentication depends on the selected EAP method. For example, EAP-TLS uses X.509 certificates and EAP-SIM uses Subscriber Identity Modules (SIM) as credentials. The specific type of credentials and corresponding EAP methods to be used with the PKM-EAP authentication is outside the scope of the IEEE 802.16 standard. However, it is required that the selected EAP method meets the mandatory criteria defined in RFC 4017 [4]. For example, the EAP-MD5 CHAP that uses passwords as credentials does not meet the criteria defined in RFC 4017 [4]. The PKM messages that transfer EAP information during authentication and reauthentication are protected by the Hashed Message Authentication Code/Cipher-based Message Authentication Code (HMAC/CMAC) tuple and messages with invalid HMAC/CMAC digests are discarded by both the BS and SSs.

The PKMv2 protocol also supports mutual RSA authentication. A client may request mutual authentication using the PKMv2 RSA-request message, which includes the client's X.509 digital certificate. The PKMv2 RSA-response message from the BS contains the BS's X.509 digital certificate, which can be used to authenticate the BS's identity. Mutual authentication can be used by itself or it can be followed by PKM-EAP authentication. If PKM-EAP authentication is also used, the mutual authentication is used only during initial network entry.

4.3 Cryptographic Methods

The format of 802.16 MAC PDU is shown in Figure 4.3. The generic MAC header is of fixed size (6 bytes) and is not encrypted. The variable size payload is

Generic MAC Header	Payload	CRC

Figure 4.3   Medium Access Control PDU format.

encrypted using one of the cryptographic methods described in the following subsections.

4.3.1   Data Encryption: DES in CBC Mode

When the DES encryption is specified for a connection, the CBC mode of the U.S. Data DES algorithm is used to encrypt the MAC PDU payloads. The CBC IV is calculated differently for the DL and UL. In the DL, the CBC is initialized with the XOR of the IV parameter included in the TEK keying information and the current frame number (right justified). In the UL, the CBC is initialized with the XOR of the IV parameter included in the TEK keying information and the frame number of the frame in which the UL-MAP granting this specific transmission was transmitted.

For the encryption of the final block of the plain text that is less than 64 bits, residual termination block processing is used. For example, consider a final block of plain text having n bits, where n is less than 64, the next-to-last cipher text block will be encrypted for a second time with DES algorithm, using the ECB mode, and the most significant n bits of the result are XORed with the final n bits of the payload to generate the short final cipher block. On the receiver's end, for decryption of the short final cipher block, the receiver encrypts the next-to-last cipher text block with the DES algorithm, using the ECB mode and XORs the most significant n bits with the short final cipher block to recover the short final clear text block.

If more than two MAC PDUs of length less than 64-bits are transmitted with the same security association (SA) on the same PHY frame, it is possible to derive the XOR of the plain texts easily. However, this is unlikely to occur in real world, as the MAC PDUs are typically larger than 64 bits. Some preventive measures such as using packing (with other SDUs or zero-length dummy SDUs) can be used to extend the size of the MAC PDU.

4.3.2   Data Encryption: AES in CCM Mode

When the AES CCM mode encryption is specified for a connection, the CCM mode of the U.S. AES algorithm specified in NIST Special Publication 800-38C, FIPS-197 is used to encrypt the MAC PDU payloads.

In this mode, the size of the MAC PDU is expanded to include the packet number (PN) and message authentication code. If the plain text contains L bytes, the encrypted frame contains a 4-byte PN, cipher text with L bytes, and a cipher text message authentication code of 8 bytes, known as the Integrity Check Value (ICV). Therefore, the total length of the encrypted frame is increased to $L + 12$ bytes. The 8-byte ICV is appended to the end of the MAC PDU payload and the encryption is performed on both the plain text payload and the appended ICV. The plain text PDU and the appended ICV are encrypted and authenticated using the active TEK, as per the algorithm specified in the CCM specification.

The PN associated with an SA is set to 1 when the SA is established and when a new TEK is installed. After each PDU transmission, the PN is incremented by 1. The SS should ensure that a new TEK is requested and transferred before the PN on either the SS or the BS reaches 0x7FFFFFFF. On UL connections, the PN is XORed with 0x80000000 prior to encryption and transmission. On DL connections, the PN is used without modifications. The PN is used by the receiver to detect replay attacks and take corrective actions.

4.3.3 Data Encryption: AES in CTR Mode

This mode requires unique initial counters and key pairs across all messages. 128-bit AES block and cipher counter block sizes are used in CTR mode. The counter is 128 bits long and is constructed from the frame number. It also requires an 8-bit rollover counter (ROC). Note that a new MAC PDU requires a new frame number and reinitialization of the counter. When the frame number reaches 0x000000 (from 0xFFFFFF), the ROC is incremented. The 8-bit ROC is inserted before the MAC PDU payload for AES-CTR encryption, that is, the ROC is the 8 MSBs of the 32-bit nonce. The ROC is not encrypted.

It is very important that the tuple value of {AES counter, KEY} not be used more than once. Therefore, a new Multicast Broadcast Service Group Traffic Encryption Key (MGTEK) must be requested and transferred before the ROC reaches 0xFF.

A 32-bit nonce is made from the concatenation of the 8-bit ROC and the 24-bit frame number. This 32-bit nonce is repeated four times to construct the initial 128-bit counter block required by the AES-128 cipher. At the most 2^{32} PDUs can be encrypted with a single MBS_Traffic_Key (MTK). The plain text PDU is encrypted using the active MTK derived from MBS Authorization Key (MAK) and MGTEK, according to the CTR mode specification. A different 128-bit counter value is used to encrypt each 128-bit block within a PDU.

The processing yields a payload that is 8 bits longer than the plain text payload.

4.3.4 Data Encryption: AES in CBC Mode

In AES, CBC mode, encryption is performed on a block-by-block basis. However, residual termination block processing is used to encrypt the final block of plain text when the final block is less than the cipher block size. Given a final block with n bits, where n is less than the cipher block size m, the next-to-last cipher text block is divided into two parts. The first part is n bits, and the other part is $m-n$ bits. The former is sent to the receiver as the final block cipher text. The final short block is padded to obtain a complete plain text block, and then encrypted with the AES algorithm in CBC mode. In the special case when the payload portion of the MAC PDU is less than the cipher block size, the most significant n bits of the generated CBC-IV, corresponding to the number of bits of the payload, are XORed with the n bits of the payload to generate the short cipher block.

The CBC IV is produced as follows. The zero hit counter is initialized to zero when the key reply message is received, and updated whenever either the PHY frame number is zero or a MAC PDU is received in a frame. The zero hit counter increases by one if the previous PHY frame number is equal to or greater than the current PHY frame number. The CBC IV is generated as the result of the AES block ciphering algorithm with the key of TEK. The plain text for the CBC IV generation is calculated with the XOR of [1] the CBC IV parameter value included in the TEK keying information [2], the 128 bits obtained by the concatenation of the 48-bit MAC PDU header, the 32-bit PHY synchronization value of the MAP that a data transmission occurs, and the XOR value of the 48-bit SS MAC address and the zero hit counter. The CBC IV is updated for every MAC PDU.

4.3.5 Encryption of TEK-128 with AES Key Wrap

As discussed before, the BS encrypts the value fields of the TEK-128 in the key reply messages it sends to a client SS. Encryption and decryption using the AES key wrap algorithm are performed as follows:

$$C,I = E_k[P]$$

$$P,I = D_k[C],$$

where $P =$ plain text 128-bit TEK, $C =$ cipher text 128-bit TEK, $I =$ Integrity Check Value, $k =$ the 128-bit key encryption key (KEK), $E_k[\ ] =$ AES key wrap encryption with key k, and $D_k[\ ] =$ AES key wrap decryption with key k.

The AES key wrap encryption algorithm accepts both a cipher text and ICV. The decryption algorithm returns a plain text key and the ICV. The default ICV in the NIST AES key wrap algorithm shall be used.

4.3.6 Calculation of HMAC Digests

The calculation of the keyed hash in the HMAC-digest attribute and the HMAC tuple uses the HMAC (IETF RFC 2104) with the secure hash algorithm SHA-1

(FIPS 180-1). The DL authentication key HMAC_KEY_D is used for authenticating messages in the DL direction. The UL authentication key HMAC_KEY_U is used for authenticating messages in the UL direction. The UL and DL message authentication keys are derived from the AK. The HMAC sequence number in the HMAC tuple is equal to the AK sequence number of the AK from which the HMAC_KEY_x was derived. When calculating the digest with this key, the HMAC sequence number in the HMAC tuple is equal to the operator shared secret sequence number. The digest is calculated over the entire MAC management message with the exception of the HMAC-digest and HMAC tuple attributes.

The HMAC sequence number in the HMAC tuple or short-HMAC tuple is equal to the AK sequence number of the AK from which the HMAC_KEY_x was derived. In the case of PKMv2, short-HMAC digest calculations include the HMAC_PN_* that should be concatenated after the MAC management message.

4.3.7 Derivation of TEKs, KEKs, and Message Authentication Keys

The BS generates AKs, TEKs, and IVs. A random or pseudo-random number generator is used to generate AKs and TEKs. A random or pseudo-random number generator may also be used to generate IVs.

The 56-bit DES keys are 8-byte (64-bit) quantities where the seven most significant bits (i.e., seven leftmost bits) of each byte are the independent bits of a DES key, and the least significant bit (i.e., rightmost bit) of each byte is a parity bit computed on the preceding seven independent bits and adjusted so that the byte has odd parity.

PKM does not require odd parity. The PKM protocol generates and distributes 8-byte DES keys of arbitrary parity, and it requires that implementations ignore the value of the least significant bit of each byte.

The keying material for two-key 3-DES consists of two distinct (single) DES keys. The 3-DES KEK used to encrypt the TEK is derived from a common AK. The KEK is derived as follows:

$$KEK = Truncate(SHA(K\_PAD\_KEK \mid AK),128)$$

K_PAD_KEK = 0x53 repeated 64 times, that is, a 512-bit string.

The function $Truncate(x,n)$ denotes the result of truncating x to its leftmost n bits, and $SHA(x|y)$ denotes the result of applying the SHA-1 function to the concatenated bit strings x and y. The keying material of 3-DES consists of two distinct DES keys. The 64 most significant bits of the KEK are used in the encrypt operation. The 64 least significant bits are used in the decrypt operation.

The HMAC authentication keys are derived as follows:

$$HMAC\_KEY\_D = SHA(H\_PAD\_D|AK)$$

$$HMAC\_KEY\_U = SHA(H\_PAD\_U|AK)$$

$$HMAC\_KEY\_S = SHA(H\_PAD\_D|\text{operator shared secret}).$$

with

$$H\_PAD\_D = 0x3A \text{ repeated } 64 \text{ times}$$

$$H\_PAD\_U = 0x5C \text{ repeated } 64 \text{ times}.$$

The construction of the KEK for use with TEK-128 keys is the same as for 3-DES KEKs except that the full 128 bits of the KEK are used directly as the 128-bit AES key, instead of the KEK being split into two 64-bit DES keys.

4.3.8 Cipher-Based MAC

A BS or SS may support management message integrity protection based on cipher-based MAC—together with the AES block cipher.

The calculation of the keyed hash value contained in the CMAC-digest attribute and the CMAC tuple uses the CMAC Algorithm with AES. The DL authentication key CMAC_KEY_D is used for authenticating messages in the DL direction. The UL authentication key CMAC_KEY_U is used for authenticating messages in the UL direction. UL and DL message authentication keys are derived from the AK.

For authentication of multicast messages in the DL, a CMAC_KEY_GD per group is used, where the group authentication key is derived from Group Key Encryption Key (GKEK). The CMAC-digest and CMAC tuple attributes are applicable only to the PKM version 2 [2]. In PKM version 2, the AK identifier (AKID) used in the computation of the CMAC value tuple is the 64-bit AKID of the AK from which the CMAC_KEY_x was derived.

The CMAC packet number counter (CMAC_PN_*) is a 4-byte sequential counter that is incremented in the context of UL messages by the SS, and in the context of DL messages by the BS. The BS will also maintain a separate CMAC_PN_* for multicast packets per each Group Security Association (GSA) and increment that counter in the context of each multicast packet from the group. For MAC messages that have no CID, for example, RNG-REQ message, the CMAC_PN_* context will be the same as used on the basic CID. If basic CID is unknown (e.g., in network reentry situation) then CID 0 should be used.

The CMAC packet number counter, CMAC_PN_*, is part of the CMAC security context and must be unique for each MAC management message with the CMAC tuple or digest. Any tuple value of {CMAC_PN_*, AK} shall not be used more than once. The reauthentication process must be initiated by BS or SS to establish a new AK before the CMAC_PN_* reaches the end of its number space. The digest is calculated over a field consisting of the AKID by the CMAC packet number counter, expressed as an unsigned 32-bit number, followed by the 16-bit connection ID on which the message is sent, followed by 16-bit zero padding (for the header to be aligned with AES block size), and followed by the entire MAC management message with the exception of the CMAC-TLV (Type-Length-Value).

The least significant bits of the digest are truncated to yield a 64-bit length digest:

CMAC value <= Truncate64 (CMAC (CMAC_KEY_*, AKID CMAC key sequence number | CMAC_PN |CID |16-bit zero padding | MAC_Management_ Message))

If the digest is included in an MPDU that has no CID, for example, a RNG-REQ message, the CID used takes the value of the basic CID. If basic CID is unknown (e.g., in network re-entry situation) then CID 0 is used.

4.3.9 Key Derivation Functions for PKM Version 2

The construction of the KEK for use with TEK-128 keys is the same as for 3-DES KEKs except that the full 128 bits of the KEK are used directly as the 128-bit AES key, instead of the KEK being split into two 64-bit DES keys.

For broadcast services, the BS encrypts the value fields of the GKEK in the key update command message for the GKEK update mode and sends the encrypted GKEK to each SS. GKEK can be encrypted with 3-DES, RSA, or AES in the following way.

First, two-key 3-DES in the encrypt-decrypt-encrypt (EDE) mode is used to encrypt the GKEK when the encryption algorithm identifier in the cryptographic suite is equal to 0x01. Encryption and decryption are performed according to

$$C = E_{k1}[D_{k2}[E_{k1}[P]]],$$

$$P = D_{k1}[E_{k2}[D_{k1}[C]]],$$

where P = plain text 128-bit GKEK, C = cipher text 128-bit GKEK, k_1 = leftmost 64 bits of the 128-bit KEK, k_2 = rightmost 64 bits of the 128-bit KEK, $E[\ ]$ = 56-bit DES ECB mode encryption, and $D[\ ]$ = 56-bit DES ECB mode decryption.

The RSA method of encrypting the GKEK is used when the encryption algorithm identifier in the cryptographic suite is equal to 0x02.

128-bit AES in ECB mode is used when the encryption algorithm identifier in the cryptographic suite equal to 0x03. Encryption and decryption are performed as:

$$C = E_{k1}[P],$$

$$P = D_{k1}[C],$$

where P = plain text 128-bit GKEK, C = cipher text 128-bit GKEK, k_1 = the 128-bit KEK, $E[\ ]$ = 128-bit AES ECB mode encryption, and $D[\ ]$ = 128-bit AES ECB mode decryption.

The 128-bit AES key wrap algorithm is used when the encryption algorithm identifier in the cryptographic suite is equal to 0x04. This 128-bit AES key wrap algorithm is defined only for PKM version 2. Encryption and decryption are performed as:

$$C,I = E_k[P],$$

$$P,I = D_k[C],$$

where P = plain text 128-bit GKEK, C = cipher text 128-bit GKEK, k = the 128-bit KEK derived from the AK, $E_k[\ ]$ = AES key wrap encryption with key k, $D_k[\ ]$ = AES key wrap decryption with key k.

4.4　X.509 Digital Certificates

4.4.1　SS X.509 Digital Certificate

The format of the X.509 version 3 digital certificate used in IEEE 802.16 is shown in Table 4.3. However, there are some restrictions on the attribute values of the certificate. The SS's certificate is required to have a lifetime greater than the operational lifetime of the SS and these certificates are not renewable. The validity period of the manufacturer's certificate must exceed that of the SS's certificate. The SS certificate validity period begins with the date of generation of the device's certificate.

SS certificates have serial numbers and are signed by a particular issuer and these serial numbers should be issued by the manufacturer in increasing order. Therefore, the certificate with the later (i.e., higher) value of the Certificate Validity (tbsCertificate.validity.notBefore) field is required to have a serial number that is greater than the serial number of a certificate with an earlier (i.e., lower) tbs Certificate.validity.notBefore field value.

All the digital certificates used in IEEE 802.16 are required to be signed with the RSA signature algorithm using SHA-1 as one-way hash function. The RSA signature algorithm is shown in Figure 4.4.

The format of the manufacturer's certificate is shown in Figure 4.5. The country name, organization name, and the common name attributes are required to be included with the specific values as shown in Figure 4.5. An OrganizationalUnit-Name attribute with a specific value of "wireless MAN" must be included in all manufacturer certificates.

Table 4.3   X.509 Version 3 Digital Certificate Format

X.509 v3 Field	Description
Version	Indicates X.509 certificate version
Serial number	Integer that an CA assigns to the certificate
Signature	OID and parameters used to sign the algorithm
Certificate issuer	Name of CA that issued the certificate
Certificate validity	Specifies when the certificate becomes active and expires
Subject	Name identifying whose public key is certified
Public key info	Contains public key information
Issuer unique ID	Optional field
Subject unique ID	Optional field
Extensions	The extension data
Signature algorithm	OID and algorithm for signing certificate
Signature value	Digital signature computed on certificate

```
sha-1WithRSAEncryption OBJECT IDENTIFIER :=
{iso(1) member-body(2) us(840) rsadsi(113549) pkcs(1) pkcs-
1(1) 5}
```

Figure 4.4   RSA signature algorithm.

```
Country Name=<Country of Manufacturer>
[StateOrProvinceName=<state/province>]
[Locality Name=<City>]
Organization Name=<Company Name>
OrganizationalUnitName=Wireless MAN
[OrganizationalUnitName=<Manufacturing Location>]
commonName=<Company Name><Certification Authority>]
```

Figure 4.5   Manufacturer certificate.

```
Country Name=<Country of Manufacturer>
Organization Name=<Company Name>
OrganizationalUnitName=<manufacturing location>
Common Name=<Serial Number>
Common Name=<MAC Address>
```

Figure 4.6   Subscriber Station certificate.

If the OrganizationalUnitName representing the actual manufacturing location is included, it must be preceded by the OrganizationalUnitName with the value "wireless MAN." The StateOrProvinceName and locality name are optional attributes that may be included. No other attributes are allowed and should not be included in the manufacturer certificate.

The SS's certificate with all the fields is shown in Figure 4.6. The MAC address field contains the SS's 48-bit IEEE MAC address. The MAC address is represented as six pairs of hexadecimal digits separated by colons, where the hexadecimal alphabets are in uppercase. The manufacturing location of the SS is specified in the OrganizationalUnitName in an SS certificate and must be the same as the one used in the manufacturer certificate. The country name, organizational name and common name attributes are required and no other attributes are allowed.

The public key Information field (Table 4.3) contains the public key algorithm identifier and the public key. The SS certificate issued by the manufacturer is expected to be stored in a permanent, write-once memory. If SSs have factory-installed RSA private/public key pairs, they are also required to have factory-installed SS certificates. If the SSs depend on internal algorithms to generate an RSA key pair, then the SSs are required to support a mechanism for installing a manufacturer-issued SS certificate following key generation. The SS software should also contain one ore more Certificate Authority (CA) certificate of the manufacturer CA(s) that signed the SS certificate. The BS establishes the identity of the SS and its services with the AK exchange as shown in Figure 4.7. The details of the authorization procedure are described in later sections.

4.4.2   BS Certificate

The BS certificate has the format shown in Figure 4.8.

4.5   PKM Version 1: Security Associations

IEEE 802.16 defines an SA as a set of security information that need to be shared between an SS and one or possibly more SSs to establish secure communication over the broadband wireless network.

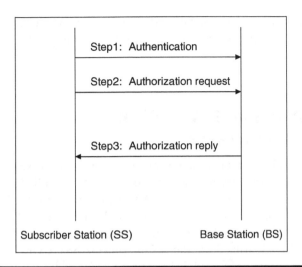

Figure 4.7 Authentication and authorization message exchanges between Subscriber Station and Base Station.

Three types of SAs (primary, static, and dynamic) are defined in 802.16. The primary security association is established during initialization of an SS. Static SAs are provisioned within the BS ahead of time. Dynamic SAs are the SAs that are established and eliminated dynamically in response to the requests for initialization and termination of specific service flows. Dynamic SAs are established by BS by issuing an SA add message. When an SS receives an SA add message, it creates a TEK state machine for each SA that is listed in the message. The SA's shared information consists of cryptographic suites, which may include TEKs and IVs, depending on the type of the cryptographic suite.

SAs are identified by security association identifiers (SAIDs). The basic CID (connection identifier) of the SS is the SAID of SS's primary SA. SS requests the SA's keying material from BS using PKM protocol messages. The keying material, which may include keys and IVs, has limited lifetime. Before the expiration of the older keys, SS may request new keying material from the BS. It is the responsibility of SS to keep track of all the expiration times of the keying material. All transport connections are mapped to an existing SA. Multicast transport connections may be

```
Country Name=<Country of Operation>
Organization Name=<Name of Operator>
OrganizationalUnitName=<WirelessMAN>
Common Name=<Serial Number>
Common Name=<BS Id>
```

Figure 4.8 Subscriber Station certificate.

mapped to any static or dynamic SA, and the secondary management connections are mapped to primary SA. The basic and primary management connections are not mapped to any SA.

4.6 PKM Version 1: Authorization and Key Exchange

An SS uses the PKM protocol to exchange authorization messages with and to obtain traffic encryption keying material from the BS. The PKM protocol follows a client-server model, where the client (SS) authenticates with the BS and requests keying material from the server (BS). The PKM establishes a shared secret key, known as the AK between BS and SS. This secret key is used for subsequent exchanges of TEKs. During the authorization process, SS presents its unique X.509 certificate which contains SS's public key and SS MAC address. The BS verifies the SS's X.509 certificate and sends the AK encrypted with public key of that SS.

4.6.1 SS Authorization and AK Exchange

After the BS authenticates a client SS's identity, then the BS provides an AK for the authenticated SS. This AK is the key from which a KEK and message authentication keys are derived and finally the authenticated SS will be provided with the identities (i.e., the SAIDs) and properties of primary and static SAs by the BS for which the SS is authorized to obtain keying information. SS periodically seeks reauthorization with the BS once the initial authorization is achieved from the BS. Authorization state machine is the one which manages the reauthorization of the SS. SS must maintain its reauthorization status to maintain fresh keying material. The detailed process of authorization and the AK exchange is described subsequently.

1. Step 1: Authentication information message

$$\text{SS} \xrightarrow{\text{Certificate (Manufacturer (SS))}} \text{BS}$$

SS begins the authorization by sending authentication information message to the BS. This authentication information message contains the X.509 certificate issued by the manufacturer of the SS. This is used by the BS to identify the manufacturer of the SS.

2. Step 2: Authorization request message

$$\text{SS} \xrightarrow{\text{Cert (SS) | Capabilities | CID}} \text{BS}$$

Table 4.4   Data Encryption Algorithm Identifiers

Value	Description
0	No data encryption
1	CBC Mode, 56-bit DES
2	CCM Mode, 128-bit AES
3	CBC Mode, 128-bit AES
128	CTR Mode, 128-bit AES for MBS
4–127 and 129–255	Reserved

Abbreviations: DES, Data Encryption Standard; AES, Advanced
Encryption Standard; CBC, Cipher Block-Chaining;
CTR, Counter; MBS, Multicast Broadcast Services.

Immediately following the authentication information message, SS sends an authorization request message. This is the request for the AK and for the SAIDs. The parameter "cert" (SS) is the certificate of the SS which sends the authorization request message. The second parameter in the authorization request message is the "capabilities." The capabilities include the cryptographic algorithms the requesting SS supports, cryptographic suite identifiers indicating packet data encryption, and packet data authentication algorithm.

Cryptographic suites: The cryptographic suite defines the set of methods or algorithms for data encryption, data authentication, and TEK encryption. The list of supported cryptographic suites in 802.16 is summarized in Table 4.2. The cryptographic suite is identified by a 24-bit integer. The most significant byte indicates the encryption algorithm and key length. The middle byte indicates the data authentication algorithm. The least significant byte indicates the TEK encryption algorithm.

The list of currently supported data encryption algorithm identifiers is shown in Table 4.4. If the value field is 0 then no data encryption is performed.

The list of supported authentication algorithm identifiers is shown in Table 4.5. If the value filed is 0 then no data authentication is used.

Table 4.5   Data Authentication Algorithm Identifiers

Value	Description
0	No data authentication
1	CCM Mode, 128-bit AES
2–255	Reserved

Table 4.6   TEK Encryption Algorithm Identifiers

Value	Description
0	Reserved
1	3-DES EDE with 128-bit key
2	RSA with 1024-bit key
3	AES ECB with 128-bit key
4	AES key wrap with 128-bit key
5–255	Reserved

Abbreviations: AES, Advanced Encryption Standard; DES,
Data Encryption Standard; ECB, Electronic
Codebook; TEK, Traffic Encryption Keys;
EDE, encrypt-decrypt-encrypt.

The list of supported TEK encryption algorithm identifiers is shown in Table 4.6. If the value field is 0 then it is reserved.

The third parameter in the authorization request message is the CID. In 802.16, all connections are identified by a 16-bit CID. When an SS initializes and registers with the BS, three management connections are established between the SS and the BS in each direction (UL and DL). The basic connection is used by the BS MAC and SS MAC to exchange short, time-critical MAC management messages. The primary management connection is used by the BS MAC and SS MAC to exchange longer, more delay-tolerant MAC management messages. Finally, the secondary management connection is used by the BS and SS to transfer delay-tolerant, standard-based management messages.

After the initial ranging is completed, an SS needs to authorize with the BS. This process is initiated by sending an authorization request message. In response to the authorization request message, BS determines the encryption algorithms and protocol it shares with the SS, and issues AK encrypted with SS's public key by validating SS's identity. The BS rejects authorization request message if none of the cryptographic suites offered by the SS are satisfactory.

3. Step 3: Authorization reply message

$$\text{BS} \xrightarrow{\text{RSA-Encrypt (PubKey (SS), AK),lifetime,SeqNo,SAID list}} \text{SS}$$

This authorization reply message is sent from the BS to SS in response to the authorization request message. The SeqNo is used to distinguish between successive generations of AKs. This SeqNo indicates the SS the AK that it has to use. SS periodically refreshes its AK by reissuing an authorization request to the BS. The active lifetime of the AK is indicated by the lifetime parameter. The reauthorization is similar to the authorization with the exception that authentication information message will not be issued by the SS. To avoid service interruptions between SS and BS, SS AKs have overlapping lifetimes. During transition periods, that is, when BS or SS changes from one AK to another, two simultaneously active AKs will be supported by both BS and SS. The fourth parameter is SAID.

4.6.2   TEK Exchange Overview

A separate TEK state machine will be started by the SS for each of the SAIDs identified in the authorization reply message. The keying material associated for each SAID is managed by the TEK state machine. The TEK state machine is responsible for periodically refreshing the keying material for SAIDs. The key reply message that is sent from BS to SS determines keying material of the BS for a specific SAID. The key reply contains TEK that is encrypted by triple DES (EDE mode), which uses two-key. Triple DES KEK is derived from AK.

■ Key Reply Message

$$\text{BS} \xrightarrow{\text{Triple-DES (TEK), Lifetime.}} \text{SS}$$

4.6.3   Encryption of TEK with 3-DES

If the TEK encryption algorithm identifier in the cryptographic suite equal to $0x01$, 3-DES will be used for encryption. The TEK that is sent to SS in the key reply message from the BS is encrypted using 3-DES. Encryption is done using two-key triple DES in the EDE mode.

The keying material for the 3-DES algorithm is derived from the KEK, which is initially derived from the AK. If the key request message from the SS indicates the key sequence number of the new active AK, then this AK will be used to decrypt KEKs otherwise the older AK will be used.

The function Truncate(x, n) truncates x to its leftmost n bits. SHA($x \mid y$) function denotes the result of applying the SHA-1 function to the concatenated bit strings x and y. The keying material of 3-DES consists of two distinct DES keys. For the encryption operation the 64 most significant bits of the KEK will be used and for the decryption operation the 64 least significant bits will be used.

Two sets of keying material will be maintained by the BS for each of the SAID. These keying materials have overlapping lifetimes to avoid disruption of service, where each generation (of keying material) becomes active halfway through the life of the predecessor and expires halfway through the life of its successor. When a BS responds with key reply message, it includes both the generations of keying material and remaining life time of each of the two sets of keying material in addition to TEK and CBC IV. The SS depends on the lifetimes of the keying material specified in the key reply message and determines when the BS will invalidate the keying material for that particular TEK and depending on this the SS will schedule key requests such that SS requests and receives new keying material before the expiration of the present keying material. TEK state machine is active as long as

1. SS is authorized and is operating in the BS's security domain, that is, SS has a valid AK and
2. SS is authorized and is allowed to participate in a particular SA, that is, BS continues to provide fresh keying material during rekey cycles.

When authorization reject is received by the SS from the BS, the parent authorization state machine stops all its child TEK state machines. During reauthorization if the static SAIDs change, individual TEK state machines can be started or stopped. Authorization and TEK state machines communicate through the passing of events and protocol messaging. The events generated by the authorization state machine are stop, authorized, authorization pending, and authorization complete. Any of these events are not targeted by the TEK state machines to the authorization machine. TEK state machine indirectly affects authorization state machine through messaging a BS sends in response to an SS's requests. SS will not start a TEK state machine for a static SA's if the SS receives an authorization reply from the BS that indicates the cryptographic suites that SS does not support.

4.6.4   Authorization State Machine

Authorization state machine contains six states and eight events. Figure 4.9 depicts the protocol messages transmitted and internal events generated for each of the model's state transitions. Any additional internal actions such as clearing or starting of timers that accompany the specific state machines are not shown in the diagram. This authorization state machine maintained by the SS is responsible for achieving authorization from the BS. Reauthorization is also achieved through this authorization state machine.

In the authorization state machine flow diagram,

a) Ovals represent states.
b) Events are represented in italics.
c) Messages are represented in normal font.
d) The labeling of state transitions is shown as: <the cause of the transition>/<The triggered messages and events by the transition>. If there are many events or messages before the slash "/", any of them can cause the transition. If there are many events or messages after the slash "/", all of the actions will take place.

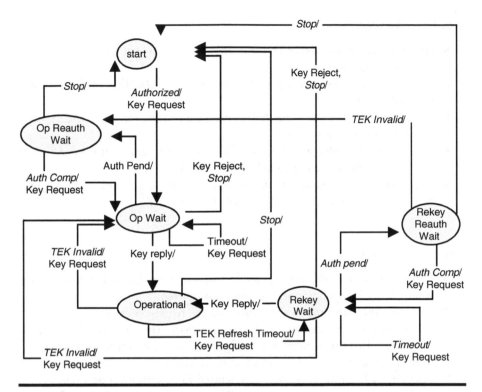

Figure 4.9   Authorization state machine.

4.6.4.1   States

- *Start*: This is the initial state of the state machine and no resources are assigned.
- *Authorize wait* (*auth wait*): SS after receiving communication established event issues authentication information and authentication request message and is waiting for reply.
- *Authorized*: SS receives authorization reply message and starts TEK state machine for each of the SAID identified in the reply message.
- *Reauthorize wait* (*reauth wait*): When SS receives an indication that authorization is no longer valid or SS was about to expire and when it receives this indication it sends auth request message to BS and is waiting for response.
- *Authorize reject wait* (*auth reject wait*): If the error code in the authorization reject message indicates that the error was not of permanent nature, SS sets a timer and goes to auth reject wait state.
- *Silent*: If the error code in the authorization reject message indicates that the error was of permanent nature, this triggers the transition into silent state.

4.6.4.2   Messages

- *Authorization request (auth request)*: This message is sent from SS to BS for requesting AK, SAIDs.
- *Authorization reply (auth reply)*: Receive AK, SAIDs. This message is sent from BS to SS.
- *Authorization reject (auth reject)*: This message is sent from BS to the SS when authorization is rejected.
- *Authorization invalid (auth invalid)*: This message is sent from BB to SS when BS does not recognize the SS being authorized.
- *Authentication information (authent info)*: This message contains SS's manufacturer's X.509 Certificate. This is sent from SS to BS.

4.6.4.3   Events

- *Communication established*: SS sends this event to the authorization FSM when the basic capability negotiation is completed.
- *Timeout*: The wait timer is timed out and the request is resent.
- *Authorization grace timeout (auth grace timeout)*: The authorization timer is timed out. This timer specifies a configurable amount of time that signals SS to reauthorize before authorization actually expires.
- *Reauthorize (reauth)*: This event is generated for reauthorization cycle.
- *Authorization invalid (auth invalid)*: This event is generated when there is a failure in authenticating key reply or key reject message. This message is also sent from BS to SS.
- *Authorization reject (auth reject)*: SS receives auth reject in response to an auth request. If the error code indicates that the error was not of permanent nature, the state machine will set a timer and transition into auth reject wait state.

4.6.5   TEK State Machine

The TEK state machine, as shown in Figure 4.10, consists of six states and nine events. BS maintains two active TEKs per SAID. The key reply message consists of both of these TEKs. BS uses older of the two TEKs to encrypt DL traffic and either older or newer TEK for decryption depending on which of the two keys SS was using. SS uses the newer of its two TEKs to encrypt UL traffic and either newer or older TEK for decryption depending on which key the BS was using. SS uses TEK state machine to synchronize TEKs with those of BS. After the expiration time of the older of the SS's TEK and before the expiration of its newer TEK, TEK state machine issues key request message for refreshing keying material.

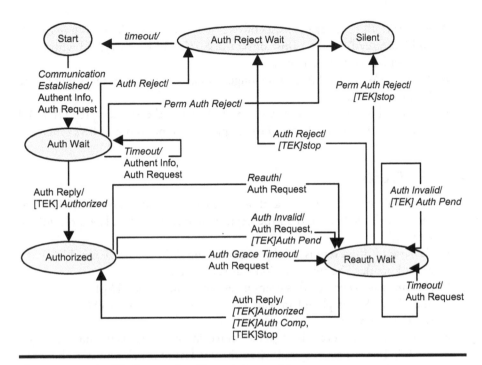

Figure 4.10   Traffic Encryption Keys state machine.

4.6.5.1   States

- *Start*: This is the initial state of FSM and no resources are assigned.
- *Operational wait (op wait)*: The TEK state machine sent its initial request (key request) for its SAID's keying material (TEK and CBC IV), and is waiting in this state for a reply from the BS.
- *Operational reauthorize wait (OP reauth wait)*: The TEK state machine is in this state if it does not have valid keying material.
- *Operational*: In this state the SS will have valid keying material for the particular SAID.
- *Rekey wait*: SS requests for update for this SAID. At this point the newer key of the two TEKs does not expire and the newer key can be used for both encryption and decryption.
- *Rekey reauthorize wait (rekey reauth wait)*: If the TEK state machine has valid traffic keying material, TEK machine will be placed in this state. Re-authorization cycle is initiated by the authorization state machine.

4.6.5.2   Messages

- *Key request*: The key request message is sent by the SS to the BS for a TEK for SAID.

■ *Key reply*: This message consists of two active sets of traffic keying materials and is sent by BS to SS. This consists of SAIDs TEKs encrypted with triple DES with keys derived from AK.
■ *Key reject*: When the SAID is no longer valid, key reject message is sent by the BS to SS's TEK.
■ *Invalid*: This message is sent by the BS to SS when the SS encrypted UL PDU with an invalid TEK, that is, TEK sequence number is out of BS's range.

4.6.5.3 Events

■ *Stop*: This event is sent by authorization FSM to an active TEK FSM to terminate TEK FSM and all the keying material from SS's key table will be removed.
■ *Authorized*: This event is sent by authorization FSM to a nonactive TEK FSM to indicate that the authorization was successful.
■ *Authorization pending*: This event is sent by authorization FSM to TEK FSM and TEK FSM is placed in wait state while authorization FSM completes reauthorization.
■ *TEK invalid*: This event is triggered on receipt of a TEK invalid message from the BS. This is also triggered when there is loss of synchronization between SS and BS.
■ *Timeout*: The retry timer is timed out. The request is retransmitted.
■ *TEK refresh timeout*: The TEK refresh timer timed out. This event signals TEK state machine to issue a new key request for refreshing keying material.

4.7 PKM Version 1: Parameter Values

The privacy configuration parameters are shown in Table 4.7. The AK lifetime is the lifetime of the AK, the BS sends to the SS. This lifetime is assigned by the BS. The minimum value of this parameter is one day and the maximum value is 70 days. The default value for the AK lifetime is seven days. The second parameter in the table is the TEK lifetime. This is the lifetime of the TEK, the BS sends to the requesting SS. The lifetime of this parameter is set by the BS. The minimum value for the TEK lifetime is 30 minutes and maximum value is seven days. The default value is 12 hours. These two parameters are assigned by the BS.

The authorization request retransmission interval from authorize wait state is specified by the authorization wait timeout and the minimum value for this parameter is 2 seconds, maximum value is 30 seconds and the default value is 10 seconds. The authorization request retransmission interval from reauthorize wait state is specified by the reauthorization wait timeout and the minimum value for this parameter is 2 seconds, maximum value is 30 seconds and the default value is 10 seconds. The time prior to authorization expiration SS begins reauthorization is

Table 4.7   Privacy Configuration Parameters

BS/SS	Name	Description	Min in Days	Default in Days	Max in Days
BS	AK lifetime	Lifetime in seconds	1	7	70
BS	TEK lifetime	Lifetime in seconds	30 minutes	12 hours	7
SS	Auth wait timeout	Retransmission interval	2 seconds	10 seconds	30 seconds
SS	Re-auth wait timeout	Retransmission interval	2 seconds	10 seconds	30 seconds
SS	Auth grace time	Time prior to expiration	5 minutes	10 minutes	35
SS	Operat wait timeout	Retransmission interval	1 second	1 second	10 seconds
SS	Rekey timeout	Retransmission interval	1 second	1 second	10 seconds
SS	TEK grace time	Time prior to expiration	5 minutes	1 hour	3.5
SS	Auth rej timeout	Delay	10 seconds	60 seconds	10 minutes

Abbreviations: BS, Base Station; SS, Subscriber Station; TEK, Traffic Encryption Keys; AK, Authorization Key.

specified by authorization grace time and the minimum value for this parameter is 5 minutes, maximum value is 35 days and the default value is 10 minutes. The key request retransmission interval from operational wait state is specified by operational wait timeout parameter and the minimum value is one second, maximum value is ten seconds and the default value for this parameter is one second. The key request retransmission interval from rekey wait state is specified by rekey wait timeout parameter and the minimum value is one second, maximum value is ten seconds, and the default value for this parameter is one second. The time at which the SS begins rekeying prior to expiration of TEK is specified by TEK grace time and the minimum value is 5 minutes, maximum value is 3.5 days, and the default value for this parameter is 1 hour. The delay before resending auth request after receiving auth reject is specified by authorize reject wait time out and the minimum value for this parameter is 10 seconds, maximum value is ten minutes, and the default value is 60 seconds. These are all the timer values that are set in the authorization state machine and the TEK state machine. The time of TEK grace time must be less than half the TEK lifetime.

4.8   PKM Version 1: Message Encodings

All PKM messages are encoded in the TLV format.

Type: The type field is of one byte. An attribute with unknown type will be ignored by the PKM server. Similarly, attributes of unknown type will be ignored by the PKM client. All the unknown attribute types will be kept in log files by both the PKM client and the server.

Length: The length of the attributes value is determined by the length filed and the value is in bytes. The length field does not contain the type and length fields.

Value: The value field contains zero or more bytes and information specific to the attributes is contained in this field. The format and length of the value field is determined by the type and length fields.

4.9   PKM Version 2: Three-Way Handshake

The AK can be derived in one of the three different ways depending on the authentication scheme used. Before the three-way handshake begins, both the BS and SS derive a shared KEK and the HMAC/CMAC keys. The PKMv2 three-way handshake sequence proceeds as follows.

1. During the initial network entry or reauthorization, the BS sends PKMv2 SA-TEK-Challenge (including a random number BS_Random) to the SS after protecting it with the HMAC/CMAC tuple. If the BS does not receive

PKMv2 SA-TEK-Request from the SS within SAChallenge-Timer, it resends the previous PKMv2 SA-TEK-Challenge up to SAChallengeMaxResends times. If the BS reaches its maximum number of resends, it initiates another full authentication or drops the SS.

2. If HO Process Optimization Bit #1 is set, indicating that PKM authentication phase is omitted during network re-entry or handover, the BS begins the three-way handshake by appending the SAChallenge tuple TLV to the RNG-RSP. If the BS does not receive PKMv2 SA-TEK-Request from the MS within SaChallengeTimer (suggested to be several times greater than the length of SaChallengeTimer), it may initiate full reauthentication or drop the MS. If the BS receives an initial RNG-REQ during the period that PKMv2 SA-TEK-Request is expected, it shall send a new RNG-RSP with another SaChallenge TLV.

3. The SS sends PKMv2 SA-TEK-Request to the BS after protecting it with the HMAC/CMAC. If the SS does not receive PKMv2 SA-TEK-Response from the BS within SATEKTimer, it must resend the request. The SS may resend the PKMv2 SA-TEK-Request up to SATEKRequestMaxResends times. If the SS reaches its maximum number of resends, it must initiate another full authentication or attempt to connect to another BS. The SS includes, through the security negotiation parameters attribute, the security capabilities that it included in the SBC-REQ message during the basic capabilities negotiation phase.

4. Upon receipt of PKMv2 SA-TEK-Request, the BS confirms that the supplied AKID refers to an AK that is available. If the AKID is unrecognized, the BS ignores the message. The BS also verifies the HMAC/CMAC. If the HMAC/CMAC is invalid, the BS ignores the message. The BS must verify that the BS_Random in the SA TEK Request matches the value provided by the BS in the SA Challenge message. If the BS_Random value does not match, the BS shall ignore the message. In addition, the BS must verify the SS's security capabilities encoded in the security negotiation parameters attribute against the security capabilities provided by the SS through the SBC-REG message. If security negotiation parameters do not match, the BS should report the discrepancy to higher layers.

5. Upon successful validation of the PKMv2 SA-TEK-Request, the BS sends PKMv2 SATEK-Response back to the SS. The message includes a compound TLV list each of which identifies the primary and static SAs, their SA identifiers (SAID), and additional properties of the SA (e.g., type, cryptographic suite) that the SS is authorized to access. In case of HO, the details of any dynamic SA that the requesting MS was authorized in the previous serving BS are also included. In addition, the BS must include, through the security negotiation parameters attribute, the security capabilities that it wishes to specify for the session with the SS (these will generally be the same as the ones insecurely negotiated in SBC-REQ/RSP). Additionally, in case of HO, for each active SA in previous serving BS, corresponding TEK, GTEK, and

GKEK parameters are also included. Thus, SA_TEK_Update provides a short-hand method for renewing active SAs used by the MS in its previous serving BS. The TLVs specify SAID in the target BS that shall replace active SAID used in the previous serving BS and also "older" TEK-parameters and "newer" TEK-parameters relevant to the active SAIDs. The update may also include multicast/broadcast Group SAIDs (GSAIDs) and associated GTEK parameter pairs. In case of unicast SAs, the TEK-parameters attribute contains all of the keying material corresponding to a particular generation of an SAID's TEK. This would include the TEK, the TEK's remaining key life-time, its key sequence number, and the CBC IV. The TEKs are encrypted with KEK. In case of group or multicast SAs, the TEK-parameters attribute contains all of the keying material corresponding to a particular generation of a GSAID's GTEK. This would include the GTEK, the GKEK, the GTEK's remaining key lifetime, the GTEK's key sequence number, and the CBC IV. The type and length of the GTEK is equal to the ones of the TEK. The GKEK should be identically shared within the same multicast group or the broadcast group. Contrary Key-Update Command, the GTEKs and GKEKs are encrypted with KEK because they are transmitted as a unicast here. Multiple iterations of these TLVs may occur suitable to recreate and reassign all active SAs and their (G)TEK pairs for the SS from its previous serving BS. If any of the SA parameters change, then those SA parameters encoding TLVs that have changed will be added. The HMAC/CMAC is the final attribute in the message's attribute list.

6. Upon receipt of PKMv2 SA-TEK-Response, an SS verifies the HMAC/CMAC. If the HMAC/CMAC is invalid, the SS ignores the message. Upon successful validation of the received PKMv2 SA-TEK-Response, the SS installs the received TEKs and associated parameters appropriately. The SS also must verify the BS's security negotiation parameters of TLV encoded in the security negotiation parameters attribute against the security negotiation parameters of TLV provided by the BS through the SBC-RSP message. If the security capabilities do not match, the SS should report the discrepancy to upper layers. The SS may choose to continue the communication with the BS. In this case, the SS may adopt the security negotiation parameters encoded in SA-TEK-Response message.

4.10   PKM Version 2: Mutual Authentication

Mutual authentication can take place in one of the two modes of operation. In the first mode, only mutual authentication is used. In the other mode, mutual authentication is followed by EAP authentication. In this second mode, the mutual authentication is performed only for initial network entry and only EAP authentication is performed in the case that authentication is needed for re-entry.

SS mutual authorization, controlled by the PKMv2 authorization state machine, is the process of:

1. The BS authenticating a client SS's identity.
2. The SS authenticating the BS's identity.
3. The BS providing the authenticated SS with an AK, from which a KEK and message authentication keys are derived.
4. The BS providing the authenticated SS with the identities (i.e., the SAIDs) and properties of primary and static SAs for which the SS is authorized to obtain keying information.

After achieving initial authorization, an SS should periodically seek reauthorization with the BS. This reauthorization is also managed by the SS's PKMv2 authorization state machine. An SS must maintain its authorization status with the BS to be able to refresh aging TEKs and GTEKs. TEK state machines manage the refreshing of TEKs. The SS or BS may run optional authenticated EAP messages for additional authentication.

The SS sends an authorization request message to its BS immediately after sending the authentication information message. This is a request for an AK, as well as for the SAIDs identifying any static security SAs that the SS is authorized to participate in. The authorization request includes:

■ A manufacturer-issued X.509 certificate.
■ A list of cryptographic suite identifiers, each indicating a particular pairing of packet data encryption and packet data authentication algorithms that the SS supports.
■ The SS's basic CID. The basic CID is the first static CID that the BS assigns to an SS during initial ranging—the primary SAID is equal to the basic CID.
■ A 64-bit random number generated in the SS.

In response to an authorization request message, a BS validates the requesting SS's identity, determines the encryption algorithm and protocol support it shares with the SS, activates an AK for the SS, encrypts it with the SS's public key, and sends it back to the SS in an authorization reply message. The authorization reply includes:

■ The BS's X.509 certificate.
■ A pre-PAK encrypted with the SS's public key.
■ A 4-bit PAK sequence number, used to distinguish between successive generations of AKs.
■ A PAK lifetime.

■ The identities (i.e., the SAIDs) and properties of the single primary and zero or more static SAs for which the SS is authorized to obtain keying information.
■ The 64-bit random number generated in the SS.
■ A 64-bit random number generated in the BS.
■ The RSA signature over all the other attributes in the auth-reply message by BS, used to assure that the authenticity of the earlier PKMv2 RSA-Reply messages.

An SS must periodically refresh its AK by reissuing an authorization request to the BS. Reauthorization is identical to authorization. To avoid service interruptions during reauthorization, successive generations of the SS's AKs have overlapping lifetimes. Both SS and BS must be able to support up to two simultaneously active AKs during these transition periods. The operation of the authorization state machine's authorization request scheduling algorithm, combined with the BS's regimen for updating and using a client SS's AKs, ensures that the SS can refresh TEK keying information without interruption.

4.11   PKM Version 2: Multicast and Broadcast

4.11.1   Multicast and Broadcast Services Support

PKMv2 supports strong protection from theft of multicast and broadcast service by encrypting MBS defined in the IEEE 802.16e amendment.

MBS requires a MBS GSA. It is the set of security information that multiple BS and one or more of its client SSs share but not bound to any MS authorization state to support secure and access-controlled MBS content reception across the IEEE 802.16 network. Each MBS capable of MS may establish a MBS SA during the MS initialization process. MBS GSAs shall be provisioned within the BS. A MBS GSA's shared information includes the cryptographic suite employed within the GSA and key material information such as MAKs and MGTEKs. The exact content of the Marin General Services Authority (MGSA) is dependent on the MGSA's cryptographic suite. As like any other unicast SAs, MBS GSA is also identified using 16-bit SAIDs. Each MS shall establish one or more MBS GSA with its serving BS. Using the PKMv2 protocol, an MS receives or establishes an MBS GSA's keying material. The BS and MBS content server must ensure that each client MS only has access to the MGSAs it is authorized to access. An SA's keying material (e.g., MAK and MGTEK) has a limited lifetime. When the MBS content server or BS delivers MBS SA keying material to an MS, it also provides the MS with that material's remaining lifetime. It is the responsibility of the MS to request new keying material from the MBS server or BS before the set of keying material that the MS currently holds expires at the MBS server or BS.

4.11.2 Optional Multicast Broadcast Rekeying Algorithm

When the Multicast Broadcast Rekeying Algorithm (MBRA) is supported, the MBRA is used to refresh traffic keying material efficiently not for the unicast service, but for the multicast or the broadcast service.

Note that an SS may get the traffic keying material before an SS is served with the specific multicast service or the broadcast service. The initial GTEK request exchange procedure is executed by using the key request and key reply messages that are carried on the primary management connection. The GTEK is the TEK for multicast or broadcast service. Once an SS shares the traffic keying material with a BS, the SS does not need to request new traffic keying material. A BS updates and distributes the traffic keying material periodically by sending two key update command messages.

A BS manages the M&B (Multicast & Broadcast) TEK grace time for the respective GSA-ID in itself. The GSA-ID is the SA-ID for multicast or broadcast service. This M&B TEK grace time is defined only for the multicast service or the broadcast service. This parameter means time interval (in seconds), before the estimated expiration of an old distributed GTEK. In addition, the M&B TEK grace time is longer than the TEK grace time managed in an SS.

A BS distributes updated traffic keying material by sending two key update command messages before old distributed GTEK is expired. The usage type of these messages is distinguished according to the key push modes included in the key update command message. The purpose of the key update command message for the GKEK update mode is to distribute the GKEK. The key update command message for the GKEK update mode is carried on the primary management connection. A BS intermittently transmits the key update command message for the GKEK update mode to each SS to reduce the BS's load in refreshing traffic key material. The GKEK is needed to encrypt the new GTEK. The GKEK may be randomly generated in a BS or an ASA server.

A BS transmits the PKMv2 group key update command message for the GTEK update mode carried on the broadcast connection after the M&B TEK grace time starts. The aim of the key update command message for the GTEK update mode is to distribute new GTEK and the other traffic keying material to all SSs served with the specific multicast service or the broadcast service. This GTEK is encrypted with already transmitted GKEK.

An SS shall be capable of maintaining two successive sets of traffic keying material per authorized GSA-ID. Through operation of its GTEK state machines, an SS shall check whether it receives new traffic keying material or not. If an SS gets new traffic keying material, then its TEK grace time is not operated. However, if it does not have that, then an SS shall request a new set of traffic keying material a configurable amount of time, the TEK grace time, before the SS's latest GTEK is scheduled to expire.

If an SS receives two valid key update command messages and shares new valid GKEK and GTEK with a BS, then that SS does not need to request a new set of traffic keying material.

If an SS does not receive at least one of the two key update command messages, then that SS sends the key request message to get a new traffic keying material. A BS responds to the key request message with the key reply message. In other words, if an SS does not get valid new GKEK or GTEK, then the GTEK request exchange procedure initiated by an SS is executed.

An SS tries to get the GTEK before an SS is served with the specific service. The initial GTEK request exchange procedure is executed by using the key request and key reply messages that are carried on the primary management connection.

A BS must be capable of maintaining two successive sets of traffic keying material per authorized GSA-ID. That is, when GKEK has been changed a BS manages the M&B TEK grace time for the respective GSA-ID in itself. Through operation of its M&B TEK grace time, a BS shall push a new set of traffic keying material. This M&B TEK grace time is defined only for the multicast service or the broadcast service in a BS. This parameter means time interval (in seconds) before the estimated expiration of an old distributed GTEK. That is, the M&B TEK grace time is longer than the TEK grace time managed in an SS.

A BS distributes updated GTEK by using two key update command messages when the GKEK has been changed, or by using one (the second) key update command message otherwise, around the M&B TEK grace time, before the already distributed GTEK expires. Those messages are distinguished according to a parameter included in that message, "key push modes." A BS transmits the first key update command message to each SS served with the specific service before the M&B TEK grace time. The first key update command message is carried on the primary management connection. A BS intermittently transmits the first key update command message to each SS to reduce the BS's load for key refreshment. The purpose of the first key update command message is to distribute the GKEK. This GKEK is needed to encrypt the updated GTEK. The GKEK is also encrypted with the SS's KEK. The GKEK may be randomly generated in a BS or an ASA server.

A BS transmits the PKMv2 group key update command message carried on the broadcast connection after the M&B TEK grace time. The aim of the second key update command message is to distribute the GTEK to the specific service group. This GTEK is encrypted with transmitted GKEK before the M&B TEK grace time.

An SS must also be capable of maintaining two successive sets of traffic keying material per authorized GSA-ID. Through operation of its GTEK state machines, an SS checks whether it receives new traffic keying material or not. If an SS gets new traffic keying material, then its TEK grace time is not operated. However, if it does not have that, then the SS requests a new set of traffic keying material for the TEK grace time before the SS's latest GTEK is scheduled to expire.

4.12 Key Usage

BS maintains all the keying information for all SAs. The PKM protocol defined in this specification is responsible for synchronizing all the keying information between the BS and the SS.

4.12.1 AK Key Lifetime

SS initiates an authorization exchange with its BS after completing initial capabilities. When the BS receives an auth request message from the unauthorized SS, it will initiate the activation of a new AK. This key is sent back in the key reply message by the BS. Thus according to the predefined AK will expire. The SS is considered unauthorized if SS fails to reauthorize before the expiration of its current AK and BS will not hold any active AKs for the SS. If SS does not reauthorize within a specified time to the BS then all the keys from the keying table will removed. The AK's active lifetime in an authorization reply message that a BS reports is reflected perfectly in implementation permits.

4.12.2 AK Transition Period on BS Side

Whenever an SS requests for an AK, the BS is ready to send an AK. The BS will be able to support two active keys for an SS. These two keys have overlapping lifetimes. AK transition period begins when the BS receives an auth request message from an SS and the BS has a single active AK for that SS. When the BS receives an auth request, the BS will activate a second AK. The key sequence number of this second AK will be greater than one than that of the previous AK. The lifetime of the second AK will be the remaining lifetime of the first AK plus the predefined AK lifetime parameter and this value is set by the BS. Thus the active lifetime of the second key which is new will remain active for one AK lifetime beyond the expiration of the first. The transition period will end with the expiration of the older key. If BS is in the middle of an SS's AK transition period, and if SS holds two active AKs, the response to auth request messages will be with the newer of the two active keys. An auth request will be triggered upon the expiration of the older AK. New key transition period begins once the older key expires.

4.12.3 BS Usage of AK

The keying material that is derived from the SS's AK will be used by the BS for verifying the HMAC-digests in the key request messages that are sent by the SS. The same keying material will also be used for calculating the HMAC-digests that a BS writes into key reply, key reject, and TEK invalid messages that are sent to the SS. The same keying material can also be used for encrypting the TEK that is sent in the key reply messages to an SS.

For the verification of the HMAC-digest in key request messages that is received from the SS, a BS will use an HMAC_KEY_U derived from one of the SS's active AKs. The key request message contains the AK key sequence number which allows the BS to determine which HMAC_KEY_U was used to authenticate the message. If the key sequence number of the AK indicates the newer of the two AKs, the BS will identify this as an implicit acknowledgment and confirms that the SS has obtained the newer of the SS's two active AKs. A BS will use an HMAC_KEY_D derived from the active AK when calculating HMAC-digests in key reply, key reject, and TEK invalid message. When sending key reply, key reject, or TEK invalid messages, if the newer key has been implicitly acknowledged, the BS will use the newer of the two active AKs. If the newer key has not been implicitly acknowledged, the BS will use the older of the two active AKs to derive the KEK and the HMAC_KEY_D. The BS will use a KEK derived from an active AK when encrypting the TEK in the key reply messages.

The BS uses the HMAC_KEY_U and HMAC_KEY_D that are derived from one of the active AKs for calculating the HMAC digests. If the newest AK has been implicitly acknowledged from the SS then the BS uses the newer of the two active AKs to derive the HMAC_KEY_D for signing messages. If there is no implicit acknowledgment of the newer key, the older of the two active AKs will be used by the BS to derive the MAC_KEY_D.

The following subsections describe the calculation of HMAC-digests, derivation of DES, authentication keys, and derivation of HMAC authentication keys.

4.12.3.1 Calculation of HMAC-Digests

The keyed hash in the HMAC-digest attribute and the HMAC tuple will be calculated by using the HMAC with the SHA-1 hash algorithm. For authenticating messages in the DL direction, the DL authentication key HMAC_KEY_D will be used and for authenticating messages in the UL direction, the UL authentication key HMAC_KEY_U will be used. Authentication keys in the UL and DL message are derived from the AK. The sequence number of the HMAC in the HMAC tuple will be equal to the AK sequence number of the AK from which the HMAC_KEY_X was derived. The HMAC digest will be calculated over the entire MAC management message.

4.12.3.2 Derivation of DES, Authentication Keys

AKs are generated by the BS. To generate AK and TEK a random or pseudo-random number generator will be used. IV are also generated by random or pseudo-random generators. FIPS 81 defines "56-bit DES keys as 8-byte (64-bit) quantities where the seven most significant bits (i.e., seven leftmost bits) of each byte are the independent bits of a DES key, and the least significant bit (i.e., rightmost bit) of each byte is a parity bit computed on the preceding seven independent bits and adjusted so that the byte has odd parity. PKM does not require odd parity. The PKM protocol generates and distributes 8-byte DES keys of arbitrary parity, and it requires that implementations ignore the value of the least significant bit of each."

4.12.3.3   Derivation of HMAC Authentication Keys

The derivation of HMAC authentication keys are shown in Table 4.7. HMAC authentication keys are derived using secure hash algorithm. AK in the table is the authorization key. This AK may be either the older AK or newer AK depending on the key sequence number in the key request message that is sent by the SS to the BS. If the key request message indicates the key sequence number of the newer AK the new AK will be used to derive HMAC keys otherwise old AK will be used to derive HMAC keys.

4.12.3.4   Encryption of AK

For encrypting the AK, RSA public key encryption algorithm will be used with the public key of the SS as the key. The protocol uses 65537 (0x010001) as its public exponent and a modulus length of 1024 bits. The PKM protocol employs the RSAES-OAEP encryption scheme (PKCS #1). RSAES-OAEP requires the selection of a hash function, a mask-generation function, and an encoding parameter string. When encrypting the AK the default selections specified in PKCS #1 is used. These default selections for the hash function are SHA-1, the empty string for the encoding parameter string and MGF1 with SHA-1 for the mask generation function.

4.12.4   AK Period on SS Side

SS maintains an AK and is responsible for obtaining authorization from BS. SS will be ready to use two of the recently obtained AKs in the following manner.

4.12.4.1   SS Authorization

AK that has been sent to SS have limited lifetime. It is the responsibility of the SS to periodically refresh the AK by issuing the authorization request message. Authorization state machine manages the scheduling of the authorization requests and the scheduling and refreshing of the keys. Before a configurable duration of time, called the authorization grace time, the SS's authorization state machine schedules the beginning of reauthorization before the latest AK expires. This authorization grace time is configured so that the SS's authorization retry period

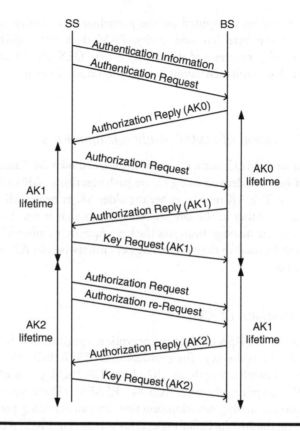

Figure 4.11 Authorization key use in the Base Station and Subscriber Station.

is sufficiently long enough to allow system delays and for providing adequate time for the SS for successful completion of an authorization exchange so that all this happens before the expiration of its most current AK. The knowledge of the authorization grace time is not necessarily known to BS. The BS will deactivate a key by tracking the lifetimes of its AKs once it has expired. The TEK management on both the BS and the SS sides is shown in Figure 4.11. The key request message that is sent from SS to BS represents the key sequence number of the key that is used by the SS. The SS uses AK for different purposes. This is described in the next session.

4.12.4.2 SS Usage of AK

SS when calculating the HMAC_digests will use the HMAC_KEY_U derived from the newer of its two most recent AKs. For authenticating key reply, key reject, and TEK reject messages the SS uses the HMAC_KEY_D derived from either newer or older AKs. The SS decrypts an encrypted TEK that is found in the key reply message with the KEK derived from either newer or older AKs. The most recent AK will be used for deriving KEK if the key request message indicates the key sequence number of the most recent AK, then this key is used for encrypting and decrypting TEKs. HMAC_KEY_U that is derived from the newer of its two most recent AKs will be used by the SS when calculating the HMAC-digests of the HMAC tuple attribute.

4.12.5   TEK Lifetime

Two sets of active TEKs per SAID will be maintained by the BS. These two sets correspond to two successive generations of keying material. The TEKs generated by the two sets will have overlapping lifetimes as determined by the TEK lifetime which is a predefined BS parameter determined by the system configuration. The key sequence number of the newer TEK will be one greater (modulo 4) than that of the older TEK. "Each TEK becomes active halfway through the lifetime of its predecessor and expires halfway through the lifetime of its successor." TEK will be no longer be used once the lifetime expires and the TEK becomes inactive. TEK parameters for the two active TEKs will be present in the key reply messages that are sent by BS to SS. The active lifetimes of TEKs' that a BS sends in a key reply message will be reflected as accurately as implementation rules and according to the time the key reply message is sent which indicates the remaining lifetimes of these TEKs.

4.12.5.1   BS Usage of TEK

The transitions of the BS between the two active TEKs, depends on whether a particular TEK is used for either DL or UL traffic. For every SAID, the BS will have transitions between active TEKs with the expiration of the older TEK, and the transition to the newer TEK for encryption will be done immediately by the BS and hence the UL transition period begins from the time the BS sends the newer TEK in a key reply message and the transition ends on the expiration of the older TEK. It is the responsibility of the SS to send key request messages for updating its keys in a timely fashion. Regardless of whether a client SS has received a copy of a TEK, the BS will make transition to a new key. Depending on the TEK that is used for DL or UL traffic, the BS uses the two active TEKs differently. The BS uses the two active TEKs of the SAID, that is, BS uses the older of the two TEKs for the encryption in the DL traffic and either the older or the newer one for decrypting DL traffic. Only the second half of that TEK's total lifetime will be used by the BS for encryption and the entire lifetime of the TEK for decryption.

4.12.5.2   SS Usage of TEK

SS maintains two successive sets of traffic keying material per authorized SAID. These keying materials have overlapping lifetimes. An SS requests a new set of

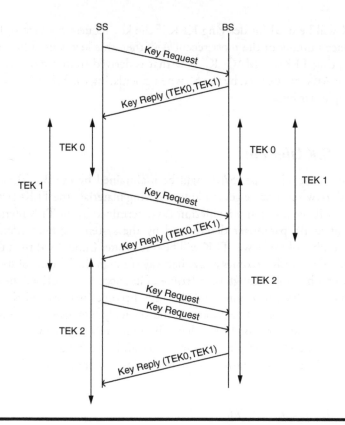

Figure 4.12 Traffic Encryption Keys lifetime in the Base Station and Subscriber Station.

traffic keying material through the operation of its TEK state machines and this is done as before by a configurable amount of time called the TEK grace time. All this happens before the SS's latest TEK expires. For each SAID that is authorized, the SS uses newer of the two TEKs for encrypting the DL channel and either newer or older of the two TEKs for decrypting the DL channel.

The left side of Figure 4.12 illustrates the SS's maintenance and usage of an SA's TEKs, the TEK that will be used to encrypt MAC PDU payloads is indicated by the shaded portion of a TEK's lifetime.

4.13 Conclusions

Multiple security threats arise in wireless networks. Security in IEEE 802.16 is considered especially important. The security sublayer in the fixed and mobile 802.16 standard provides subscribers with encryption and authentication services across the broadband wireless network. The security sublayer employs an authenticated client/server key management protocol. The technology provides vendors with a number of options, which is expected to stimulate a considerable amount of further research.

Acknowledgment

The authors would like to thank Joshua Wright of Aruba Networks for his thorough review of this chapter and invaluable comments which greatly improved the quality.

References

1. IEEE Std 802.16-2004, "IEEE Standard for Local and Metropolitan Area Networks, part 16, Air Interface for Fixed Broadband Wireless Access Systems," IEEE Press 2004.
2. IEEE Std 802.16e-2005, "IEEE Standard for Local and Metropolitan Area Networks, part 16, Air Interface for Fixed and Mobile Broadband Wireless Access Systems," IEEE Press 2005.
3. "WirelessMAN: Inside the IEEE 802.16 Standard for Wireless Metropolitan Area Networks," C. Eklund, R. B. Marks, S. Ponnuswamy, K. L. Stanwood, and N. J. M. Van Waes, IEEE Standards Information Network/IEEE Press (May 15, 2006).
4. RFC 4017, "Extensible Authentication Protocol (EAP) Method Requirements for Wireless LANs," D. Stanley, J. Walker and B. Aboba.
5. RFC 3748, "Extensible Authentication Protocol (EAP)," B. Aboba et al.

Chapter 5

Mobility Support and Conservation of Terminal Energy in IEEE 802.16 Wireless Networks

Sayandev Mukherjee, Kin K. Leung, and
George E. Rittenhouse

The IEEE 802.16-2004 standard (formerly called 802.16d) has been proposed to provide last-mile connectivity to fixed locations by radio links. Despite this original objective, we study in this paper whether mobility can be supported by the 802.16-2004 network without any change in the specification. Mobility enhancements are considered in a later standard (IEEE 802.16e). However, we anticipate that 802.16-2004 devices will be deployed in the field before the 802.16e standard does. Thus, our proposed techniques can be useful independent of the new 802.16e standard. Mobility capability involves two main issues: connection handoff and correct reception for moving terminals. We find that seamless connection handoff can be achieved within the 802.16-2004 standard by: (*i*) applying some of the existing functionalities defined for the terminal initialization process, (*ii*) devising a set of protocols for message exchanges for handoff, and (*iii*) forwarding some of the operational parameters from the current Base Station (BS) to a new one via the backhaul network, instead of over the radio link. As for reception at moving terminals, our analysis of bit-error rate (BER) for the 802.16-2004 orthogonal frequency division multiplexing access (OFDMA) mode shows that under typical radio conditions, the 802.16-2004 link can provide satisfactory error performance for terminal speed up to tens of kilometers per hour. As a result we show that the current 802.16-2004 standard with our proposed technique can support user mobility.

5.1 Introduction

Wireless local area networks (WLANs) based on the IEEE 802.11 standards have been widely deployed and used in airports, offices, and homes. Building on this success, the IEEE 802.16 standard [1–3] approved in 2001 specifies the air interface and Medium Access Control (MAC) protocol for wireless metropolitan area networks (MANs). The idea there is to provide broadband wireless access to buildings through external antennas communicating with radio BSs. The wireless MAN thus offers an alternative to fiber-optic link, cable modem, and digital subscriber loop. Using the new standard, home and business users can be connected via radio links directly to telecommunication networks and Internet.

To overcome the disadvantage of the line-of-sight requirement between transmitters and receivers in the 802.16 standard, the 802.16a standard was approved in 2003 to support non-line-of-sight links, operational in both licensed and unlicensed frequency bands from 2 to 11 GHz, and subsequently revised to create the 802.16d [4] standard. With such enhancements, the 802.16d standard (now called 802.16-2004) has been viewed as a promising alternative for providing the last-mile connectivity by radio link. As a result, many large and small companies are actively developing and testing 802.16-2004 products. However, the 802.16-2004 specification was devised primarily for fixed wireless users. The 802.16e committee [5] was subsequently formed with the goal of extending the 802.16-2004 standard to support mobile terminals.

The primary objective of this chapter is as follows. Although the 802.16-2004 standard is devised for fixed terminal locations, we explore whether the existing specification itself, without any changes or modifications, can be applied to support terminal mobility. Our results in this paper reveal that it is indeed possible for 802.16-2004 without changes to support mobility. There are two aspects of the mobility support. First, the quality of service (QoS) requirements between a mobile terminal and its BS should be satisfied, while the terminal is moving within the coverage area of the BS. Secondly, when the terminal moves from one BS to the next, the network should be capable of handing off the connection from the original BS to the new one, with an objective of minimizing data loss and delay in the handoff process. Correspondingly, we establish the feasibility of mobility support for the 802.16-2004 standard by: (*i*) devising a set of protocols for exchanges of signaling messages for connection handoff from a BS to a neighboring one, and (*ii*) showing reasonable performance in terms of BER under typical radio and user-mobility conditions.

Besides providing new insights into whether the existing 802.16-2004 standard with its original intent to serve fixed locations can indeed support mobility, this work may also have significant commercial implications. First, the 802.16e equipment for mobile environments will not be widely available for at least a couple of years. On the other hand, the 802.16-2004 products will become commonly available very soon (e.g., Intel has promised such with their "Rosedale" chip). Therefore, service providers could start to realize revenue right away by applying the techniques in this chapter to support mobility using existing 802.16-2004 standard. Furthermore, as 802.16-2004-enabled devices will be widely available soon, our proposed techniques can be applied to support mobility capabilities for the "legacy" 802.16-2004 devices, regardless of the final acceptance of the new 802.16e standard. As user mobility now causes the important issue of battery life, which does not exist for fixed wireless networks where power supply is "unlimited," we also propose here a scheme that implements sleep and wakeup for 802.16-2004 devices without any change to the standard to save battery energy. This may be seen as the complementary development to the mobility support proposal, without which service to mobile users will be constrained.

The rest of this chapter is organized as follows. Section 5.2 discusses the objective of connection handoff, application of certain functionalities defined in the 802.16-2004 standard for connection handoff, and handoff protocols. We also identify an existing message in the 802.16-2004 standard that can be used to enable handoff. In Section 5.3, we propose a sleep and wakeup scheme that also does not use new messages. Then, we analyze and show in Section 5.4 the feasibility of the 802.16-2004 physical (PHY) layer for supporting hard handoff and mobility, and the feasibility of the proposed sleep/wakeup mechanism to prolong terminal battery life. We present our conclusions in Section 5.5.

5.2 Protocols for Connection Handoff

5.2.1 Handoff Objective and Mobility Management

As the quality of an established radio link between a Subscriber Station (SS) (or terminal) and its BS deteriorates due to mobility, the objective of handing off the connection to a neighboring BS is to maintain the Internet Protocol (IP) connectivity between the SS and the corresponding host. A major goal is to minimize packet loss and delay induced by the handoff process. As the 802.16-2004 standard defines only the PHY and MAC layers, without loss of generality, suppose that the network under study employs the Hierarchical Mobile IP (HMIP) algorithm [6]

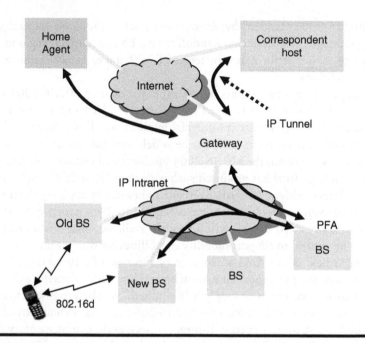

Figure 5.1 Hierarchical mobile Internet Protocol for 802.16-2004 network. (From K. K. Leung, S. Mukherjee, and G. E. Rittenhouse, *Proc. 2005 IEEE Wireless Communications and Networking Conference,* WCNC 2005. With permission.)

for micromobility management. (Similar observations apply to other mobility management algorithms such as [7] and [8].) Using the common terminology for mobile networks, Figure 5.1 shows the architecture of the HMIP for the 802.16 network under consideration. Specifically, one router is designated the Primary Foreign Agent (PFA) and serves as the "anchor point" for each SS (or connection). That is, data from and to a given SS always goes through the corresponding PFA. In addition, the PFA also keeps track of the operational parameters for the 802.16-2004 connections associated with the SS. As shown in the figure, the communication path consists of multiple IP tunnels and packets are forwarded by tunneling.

5.2.2   Initialization Process

As our objective is to support mobility without standard change, we have to use the features and protocols defined in the existing 802.16-2004 standard. We observe that in the most basic sense, handoff is to tear down the existing connection with the current BS and to set up a new connection with a neighboring BS with better link quality. Let us ignore the delay in setting up the new connection for a moment. The key functionalities for handoff are quite similar to the initialization process of a SS when registering with a BS upon power up. This is the starting point of our approach. Namely, we attempt to reuse some of the functionalities of the initialization process defined in the 802.16-2004 standard to assist connection handoff. Toward this goal, it is instructional to first review the initialization process. Then, we identify a set of required functionalities for connection handoff.

A schematic diagram of steps in the initialization process is given in Figure 5.2, which is a simplified version of Figure 55 in Section 6.3.9 of the draft standard [4]. In the first step of the process, an SS begins scanning its frequency list to identify an operating channel (or it may be programmed to log on with a specified BS). After deciding on the channel to attempt communication, the SS tries to synchronize to the downlink (DL) transmission by detecting the periodic frame preambles. Once the PHY layer is synchronized, the SS in step 2 looks for the periodically broadcast downlink channel descriptor (DCD) and uplink channel descriptor (UCD) messages, from which the SS learns the modulation and forward-error-control information for the chosen channel.

With the channel parameters known, the SS identifies a transmission opportunity from the uplink (UL) medium access protocol (MAP) to send ranging message(s) to the target BS. Based on the range-response message from the BS, the SS can adjust its transmission power and timing. Furthermore, the message also provides the SS with the basic and primary management connection identifiers (CIDs). After the ranging process is completed, the SS and BS exchange two messages to inform each other of their capabilities.

The next step is for the SS to go through the authentication procedure and exchange of encryption keys with the BS. The step involves several messages

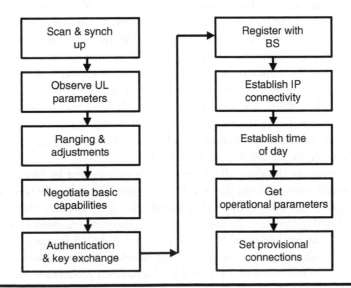

Figure 5.2 Initialization steps for 802.16-2004. (From K. K. Leung, S. Mukherjee, and G. E. Rittenhouse. *Proc. 2005 IEEE Wireless Communications and Networking Conference*, WCNC 2005. With permission.)

exchanged between the SS and BS. It starts with the SS sending its X.502 digital certificate (MAC address and SS public key), cryptographic algorithm, and basic CID to the BS. At the end of the step, both the SS and BS agree upon the authorization and traffic-encryption keys and their associated lifetimes.

In the registration step, the SS sends the BS a request message to register with the network. The BS returns a response message to indicate success or failure of the registration and, if successful, a secondary management CID. Then, the SS acquires an IP address and related parameters via dynamic host communication protocol (DHCP). In the next step, the SS sends a request for time and receives a response from a timeserver. The DHCP server also provides the address of the Trivial File Transfer Protocol (TFTP) server from which the SS can obtain a configuration file containing operational parameters. As a final step, connections are set up for service flows between the SS and BS. There are alternative ways to set up the connections. One way is for the BS to send a dynamic service addition (DSA) message to the SS. The request message contains service flow IDs, possibly CIDs and their QoS parameters. The connection setup is completed after the SS returns a DSA response to the BS and the BS sends an acknowledgment.

5.2.3   Functionalities for Connection Handoff

We obtain the functionalities required by connection handoff by eliminating unnecessary steps in the initialization process. As a result, the schematic diagram in Figure 5.2 can be reduced to Figure 5.3 for connection handoff.

Let us discuss why the functionalities (with over-the-air message exchanges) in Figure 5.2 are sufficient for connection handoff. First, it is assumed that the current BS and the new BS involved in the handoff have identical capabilities, so the negotiation of basic capabilities in step 4 in Figure 5.2 becomes unnecessary. User reauthentication can be achieved by exchange of control messages in the backhaul network. In addition, encryption keys and their associated parameters can be forwarded from the current BS to the new BS also via the backhaul network. Thus, messages exchanged over the radio link for steps 5 and 6 can be avoided. (How authentication and forward of encryption keys can be done via the backhaul network is discussed in the following subsection.)

Furthermore, as the same IP connectivity is maintained by use of HMIP in spite of handoff, one can avoid the need for re-establishing a new IP connection. As the existing IP connection remains unchanged, there is no need for the SS to receive new operational parameters. In addition, as it is reasonable to assume that BSs are synchronized, say by the Global Positioning System (GPS), it is unnecessary for the SS to re-establish time of day as part of the handoff process. Based on all these observations, the functionalities required by the handoff process are thus obtained, as shown in Figure 5.3.

It is worth noting that by comparing Figures 5.2 and 5.3, the handoff procedure actually represents a "short" initialization process. This not only enables handoff

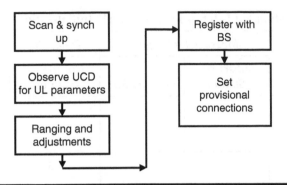

Figure 5.3 Functionalities for connection handoff. (From K. K. Leung, S. Mukherjee, and G. E. Rittenhouse, *Proc. 2005 IEEE Wireless Communications and Networking Conference,* **WCNC 2005. With permission.)**

to reuse existing functionalities but also helps keep the handoff latency satisfactorily low.

We now estimate the latency for the handoff functionalities. Table 5.1 shows the messages involved in the functionalities and their estimated latency by assuming: (*i*) Seven milliseconds per frame, which is a medium frame length, (*ii*) transmission of DCD and UCD every five frames, and (*iii*) messages can be processed and responding messages can be sent in the next frame. Note that delay incurred in channel synchronization and observation of UL parameters can be reduced if a second radio chain is used to perform the task while the first one continues its normal operations.

5.2.4 Handoff Protocol and Message Exchanges

Figure 5.4 shows the sequence of message exchanges for connection handoff. We note that as a SS can stay silent (with no transmission) at times, its BS may not recognize the need of handoff when the SS moves away for the BS. So SS initiated handoff is more appropriate than that initiated by BS. When a SS realizes a need for handoff (e.g., by checking error rate for the MAPs periodically broadcast from BS on the DL or by measuring the received signal strength), it sends a handoff request (HO-REQ) to its current BS (denoted as the old BS). In turn, the BS returns with a handoff acknowledgment (HO-ACK) message to signify that the SS can start the handoff process. It is important to note that both HO-REQ and HO-ACK messages are not defined in the 802.16-2004 standard. We include them here mainly to illustrate the handoff protocol and discuss later how one can replace these messages by an existing one defined in the standard.

Soon after the old BS responds to the SS's request for handoff, the old BS sends the Backhaul Network message 1 (BN-MSG1) to inform the PFA, which is the "anchor" point for the SS, of the MAC address, CIDs, encryption keys, and other

Table 5.1   Estimated Latency for the "Short" Initialization

Functions	Message Exchanged		Delay (for 7 ms per Frame)	
	Subscriber Station	Base Station	Number of Frames	msec
Synch up with downlink channels	–	–	5	35
Observe uplink parameters	–	–	5	35
Ranging and adjustment	2x RNG-REQ	2x RNG-RSP	4	28
Registration	REG-REQ	REQ-RSP	2	14
Establish connections	DSA-RSP	DSA-REQ, DSA-ACK	3	21
Total handoff latency			19	133

Source: From K. K. Leung, S. Mukherjee, and G. E. Rittenhouse, *Proc. 2005 IEEE Wireless Communications and Networking Conference*, WCNC 2005. With permission.

service parameters associated with the SS. Upon receiving the MSG1, the PFA forwards BN-MSG2 messages, which contain information about the SS's MAC address, connections, and operational parameters, via the backhaul network to alert all BS's surrounding the old BS, to look out for the possible handoff of the SS. This list of neighboring BSs, which are the likely candidates for handoff, is maintained at the PFA, and is analogous to the neighbor list in code division multiple access (CDMA) systems.

Following the reception of the HO-ACK message, the SS proceeds to execute the functionalities in Figure 5.3. That is, it scans and synchronizes with a new channel of a neighboring BS (denoted as the new BS in the diagram). Then, it obtains the UL transmission parameters, completes the ranging and adjustment procedure, registers and sets up provisional connections with the new BS. Once the "short initialization process" is completed, the new BS sends the BN-MSG3 to inform the PFA of the completion of the handoff. In turn, the PFA sends the BN-MSG4 to reset PHY and MAC associated with the SS on the old BS. As the new connections are established between the SS and the new BS, the PFA starts to tunnel data to the new BS for forwarding to the SS.

Before continuing, we note that there is a key delay requirement for the handoff protocol to work properly. That is, the BN-MSG2 sent from the PFA must be received and processed by all BSs surrounding the old BS before the first ranging

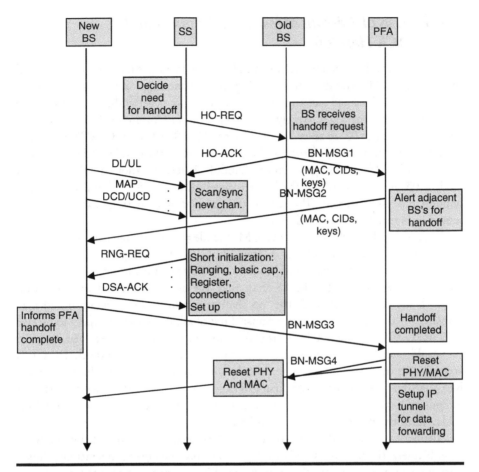

Figure 5.4 Handoff protocol. (From K. K. Leung, S. Mukherjee, and G. E. Rittenhouse, *Proc. 2005 IEEE Wireless Communications and Networking Conference,* **WCNC 2005. With permission.)**

(RNG-REQ) message from the SS arrives. This is so because without receiving the BN-MSG2 message, the neighboring BSs will not be aware of the handoff, and thus follow the rest of the steps for the normal initialization process, instead of those of the "short" process for handoff. (On the other hand, the SS knows that it has to follow the short process because it has been told to do so by receiving the HO-ACK message from the old BS.) As the scanning and synchronization with a new channel may take at least tens of milliseconds to complete, the delay requirement does not appear to be a stringent one. Rather, with a typical high-speed IP backhaul network, it is expected that the BN-MSG2 message can reach and be processed by the neighboring BSs within a couple of tens of milliseconds, which should be short in comparison with the delay incurred in channel scanning and synchronization.

5.2.5 Use of Existing Message to Request and ACK Handoff

As mentioned earlier, the HO-REQ and HO-ACK messages have not been defined in the 802.16-2004 standard and defining the new messages in the standard is not our goal either. To avoid a change to the standard, we observe that it is possible to reuse an existing message, namely the Deregistration Command (DREG-CMD) with action code of 03, to serve the place of the HO-REQ and HO-ACK messages. That is, when the SS initiates the handoff, it sends a DREG-CMD (code = 03) message to its BS. If the BS agrees to the handoff, it returns another DREG-CMD (code = 03) to the SS. When the latter is received by the SS, it signifies that the handoff process is started. The rest of the protocol and message exchanges presented in Figure 5.4 are carried out.

We now explain why the DREG-CMD (code = 03) message can be applied as such. The standard specifies [5] that "the DREG-CMD message shall be transmitted by the BS on an SS's basic CID to force the SS to change its access state. Upon receiving a DREG-CMD, the SS shall take the action indicated by the action code." If the action code is 03, the "SS shall return to normal operation and may transmit on any of its active connections." First of all, BS does not expect to receive the DREG-CMD (code = 03) message from its SSs. If it is indeed received, how the BS would interpret the message has not been specified in the standard. Thus, it is acceptable if the BS chooses to interpret the message as a request for handoff (HO-REQ). After the SS sends the first DREG-CMD (code = 03) message, the SS intends to begin a handoff, thus has a context to interpret the returned DREG-CMD from the BS as an ACK (HO-ACK).

The choice of the DREG-CMD (code = 03) message has an additional advantage. Namely, the message simply asks the SS to resume normal operations, thus it does not cause any adverse effects to the SS if it does not interpret the message in such a special way for supporting handoff. Furthermore, the message also enables correct operations for mixed SSs and BSs with or without the new handoff capability. For example, suppose that the SS has the handoff capability, but its BS does not. In this case, after receiving the first DREG-CMD (code = 03) from the SS, the BS will not send the second DREG-CMD (code = 03) to acknowledge (or approve) the handoff. Without the returned DREG-CMD, the SS simply continues its operations as defined in the original standard. In short, by initiating the handoff process from the SS, and by reusing the DREG-CMD (code = 03) message, we ensure that there are no problems arising from the misinterpretation of a message that arrives at an unexpected time due to a failure of synchronization, for example.

To prevent "ping-ponging" of an SS between an old and new BS, we propose the usual solution of a hysteresis threshold, such that a handoff will only be requested by the SS if the received signal strength from the new BS exceeds that from the old BS by at least this threshold. However, as the handoff scheme uses a "short" version of the initialization process, and in particular omits the authentication and key exchanges and request/grant of connection IDs (which are retained by the old BS and transmitted over the backhaul to the new BS), it is possible for the SS to abort the handoff at any stage before the MAC and PHY are reset at the old BS with BN-MSG4 in Figure 5.4, simply by sending another DREG-CMD (code = 03) to the old BS.

With the protocol designed for supporting terminal mobility in the 802.16-2004 networks, mobile terminals now can no longer enjoy the "unlimited" supply of power as in the fixed wireless networks. In the following, we propose and study a mechanism to conserve battery energy for terminals in the 802.16-2004 networks.

5.3   Energy Conservation via Sleep Mode

5.3.1   Need for "Sleep" and "Wakeup" Modes

The simplest way to save energy and thus prolong the battery life of a SS is to put the SS to "sleep" (i.e., kill all processes running on the SS except for the minimum required to sustain the connection) when it is not involved in any communications. During the sleep period, the SS will not transmit, but "listen" to the channel occasionally to maintain connectivity. This feature is not specified in the 802.16-2004 standard, as that standard was proposed for stationary SSs and it was assumed that power supply for the SSs would not be a critical problem.

Clearly, an SS in the sleep mode requires a complementary mechanism for "waking up" so that it can resume transmitting or receiving. The signaling message that accomplishes this is usually called the "paging" signal. Again, the 802.16-2004 standard does not include a paging signal either.

5.3.2   Message Exchanges to Enter Sleep Mode

We propose the following sequence of steps before an SS enters the sleep mode.

Based on a lack of traffic on the DL and UL, the SS decides that it should enter the sleep mode. Thus, the decision to enter the sleep mode is SS-initiated. The algorithm that the SS applies to arrive at this decision is arbitrary, and can be specific to that SS alone while being unknown to the BS.

The SS uses a CID belonging to one of its current sessions to send a bandwidth request (BR) message to the BS. This is the mechanism for requesting additional bandwidth specified in the current 802.16-2004 standard. As it is very unlikely for a SS to specify 0 bytes in the BR message (i.e., requesting zero additional bandwidth), we propose for the BS to interpret such a message from the SS as a request for permission to enter the sleep mode. A key advantage of such use of BR message with 0-byte request is that if the BS is not enabled to support this sleep/wakeup function, then a request for an additional bandwidth of zero bytes will simply be ignored or discarded, thereby causing no changes to the current session.

If the BS is capable of supporting sleep/wakeup in the SS, then it includes a pre-specified Uplink Interval Usage Code (UIUC) that serves as an acknowledgment to the SS that it is allowed to enter into the sleep mode. The specific UIUC is known and hard-wired in the BS and SS.

On receipt of this acknowledgment message from the BS, the SS enters the sleep mode after a fixed time interval (measured in the number of frames), which is set by the network operator and assumed known to both the BS and the SS. This obviates the need to transmit this interval defining the sleep start time from BS to SS or vice versa, thereby eliminating the need to change the standard to accommodate a message that does so.

During the sleep mode, the SS maintains frame synchronization (this is one of the few processes that are maintained during the sleep mode). Further, the SS decodes DL information periodically to check whether the SS is being "paged" by the BS (see the description of the "wakeup" subsequently). The period (e.g., once every five frames) is predefined by the network operator and hard-wired in the SS. Battery energy is conserved as the SS decodes data occasionally.

Note that from the perspective of the BS, the SS is treated just the same as if it were not in the sleep mode, with the exception that the BS does not transmit data to a sleeping SS without first ensuring that the SS has been waken up. The session parameters associated with a sleeping SS are retained.

The sleep period is not indefinite, but continues only for a finite number of frames, after which the SS wakes up by default if it has not already been waken up by a paging message sent from the BS, or by the arrival of data at the SS intended for transmission on the UL. This maximum sleep duration is also predetermined, fixed, and set by the network operator and assumed to be known to both BS and SS.

5.3.3   Message Exchanges to Wakeup a Sleeping SS

5.3.3.1   SS Wakes Up on Its Own

If the SS has any data to send on the UL to the BS, it simply wakes up by reviving all processes that were running before it entered the sleep mode, and then transmitting the data just as it would have if it had never entered the sleep mode. As the BS retained all session parameters when the SS first entered the sleep mode, the BS is ready to receive the data and does so. The arrival of this data from the SS alerts the BS to the fact that the "sleeping" SS has now woken up by its UL transmission. The BS then treats the SS as if it is no more in sleep mode.

5.3.3.2   BS Wakes Up the SS via a Paging Message

The BS wakes up a sleeping SS by transmitting a "paging" message during one of the periodic frames that is received and decoded by the SS. A specific "paging" message is not supported by the current 802.16-2004 standard. However, the standard permits network operators to use several UIUCs to define modulation and coding rates (burst profiles) for both the orthogonal frequency division multiplexing (OFDM) and OFDMA modes of operation. The network operator may therefore reserve one of these UIUCs for "paging." The steps involved in waking up a sleeping SS to receive DL data is given next.

In the UL-MAPs transmitted by the BS over a number of consecutive frames, the BS specifies the CID of the SS scheduled for wakeup, employing the UIUC designated for "paging" purposes with the most robust modulation and coding scheme available. The reason for this is that the SS is only receiving and decoding periodic frames during sleep mode. Further, as the sleeping SS does not transmit anything on the UL, the BS has no information about DL channel quality, and cannot tailor its coding scheme accordingly. Thus, if the DL channel suffers degradation, the SS may not receive the BS transmission, so robust modulation/coding with repetition maximizes the chance of the SS receiving the paging "message."

Upon receiving the UIUC information in the UL-MAP, the SS exits the sleep mode. To confirm with the BS that the SS has indeed exited the sleep mode, the SS transmits another bandwidth request message with 0 bytes in the BR field. The receipt of this message by the BS is interpreted by the BS as an acknowledgment by the SS that it has now "woken up" and resumed normal operation.

5.4   Feasibility of Supporting Hard Handoff

In this section, we show that the PHY layer of 802.16-2004 standard can support terminals moving with moderate speed. In particular, we present a simple analysis leading to an expression for the BER on a wireless link between a BS and an SS with the OFDMA air interface as specified in 802.16-2004. It is reasonable to assume [9] that such a link is limited by intercarrier interference (ICI), rather than by interference between OFDMA users. This is because the latter is averaged over multiple users and, for universal or low frequency reuse, may be taken to be small and relatively constant overtime, and thus absorbed into the Gaussian thermal noise. (On the other hand, for the OFDM with time division multiple access (TDMA) PHY mode, this assumption may be problematic for systems with high frequency reuse. This is so because the lack of fast power control and the bursty nature of interference in a TDMA system implies that the out-of-cell interference is more accurately modeled by a log-normal distribution [10].) In addition, we assume no fast or soft handoff (so that the terminal of interest remains supported by the given BS over the duration of the following analysis). Further, we assume no or slow power control, so that the transmitted symbol energy stays the same over the time interval of interest.

The link is subject to both fast and slow fading, the latter assumed to be almost unchanged over the duration of observation and hence absorbed into the average symbol energy at the receiver. The fast fading is assumed to be Rayleigh, given by the Clarke-Jakes model [11]. Then, it can be shown [9] that the average received symbol energy-to-noise ratio is given by

$$\bar{\gamma}_s = \cfrac{1}{1 - \cfrac{1}{N^2}\left[N + 2\sum_{i=1}^{N-1}(N-i)J_0\left(2\pi f_m T_s i\right)\right] + \cfrac{NT_s}{E_s/N_0}}$$

where N is the number of OFDM subcarriers, T_s is the duration of each M-ary Quadrature Amplitude Modulated (QAM) symbol transmitted on a subcarrier, N_0 is the noise power, E_s is the average transmitted symbol energy, and $f_m = f_c(v/c)\pi$ is the Doppler frequency, where f_c is the carrier frequency, the terminal speed, and c the speed of light. The corresponding average received bit energy-to-noise ratio is given by $\bar{\gamma}_b = \bar{\gamma}_s/\log_2 M$, or

$$\bar{\gamma}_b = \cfrac{1/\log_2 M}{1 - \cfrac{1}{N^2}\left[N + 2\sum_{i=1}^{N-1}(N-i)J_0\left(2\pi f_m T_s i\right)\right] + \cfrac{NT_s}{\log_2 M}\left(\cfrac{1}{E_b/N_0}\right)}$$

for M-ary QAM modulation [e.g., $M = 4$ for Quadrature Phase Shift Keying (QPSK) and $M = 16$ for 16-QAM]. Note that $E_b = E_s/\log_2 M$ is the average transmit energy per bit.

We assume symbol-by-symbol detection at the receiver. Let $P_b(\gamma_b)$ be the probability of BER when the received bit energy-to-noise ratio is γ_b. Then we have

$$P_b = \int_0^\infty P_b(\gamma) f_{\gamma_b}(\gamma) d\gamma \tag{5.1}$$

where $f_{\gamma_b}(\gamma)$ is the probability density function (pdf) of the bit energy-to-noise ratio under the chosen fading model. For the case of Rayleigh fading, we have

$$f_{\gamma_b}(\gamma) = \frac{\exp\left(-\gamma/\bar{\gamma}_b\right)}{\bar{\gamma}_b}, \quad \gamma \geq 0. \tag{5.2}$$

Finally, we make the assumption that the ICI may be approximated by additive white Gaussian noise (AWGN). As shown in [9], this approximation is very accurate for $N = 256$ and virtually exact for $N \geq 1024$. This approximation allows for the reuse of well-known expressions for the probability of bit-error on an AWGN channel. For QPSK modulation, the exact expression for bit error probability is available:

$$P_b(\gamma_b) = Q\left(\sqrt{\gamma_s}\right) = Q\left(\sqrt{2\gamma_b}\right) \tag{5.3}$$

whereas for general *M*-ary QAM, we only have the approximation (which applies to Gray coding)

$$P_b(\gamma_b) \approx \frac{P_M(\gamma_s)}{\log_2 M} \tag{5.4}$$

where P_M is the symbol-error probability. For example, the symbol-error probability for 16-QAM is

$$P_M(\gamma_s) = 3Q\left(\sqrt{\frac{\gamma_s}{5}}\right)\left[1 - \frac{3}{4}Q\left(\sqrt{\frac{\gamma_s}{5}}\right)\right]. \tag{5.5}$$

Let us consider as an example the OFDMA mode in the Multi-channel Multiport Distribution Systems (MMDS) band (carrier frequency $f_c = 2.6$ GHz), with a bandwidth of 12 MHz. As specified in the 802.16-2004 standard, the number of subcarriers is $N = 2048$ and the OFDM useful symbol period is $NT_s = 149.33\ \mu s$. The raw bit rate for rate 1/2 coding is about 10 Mb/sec for QPSK, and about 20 Mb/sec for 16-QAM when the cyclic prefix duration is 1/32 of the useful symbol period. For this choice of parameters, in Figure 5.5a, we plot the BER as obtained from Equations 5.1 and 5.2 for QPSK modulation ($P_b(\gamma)$ given by Equation 5.3) as a function of the terminal speed *V* for several choices of the average bit energy to noise ratio E_b/N_0. In Figure 5.5b, we repeat the plots for the case of 16-QAM modulation ($P_b(\gamma)$ given by Equations 5.4 and 5.5). For QPSK, the lack of dependence of BER on E_b/N_0 supports the validity of the approximation that the system is ICI-limited and not noise-limited, while the accuracy of this approximation is lower for the 16-QAM case. Note that with QPSK modulation, we can maintain a BER of 0.002 percent or less for terminal speeds up to 40 km/h, and with 16-QAM, we can maintain this BER for terminal speeds up to about 10 km/h. Thus, terminal mobility can be supported for these moderate speeds. It is important to note that these results correspond to cases without coding. Therefore, it is expected that higher terminal speeds can be supported when coding techniques are used.

Next, let us consider the handoff latency requirement. In a practical system, SSs anywhere in a cell could see the dominant pilot change or a new BS enter the set of candidates for handoff. The frequency of this event, which could potentially trigger a HO-REQ, depends on the speed of the SS. For slowly-moving SSs, the channel quality to the old BS does not change rapidly, so the maximum latency possible for the handoff is given by the decorrelation distance of the shadow fading from the old BS, which are several tens or even hundreds of meters. For slowly-moving SSs, the time taken to cover this distance is much larger than the time taken to complete the handoff. For a fast-moving SS in the interior of the cell, the channel quality changes rapidly over a short time interval. Thus, the introduction of a time hysteresis for the HO-REQ, that is, requiring that a new BS should have a better channel for at least some length of time

(a)

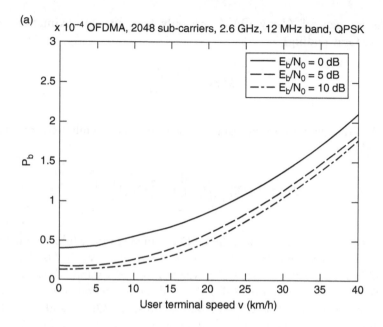

(b)

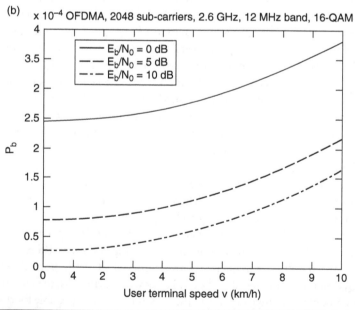

Figure 5.5 **Bit-error rate versus terminal speed for an MMDS orthogonal frequency division multiplexing access link with Rayleigh fading, 2048 subcarriers, 2.6-GHz carrier frequency, and 12-MHz bandwidth for: (a) QPSK and (b) 16-QAM modulation. (From K. K. Leung, S. Mukherjee, and G. E. Rittenhouse, *Proc. 2005 IEEE Wireless Communications and Networking Conference*, WCNC 2005. With permission.)**

before a HO-REQ is initiated, is likely to remove the need to perform handoffs arising from the majority of such events. In general, for a fast-moving SS, the time hysteresis requirement is likely to be met when the SS is moving toward the periphery of the cell and into a new cell. We now focus on this case.

Assume that the cell radius (covered and served by one BS) is 1 km and that there is a 5 percent overlap in the coverage area of two adjacent cells. Modeling these cells as circles of radius 1 km each, with distance d km between the bases, the fractional area of overlap is given by

$$\frac{2}{\pi}\left[\cos^{-1}\left(\frac{d}{2}\right)-\left(\frac{d}{2}\right)\sqrt{1-\left(\frac{d}{2}\right)^2}\right]$$

which is 5 percent when $d=1.75$ km. This in turn implies that the overlapping area extends 250 m into each cell.

Finally, consider a user terminal moving from one cell to the next in a straight line with constant velocity. To simplify the analysis, let us assume that if the terminal has not been handed off to the adjacent cell (BS) when it has passed the common overlapping coverage area of the current cell, then the connection is dropped. Then the maximum total distance over which the handoff must be completed is 250 m. For the maximum supported mobile speed of 40 km/h (with a BER of 0.002 percent or less and QPSK modulation) we see that the handoff must be completed within 22.5 sec. Such a requirement can be easily met when compared with latency estimates in Table 5.1 plus the typical delay in tens of milliseconds incurred in the message exchanges in Figure 5.4.

Consider now the probability that the BS is unable to wakeup a sleeping SS with the "paging" UIUC (due to channel degradation) in the maximum possible latency interval available to complete a handoff. Recall that the SS only receives and decodes one frame every N_f frames. If the frame duration is T_f, then each failed paging attempt represents an additional delay of $N_f T_f$ in waking up the SS. Assuming independent fading conditions over successive frames when the sleeping SS receives and decodes, we see that the number of attempts N_w made by the BS to wakeup the SS before it is successfully awakened is a geometric random variable: $P\{N_w = n\} = q^{n-1}(1-q)$, $n = 1, 2, \text{K}$, where q is the probability that the paging UIUC is incorrectly received at the SS on any single attempt.

Let $P_b(\gamma_b)$ be the BER when the received bit energy-to-noise ratio is γ_b. Now, the UIUC is four bits in length. As the bit energy-to-noise ratio is assumed the same over the duration of the UIUC, we have

$$q = 1 - \int_0^\infty \left[1 - P_b(\gamma)\right]^4 f_{\gamma_b}(\gamma) d\gamma. \tag{5.6}$$

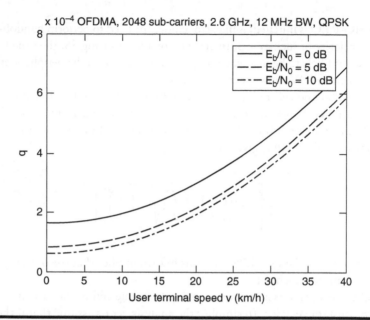

Figure 5.6 **Plot of q as given by Equation 5.6 versus mobile speed v for the system parameters as in Figure 5.5a. (From S. Mukherjee, K. K. Leung, and G. E. Rittenhouse, *Proc. 2005 IEEE Pacific Rim Conf. on Communications, Computers and Signal Processing*, PACRIM 2005. With permission.)**

Note that our proposed paging scheme requires the modulation to be QPSK. In Figure 5.6, we plot q vs. the mobile speed v for the same three choices of E_b/N_0 as in Figure 5.5.

We have shown above that for cells with radius 1 km, 5 percent overlap between adjacent cells, and mobile speed of 40 km/h (the maximum speed that can be supported with a BER of 0.002 percent or less and QPSK modulation), the handoff must be completed within 22.5 sec, which considerably exceeds the estimate of 19 frames or 133 milliseconds for exchanging the messages to perform the handoff. Even if we conservatively allocate only ten seconds to waking up a sleeping SS, with frames of duration $T_f = 7$ msec and the SS decoding one frame out of every $N_f = 5$, we have a maximum of $10s/N_f T_f = 286$ attempts to wakeup the sleeping SS within the given time frame. Thus the probability of dropping the session during handoff because of not being able to wakeup the SS in time is $P\{N_w \geq 286\} = q^{286-1}$, which is negligibly small for the range of speeds shown in Figure 5.6.

5.5 Conclusion

We have studied in this chapter that the 802.16-2004 standard can intrinsically support terminal mobility without any change in the specification, although its original intent is to provide the last-mile connectivity to fixed locations. Specifically, we have shown that mobility capability can be achieved for the 802.16-2004 by: (*i*) applying some of the existing functionalities defined for the initialization process, (*ii*) devising a new set of protocols for connection handoff, and (*iii*) forwarding some of the operational parameters from the current BS to the handoff BS via the backhaul network, instead of over the radio link. Our link-performance study shows that under typical radio conditions, the 802.16-2004 link can provide satisfactory bit-error performance for terminal speed up to tens of kilometers per hour.

We have also proposed protocol and control mechanisms to conserve battery energy via sleep and wakeup modes for mobile SSs in IEEE 802.16-2004 networks. A simple analysis demonstrates that the probability that the BS cannot successfully communicate with a sleeping SS to wake it up (to perform a handoff) before the current channel degrades so much that the session is dropped, is negligibly small for the low and moderate mobile speeds for which the link BER has also been shown to be acceptably small. Combining protocols for mobility support and conservation of battery energy for mobile SSs, our results have revealed that the 802.16-2004 standard can be used to support services in mobile environments, although the standard was originally devised for fixed wireless applications. As the 802.16-2004 devices will be widely available in the near future, our proposed techniques can be applied to enable mobility capabilities for the "legacy" devices, regardless of the final acceptance of the new 802.16e standard.

Acknowledgments

The authors would like to thank T. E. Klein and H. Viswanathan for their helpful discussions.

References

1. A. T. Campbell, J. Gomez, S. Kim, A. G. Valko, C.-Y. Wan, and Z. R. Turanyi, "Design, Implementation, and Evaluation of Cellular IP," *IEEE Pers. Commun. Mag.*, 7:4 (2000), 42–49.
2. C. Eklund, R. B. Marks, K. L. Stanwood, and S. Wang, "IEEE Standard 802.16: A Technical Overview of the WirelessMAN Air Interface for Broadband Wireless Access," *IEEE Commun. Mag.*, 40:6 (2002), 98–107.
3. B. Fong, N. Ansari, A. C. M. Fong, and G. Y. Hong, "On the Scalability of Fixed Broadband Wireless Access Network Deployment," *IEEE Commun. Mag.*, 42:9 (2004), S12–S18.
4. E. Gustafsson, A. Jonsson, and C. E. Perkins, "Mobile IPv4 Regional Registration," Internet-draft, draft-ietf-mobileip-reg-tunnel-09.txt, June 2004.
5. IEEE 802.16d, "Draft IEEE Standard for Local and Metropolitan Area Networks—Part 16: Air Interface for Fixed Broadband Wireless Access Systems," May 13, 2004.

6. IEEE 802.16e committee, http://ieee802.org/16/tge.

7. S. Das and H. Viswanathan, "On the Reverse Link Interference Structure for Next Generation Cellular Systems," *Proc. IEEE Global Telecommunications Conference 2004 (Globecom'04)* (Dallas, TX, 2004), pp. 3068–3072.

8. W. C. Jakes, *Microwave Mobile Communications*. Piscataway, NJ: IEEE Press, 1995.

9. I. Koffman and V. Roman, "Broadband Wireless Access Solutions Based on OFDM Access in IEEE 802.16," *IEEE Commun. Mag.*, 40:4 (2002), 96–103.

10. R. Ramjee, K. Varadhan, L. Salgarelli, S. R. Thuel, S.-Y. Wang, and T. LaPorta, "HAWAII: A Domain-Based Approach for Supporting Mobility in Wide-Area Wireless Networks," *IEEE/ACM Trans. Networking*, 10:3 (2002), 396–410.

11. G. L. Stüber, *Principles of Mobile Communication*, 2nd edn. Boston, MA: Kluwer Academic Publishers, 2002.

12. K. K. Leung, S. Mukherjee, and G. E. Rittenhouse, *Proc. 2005 IEEE Wireless Communications and Networking Conference*, WCNC 2005.

13. S. Mukherjee, K. K. Leung, and G. E. Rittenhouse, *Proc. 2005 IEEE Pacific Rim Conf. on Communications, Computers and Signal Processing*, PACRIM 2005.

Chapter 6

On the Best Frequency Reuse Scheme in WiMAX

S.-E. Elayoubi, O. Ben Haddada,
and B. Fourestié

In this chapter, we present and compare the frequency reuse schemes proposed in the literature for WiMAX cellular systems. These schemes are conceived to reduce intercell interference, especially at cell edges, and include the classical reuse three scheme in addition to novel hybrid frequency allocation schemes. These later define a mix of reuse one at cell centers and reuse three at cell edges, or a power/frequency scheduling that consists in allocating all available frequencies to each cell with power control to reduce interference at cell edge. Our results compare these schemes and show that the power/frequency scheduling scheme outperforms all other ones by confining intercell interference and optimally using frequency resources.

6.1 Introduction

When a frequency reuse of one is used (i.e., all sites have been assigned the entire frequency band), intercell interference becomes the limiting factor. In fact, due to heavy cochannel interference (CCI), users at cell edges may suffer degradation in connection quality. This cell edge interference problem has recently been addressed by appropriately configuring frequency usage without resorting to traditional frequency planning methods. In a frequency reuse three scheme, each cell is allocated a third of the frequency band, and a three-cell pattern is used. If intercell interference is diminished, capacity is also reduced as a cell can only use a third of the total frequency resources.

A mix of frequency-reuse one and three schemes has then been proposed to avoid interference at cell edges. This consists in dividing the frequency band into two subbands: a frequency-reuse one subband, allocated to users at cell center, and a frequency-reuse three subband, allocated to cell-edge users [1]. This scheme indeed decreases interference, but also reduces peak data rates as the entire frequency band is not allocated to each cell.

When using this fractional frequency allocation and to overcome the loss of capacity caused by the partial use of the frequency band in each cell, some manufacturers propose to implement fractional reuse as part of the scheduling decision [2]. The idea is to allocate cell-edge frequencies in adjacent cells with lower power to limit the interference. This can be viewed as a power/frequency scheduling technique based on the path loss of the user.

These frequency allocation techniques have been proposed without any performance analysis. In this paper, we analyze and compare four different frequency allocation schemes: reuse one, reuse three, a mix of reuse one and three, and power/frequency scheduling. We consider a system carrying elastic (FTP-like) traffic and evaluate the capacity taking into account that the modulation is chosen depending on the Signal to Interference plus Noise Ratio (SINR). Note that the comparison between the different scenarios is based on a basic interference simulator, combined with an analytical capacity calculation. In fact, the simulator considers a regular hexagonal cellular network and calculates the mean useful throughput per subchannel for various numbers of collisions, in each frequency planning scenario. This throughput is then used to calculate the capacity using classical queueing theory methods.

Our numerical results show the following results:

1. The reuse one scheme achieves high cell throughput, however it suffers from very low cell-edge performance.
2. The reuse three scheme decreases severely the overall throughput because only one-third of the capacity is used in each cell.
3. A reuse one at cell centers combined with a reuse three at cell edges can achieve an acceptable compromise between overall throughput and cell-edge performance.
4. Finally, if a power/frequency scheduling is implemented at the Base Station (BS), we can achieve the best performance by increasing both overall and cell-edge throughputs.

The remaining of this chapter is organized as follows. We first recall in Section 6.2 the frequency allocation schemes in mobile WiMAX. In Section 6.3, we show how to calculate the mean throughput in the cell. Section 6.4 presents a general methodology for the calculation of the capacity in WiMAX and shows the impact of the interference on the performance. Section 6.5 analyzes the classical frequency reuse schemes, while Sections 6.6 and 6.7 present the hybrid schemes. Section 6.8 eventually concludes the chapter.

6.2 Frequency Allocation in Mobile WiMAX

In orthogonal frequency division multiple access (OFDMA), the frequency band is divided into subcarriers. These subcarriers are grouped into sets, called subchannels in WiMAX, which are allocated to users. We will consider in our numerical applications the case of an Fast Fourier Transform (FFT) size of 1024 subcarriers, with an intercarrier spacing of 7.8 kHz.

Two allocation modes are possible: distributed and adjacent.

6.2.1 Distributed Allocation

In the distributed subcarriers allocation, full channel diversity is obtained by distributing the allocated subcarriers to subchannels using a permutation mechanism. This mechanism is designed to introduce frequency diversity, thus minimizing the performance degradation due to fast fading which is characteristic of mobile environments. In addition to that, WiMAX standards [3,4] specify two different distributed allocation modes: the fully used subchannelization (FUSC) mode where all subcarriers are used to form subchannels in each cell, and the partially used subchannelization (PUSC) mode where the frequency band is divided into three segments.

For illustration, with an FFT size of 1024 and after reserving the pilot and guard subcarriers, a FUSC allocation will correspond to 16 subchannels of 48 data subcarriers each, whereas a PUSC allocation will correspond to 30 subchannels, each containing 24 data subcarriers. Note that assigning subcarriers to subchannels in PUSC is a bit complicated, as it employs two permutations:

- An outer permutation divides the subcarriers into six major groups of clusters using a specific renumbering sequence.
- An inner permutation operates separately on each major group, distributing subcarriers to subchannels within the group and is based on the FUSC permutation with distinct parameters for the odd and even major groups.

This is illustrated in Figure 6.1, where two groups are assigned to one segment corresponding to a sector of the cell. Note that a segment can also be allocated to a cell in an omni-directional setting.

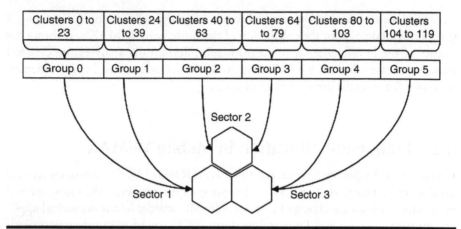

Clusters 0 to 23	Clusters 24 to 39	Clusters 40 to 63	Clusters 64 to 79	Clusters 80 to 103	Clusters 104 to 119
Group 0	Group 1	Group 2	Group 3	Group 4	Group 5

Figure 6.1 Construction of groups and segments in the partially used subchannelization allocation mode.

6.2.2 Adjacent Allocation

This method uses adjacent subcarriers to form subchannels. It corresponds to the WiMAX AAS (Advanced Array Systems) mode, designed to support Multiple-Input Multiple-Outpt (MIMO) techniques and adaptive modulation. Note that, to achieve a frequency diversity, mobiles using adjacent allocation may hop rapidly between different subchannels during their communication times.

6.3 Calculation of the Mean (Instantaneous) Throughput of the Cell

6.3.1 What Is Interference in OFDMA?

In the presence of multi-path propagation, code division multiple access (CDMA) codes, used in 3G networks, loose their orthogonality property, leading to Inter-Symbol Interference (ISI). On the contrary, the OFDM technology, used in WiMAX systems, multiplexes the data over a large number of subcarriers that are spaced apart at separate frequencies. This modulation scheme provides orthogonality between subcarriers which simplifies the detection and eliminates the intracell interference [5].

However, when cellular networks are designed using OFDMA technology, intercell interference appears as the limiting problem. In the downlink for instance, intercell interference occurs at a mobile station when a nearby base station transmits data over a subcarrier used by its serving base station, as illustrated in Figure 6.2. This is called collision and, depending on the number of interfering

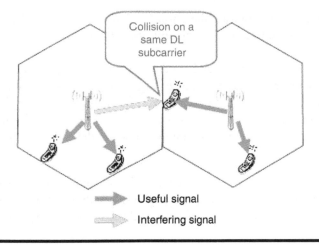

Figure 6.2   Intercell interference in WiMAX.

base stations, we can have more than one collision at the same subcarrier. As the frequency is allocated in WiMAX on the basis of subchannels, each consisting of several subcarriers, different scenarios are possible:

- In the case of adjacent allocation, when a collision occurs, all the subcarriers of the subchannel are involved. Frequency hopping is then necessary to distribute the interference between users.
- For distributed allocation, frequency diversity is ensured when constructing the subchannels, thus leading to an averaged interference between calls.

However, authors in Ref. [6] showed that the number of collisions is independent of the allocation mode, and is always distributed following a hyper-geometric distribution when the system is homogeneous.

6.3.2   Why Is Interference Limiting?

In WiMAX, Adaptive Modulation and Coding (AMC) will be used. The choice of the modulation depends on the value of SINR (also called C/I) through the perceived Bloc Error Rate (BLER): the most efficient modulation that achieves a BLER larger than say 10^{-1} is used. For each SINR value, this leads to a couple of values $(e, BLER)$, where e is the efficiency of the used modulation (e.g., e is equal to 1 bit/symbol for QPSK 1/2 and to 5 bits/symbol for 64 QAM 5/6). These values are determined by link-level curves $e(C/I)$ and BLER(C/I), available in the literature [7]. When interference increases, the SINR decreases, and the BLER increases, forcing the base station to use a more robust (less efficient) modulation. This may have two negative impacts, depending on traffic characteristics:

1. If the connection corresponds to a real time call (e.g., a video conference), changing the modulation because of large interference may result in degrading the quality of the image, or even in completely dropping the call.
2. If the connection corresponds to an elastic call (e.g., FTP-like data transfer), a lower efficiency modulation results in a lower throughput and a large transfer time.

In the remainder of this chapter and for the sake of comparison, we will consider a WiMAX system, in the downlink, carrying elastic traffic and study the impact of frequency reuse schemes on the interference and the system quality of service (QoS).

6.3.3 SINR Calculation

When calculating the SINR, we must take into account the geometric disposition of the interfering cells and the propagation conditions. These latter are the distance between the transmitter and the receiver, the shadowing and the frequency-selective fading. However, as in OFDMA the data is multiplexed over a large number of subcarriers that are spaced apart at separate frequencies, the channel consists of a set of parallel, flat and non-frequency selective fading, channels [5]. The received signal is then only impacted by distance and slow fading, and the SINR is given by:

$$\frac{C}{I} = \frac{P/L_0}{\sum_i (P/L_i) + N_0} \tag{6.1}$$

where P is the power emitted by the base station to each user, N_0 is the background noise, and L_i is the path loss between interfering base station i and the corresponding mobile. Note that the path loss between a transmitter and a receiver depends on the distance between them, in addition to a shadowing variable resulting from the obstacles between them:

$$L = K_1 \log(\text{distance}) + K_2 (\text{frequency}) + \text{shadowing}$$

where K_1 is a constant and K_2 is another constant that depends on the frequency band.

For a seek of simplicity, we will neglect in our simulations the impact of the shadowing. A more complete mathematical model detailing the shadowing factor can be found in Ref. [6].

6.3.4 Mean Throughput Calculation

Let $\bar{D}$ be the instantaneous throughput of a typical call in the cell. This throughput depends, in addition to the offered bandwidth by subcarrier W and the number M

of subcarriers by subchannel, on the efficiency of the used modulation and the BLER. This relationship is given by:

$$\bar{D} = E[MWe \times (1 - BLER)]. \tag{6.2}$$

Note that M is equal to 48 in FUSC and 24 is PUSC.

As stated here, the BLER depends on the physical layer characteristics (used modulation and path loss) and on the amount of interference. It is then correlated with the efficiency. This gives:

$$e \times (1 - BLER) = e\left(\frac{C}{I}\right) \times \left(1 - BLER\left(\frac{C}{I}\right)\right) = f\left(\frac{C}{I}\right).$$

However, the function $f(C/I)$ is not known and depends on two parameters: the position of the user in the cell (impacting its path loss), and the number of collisions that occur between the subchannel allocated to the user and subchannels allocated in adjacent cells.

To calculate the mean throughput, we then average the function $f(C/I)$ following these two parameters:

1. We first fix the number of collisions by fixing the vector of collisions **X**, where **X** is a vector or zeros and ones whose dimension is equal to the number of interfering cells and whose elements correspond each to an interfering cell. The value 1 signifies that collision occurs with the corresponding cell.
2. We determine, for each point of the cell at distance $\vec{r}$ from the base station, the throughput that is achieved with the vector of collisions **X**:

$$D(\mathbf{X},r) = MWf \frac{C}{I}(\mathbf{X},\vec{r}).$$

For each vector **X**, the mean throughput $D(\mathbf{X})$ can then be obtained by integrating the function $D(X,\vec{r})$ over the cell surface.

In our simulations, we then first obtain the mean throughput $\bar{D}$. This is done by dividing the cell into a grid and calculating, at each point, the intercell interference, the SINR and the resulting throughput. We consider a classical reuse.

6.4 General Methodology for the Calculation of the Capacity

6.4.1 Why the Mean Throughput Is Not the Right Measure?

The simplest way to calculate the capacity of the system is to consider that users that arrive to the cell will have a throughput that is equal to the mean throughput calculated before. This methodology has been considered in several systems like 3G ones [8], and is justified when real time calls are considered, knowing that the sojourn time of these calls is independent from its position in the cell.

However, when elastic calls are considered like in this study, users that are far from the base station have a larger influence on the cell capacity than cell-center users, and cannot be granted the same weight in the calculations. In fact, an elastic call stills active in the cell until downloading a file of given size, and charges then the cell for a time that is proportional to the interference it receives. We show next how to find the exact contribution of each user to the overall load of the cell.

6.4.2 Harmonic Mean

In Ref. [9], the authors faced the same problem when evaluating the capacity of a High Speed Downlink Packet Access (HSDPA) system. The solution they propose is to use the harmonic mean instead of the geometric mean when evaluating the capacity. In fact, let X be the random variable representing the sojourn time of a call that is allocated one subchannel. It is obvious that, in addition to the size of the file to be transferred, the value of the sojourn time of a user depends on the instant throughput that he receives. The mean service time of a call can then be calculated by a double integration over the cell surface and the file size. Note that, these two parameters (position and file size) being independent, the mean service time is given by:

$$E[X] = \frac{Z}{MW} E\left[\frac{1}{f(C/I)}\right]$$

which corresponds to the calculation of the harmonic mean of the throughput over the cell surface:

$$\tilde{D} = \frac{MW}{E[\frac{1}{f(C/I)}]}.$$

6.4.3 Markovian Model

To evaluate the performance using this mean throughput, we use queueing theory models. We consider a system with Poisson arrivals of rate λ, where each new call is either accepted and granted a subchannel until ending the transfer of a file of mean size $Z = 2.34$ Mbytes, or blocked. Note that if no subchannels are available, the call is blocked (the number of subchannels, denoted by N, is equal to 16 in FUSC and 32 in PUSC). This system can be modeled as a classical loss network which has a product form solution as described in Ref. [10]. We can then calculate the steady-state probabilities of U, the number of users in the cell by:

$$\pi(U) = \frac{1}{\sum_{U \leq N} \frac{(\lambda \tilde{D}/Z)^U}{U!}} \frac{(\lambda \tilde{D}/Z)^U}{U!}.$$

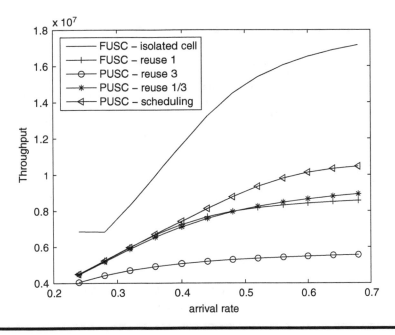

Figure 6.3 **Overall cell throughput for different frequency planning schemes.**

Using this model, we can illustrate the interference impact on the system. We plot in Figure 6.3 the cell throughput when it is isolated (with no interference) compared with its throughput with a frequency reuse one (obtained by a FUSC subchannel allocation in each cell). We first observe that, when the arrival rate of new calls increases, the throughput increases as more calls are connected. However, the intercell interference in the reuse one scheme makes less efficient modulations to be used leading to lower throughputs.

Furthermore, Figure 6.4 plots the throughput of a cell-edge user. A large degradation of this throughput is observed, and for large loads, it attains a limit of 0.15 Mbits/s, compared with 0.6 Mbits/s achieved when no interference is considered. Limiting the intercell interference is thus crucial for cell-edge user performance.

6.5 Classical Interference Mitigation: Reuse One with Tri-Sectored Cells and Reuse Three Schemes

The classical interference avoidance scheme is obtained by dividing the frequency band into three equal subbands and allocate the subbands to the cells so that adjacent cells always use different frequencies. This scheme, called reuse three scheme and illustrated in Figure 6.5, is possible using the PUSC mode. The underlying idea is to allow interference only from cells located in ring 2, leading thus to low interference.

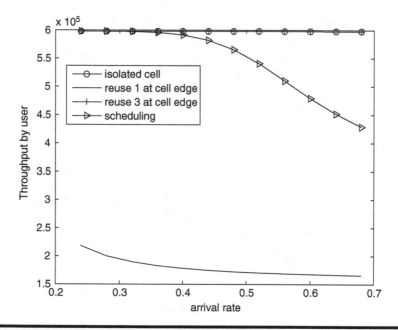

Figure 6.4 Throughput of a cell-edge user for different frequency planning schemes.

When analyzing the system, we can use the same methodology described for reuse one systems, with the difference that interference originates from farther cells. As expected, the interference at the cell edge is substantially decreased and the

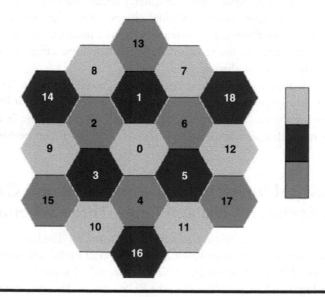

Figure 6.5 Reuse three scheme: interfering cells are in ring 2.

throughput of cell-edge users increases compared with a reuse one scheme, and is almost equal to the case with no interference. This is illustrated in Figure 6.4. However, this comes at a cost. The overall cell throughput plotted in Figure 6.3 is severely affected. This large capacity loss is due to the fact that only one-third of the resources are used in each cell, and the decrease in the interference cannot balance this loss. Innovative solutions are then to be found to decrease interference at cell edges without, or with less, capacity loss.

6.6 Fractional Reuse: A Hybrid Mix of Reuse One and Reuse Three Schemes

A hybrid solution between reuse one and reuse three schemes has been proposed [1]. The idea is to use a frequency reuse of one at the cell centers where interference is low, and a frequency reuse of three at the cell edges where users are more subject to interference. This is illustrated in Figure 6.6 and called fractional frequency reuse. This frequency allocation mode is possible in WiMAX using the PUSC mode. In fact, each segment in PUSC is decomposed into two groups, resulting in six different groups: three even groups of six subchannels each and three odd groups of four subchannels each. All even groups can thus be allocated to the cell centers, whereas only one odd group is allocated to cell-edge users. This results in the loss of two odd groups (eight subchannels); to compare with the loss of two segments, equivalent to 20 subchannels when reuse three is used.

When using this fractional reuse scheme, upon the arrival of a user, it is allocated a subchannel within the frequency band that corresponds to its position in

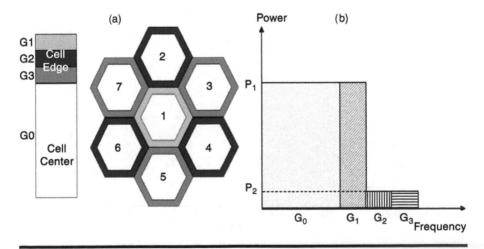

Figure 6.6 (a) Fractional frequency allocation scheme where a reuse three scheme is used at cell edges and (b) power/frequency scheduling with reduced power in cell 1 at the frequencies used for cell-edge users in the cells 2–7.

the cell. As the location of the mobile cannot be precisely known, the choice is based on the path loss: a threshold on the path loss is fixed and terminal equipments with a path loss larger than this threshold are assigned a subchannel within the frequency reuse three bandwidth.

When analyzing the performance of this scheme, we must take into account that the origins of collisions depend on the position of the user in the cell: cell-center users may receive interference from adjacent cells (cells located in ring 1 around the target cell), while cell-edge users are interfered by signals originating from farther cells (in ring 2). This results in a cell-edge user throughput comparable with that obtained with reuse three. However, even if only 22 subchannels are used in each cell, compared with 30 subchannels in a classical reuse one scheme, the overall cell throughput plotted in Figure 6.3 is comparable and even better from the reuse one case. This is due to the elimination of the cell-edge users with largest download times that generate high loads. This scheme can then be considered as a simple solution to combat interference without loosing capacity.

Note that, if a geometric mean were considered in the calculations instead of the harmonic mean, the conclusion would have been different as the elasticiy of the calls would not be taken into account. In fact, we plot in Figure 6.7 the overall cell through-put for the reuse one and fractional reuse one to three schemes when using the geometric mean in the capacity calculations, and the tendancy is inverted (reuse one better).

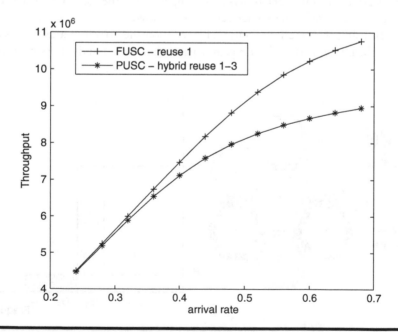

Figure 6.7 Overall cell throughput when using the geometric mean in the capacity calculations. FUSC, fully used subchannels; PUSC, partially used subchannels.

6.7 Power/Frequency Scheduling

Even if the overall cell throughput is large in the hybrid frequency allocation scheme, there is still a loss of subchannels compared with the reuse one scenario. To overcome this problem, a proposed solution is to use a power control on some frequency bands to limit interference at the cell edges. In this context and referring to Figure 6.6a, only cell 1 is allowed to transmit with full power using the "G_1" part of the spectrum whereas cells 2–7 are allowed to transmit in this part of the spectrum using only a reduced power. This is illustrated in Figure 6.6b. This will reduce the downlink interference seen by cell-edge users served by cell 1 compared with a classical reuse one scheme. The radio resources used for transmission to users equipments in a cell are controlled by the scheduler in the base station and fractional reuse can therefore be implemented as part of the scheduling decision. Fractional reuse can thus simply be seen as constraints to the scheduler.

This scheme has been proposed for the long-term evolution of 3G systems (3G LTE) systems [11], but we propose to extend it to WiMAX. This power/frequency scheduling is possible in WiMAX, as for fractional reuse, using the even and odd groups in the PUSC mode, with the difference that all groups are used in each cell with different powers. Only 22 subchannels are used with full power (18 subchannels for cell-center users and 4 for cell-edge ones). The remaining eight subchannels are allowed to be used within the cell center with a reduced power ($P_2 = P_1/R$, with $R > 1$), only when the 18 subchannels assigned for cell center are occupied.

To analyze the performance of the system, we must first characterize the interference for the different kinds of subchannels in cell 1.

1. For the 18 subchannels dedicated for the cell-center users (the even groups), interference comes for all cells (in rings 1 and 2) as a reuse one is considered.
2. For the odd group allocated to cell-edge users in cell 1 (say the group numbered 1), interference comes from center-cell users in adjacent cells with reduced power P_2, and for cell-edge users in farther cells (in ring 2) with an original full power P_1.
3. For the remaining two odd groups (groups 3 and 5) allocated to cell-center users with power P_2 when the odd groups are fully occupied, they may receive three kinds of interferences. Signals arriving from cells in ring 2 are emitted with power P_2, whereas signals arriving from adjacent cells can be emitted with power P_1 or P_2, depending on the position of the corresponding users (at cell edge or at cell center).

Once the interference is characterized, the system can be analyzed as three different queues corresponding to the three aforementioned kinds of subchannels. Note that the third queue (corresponding to the eight subchannels of groups 3 and 5) receives only calls that are blocked in queue 1.

Figure 6.3 plots, again, the cell throughput when using power/frequency scheduling with $P_2 = P_1/10$ ($R = 10$). The PUSC power/frequency scheduling scheme outperforms the FUSC reuse one and the hybrid reuse one-three schemes. This is due to the intelligent interference mitigation by assigning full power to users who do not create high interference. However, there is a degradation of the performance at cell edge, illustrated in Figure 6.4. This degradation is still acceptable as a relatively high bit rate is guaranteed at cell edge (0.44 Mbits compared with only 0.15 Mbits in a reuse one scheme and 0.6 Mbits in a reuse three one). We then conclude that this fractional power/frequency scheduling scheme is the more suitable as it achieves a high-cell throughput with an acceptable cell-edge performance.

6.8 Conclusion

In this chapter, we studied and compared different frequency reuse schemes in WiMAX cellular systems, namely a full reuse one allocation, a reuse three allocation, static and dynamic mixes of reuse one and three schemes, and power/frequency scheduling. We considered a cellular system with elastic traffic and considered as performance measures the overall cell throughput and the cell-edge user throughput. When calculating the capacity, we considered the intercell interference and its impact on the throughput through the usage of adaptive modulation.

Our numerical results show that a partial frequency reuse increases cell-edge performance substantially, with a comparable at the cost of lower overall capacity compared with a reuse one scheme. However, a mix of reuses one and three outperforms a classical reuse three scheme by achieving better cell throughput and is thus preferable.

Finally, we show that a power/frequency scheduling, consisting of using all the frequency bands at each cell with an intelligent power allocation to mitigate interference, achieves high cell throughput with an acceptable cell-edge performance. It can then be considered as the best compromise between the different proposed frequency planning schemes.

References

1. WiMAX Forum, *Mobile WiMAX Part I: A Technical Overview and Performance Evaluation*, 21 February 2006.
2. IEEE C802.16e-04/453r2, *Add Sub-Segment to the PUSC Mode*, Huaweiœ—6/11/2004.
3. IEEE 802.16-2004, *Part 16: Air Interface for Fixed Broadband Wireless Access Systems*, IEEE Standard for Local and Metropolitan Area Networks, October 2004.
4. Draft 802.16e/D9, *Part 16: Air Interface for Fixed and Mobile Broadband Wireless Access Systems*, IEEE Standard for Local and Metropolitan Area Networks, June 2005.
5. M. Wennstrom, *On MIMO Systems and Adaptive Arrays for Wireless Communication*, Ph.D. dissertation, Uppsala University, 2002.

6. S-.E. Elayoubi and B. Fourestié, On Frequency Allocation Schemes in 3G LTE Systems, in *Proceedings of IEEE PIMRC*, 2006.
7. 3GPP 25.814 V1.2.3, *Physical Layer Aspects for Evolved UTRA*, May 2006.
8. I. Koukoutsidis, E. Altman, and J. M. Kelif, *A Non-Homogeneous QBD Approach for the Admission and GoS Control in a Multiservice WCDMA System*, INRIA Research Report No. RR-5358, 2004.
9. T. Bonald and A. Proutière, *Wireless Downlink Data Channels: User Performance and Cell Dimensioning*, ACM Mobicom'03, San Diego, September 2003.
10. F. Kelly, Loss Networks, *Annals of Applied Probability*, 1991, 1, 319–378.
11. 3GPP, R1-050507, Huawei, *Soft Frequency Reuse Scheme for UTRAN LTE*, 2005.

Chapter 7

Support for QoS in IEEE 802.16 Point-to-Multipoint Networks: A Simulation Study

Claudio Cicconetti

The IEEE 802.16 is a mature standard for fixed broadband wireless access (BWA), where a single Base Station (BS) coordinates the access to the wireless medium of many Subscriber Stations (SSs) in a frame-based centralized manner. The standard specifies that the BS is responsible for providing the traffic flows of SSs with quality of service (QoS), in terms of a set of negotiated parameters. To this aim several mechanisms are defined at the Medium Access Control (MAC) layer. However, the standard does not specify the procedures that should be employed by the BS and SSs so as to enforce the negotiated level of QoS. In this work the mechanisms available for QoS support are reported in details, in the context of QoS architecture of IEEE 802.16. Furthermore, the approaches that have been proposed in the literature so far for QoS support are reviewed. Due to high level of complexity of the IEEE 802.16, most proposed solutions have been evaluated using simulation, whose results are summarized here. Finally, a simulation analysis is carried out, so as to

149

evaluate the performance of different multimedia applications, with varying number of users and MAC frame duration.

7.1 Introduction

During the last years we have witnessed a rapid growth of the interest in wireless technologies to provide last-mile broadband access to the Internet [1]. On the one hand, this is due to the staggering developments in the field of radio frequency communications, which allow for increasing transmissions rates at decreasing production costs. On the other hand, users have become more accustomed to broadband access, thus rendering attractive market segments formerly not explored, such as that of rural or low population areas [2]. This eventually results in the spreading of novel applications, such as voice over IP (VoIP), which usually have stringent requirements of QoS.

The IEEE 802.16 is establishing itself as one of the leader technologies in the context of fixed BWA [3], as corroborated by the huge number companies that have joined the WiMAX Forum [4] since it was formed in June 2001 to promote the adoption of IEEE 802.16 compliant equipment by operators of BWA systems. While in the revision of 2004 [5] several air interfaces have been added to the original single-carrier (SC) profile, which make the IEEE 802.16 well-suited for varied wireless environments, the core of the MAC layer with regard to QoS support was barely modified. In fact the standard already included native support for QoS at the MAC layer since its first version has been published in 2001, with several mechanisms to support different types of applications, which are classified by the standard into four scheduling services. However, the standard does not specify mandatory nor informative algorithms to actually provide QoS support by means of these mechanisms. This allows any manufacturer to implement its own optimized proprietary algorithms, thus gaining a competitive advantage over rivals.

This chapter is organized as follows. In Section 7.2 we briefly review the IEEE 802.16 standard, both MAC and physical (PHY) layers, and introduce the notation that will be used throughout this work. In Section 7.3 we describe the QoS architecture of IEEE 802.16 first. Then we review the solutions that have been put forward in the literature to support QoS. Finally, the performance of different multimedia applications with varying offered load and frame duration is evaluated through simulation in Section 7.4. Conclusions are drawn in Section 7.5.

7.2 IEEE 802.16

The IEEE 802.16 specifies the data and control plane of the MAC and PHY layers, as illustrated in Figure 7.1. More specifically, the MAC layer consists of three

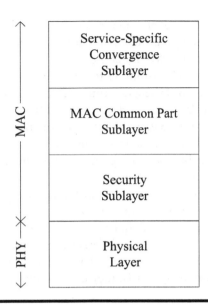

Figure 7.1 Scope of the IEEE 802.16 standard: data/control plane.

sublayers: the service-specific convergence sublayer (SSCS), the MAC common part sublayer (MAC CPS), and the security sublayer. The SSCS receives data from the upper layer entities that lie on top of the MAC layer, for example, bridges, routers, hosts. A different SSCS is specified for each entity type, including support for asynchronous transfer mode (ATM), IEEE 802.3, and Internet Protocol version 4 (IPv4) services. The MAC CPS is the core logical module of the MAC architecture, and is responsible for bandwidth management and QoS enforcement. Finally, the security sublayer provides SSs with privacy across the wireless network, by encrypting data between the BS and SSs.

This section reports the basic IEEE 802.16 MAC CPS and PHY layer functions so as to introduce the notation that will be used in the rest of this work. The interested reader can find all the details of the IEEE 802.16 specifications in the standard document [5].

7.2.1 MAC Layer

The IEEE 802.16 standard specifies two modes for sharing the wireless medium: point-to-multipoint (PMP) and mesh. With PMP, the BS serves a set of SSs within the same antenna sector in a broadcast manner, with all SSs receiving the same transmission from the BS. Transmissions from SSs are directed to and centrally coordinated by the BS. On the other hand, in mesh mode, traffic can be routed through other SSs and can occur directly among SSs. As access coordination is distributed among the SSs, the mesh mode does not include support to parameterized QoS, which is needed by multimedia applications with stringent requirements. In this study we focus on the PMP mode alone.

In PMP mode uplink (UL) (from SS to BS) and downlink (DL) (from BS to SS) data transmissions occur in separate time frames. In the DL subframe the BS transmits a burst of MAC payload data units (PDUs). As the transmission is broadcast all SSs listen to the data transmitted by the BS. However, an SS is only required to process PDUs that are addressed to it or that are explicitly intended for all the SSs. In the UL subframe, on the other hand, any SS transmits a burst of MAC PDUs to the BS in a time division multiple access (TDMA) manner. DL and UL subframes are duplexed using one of the following techniques, as shown in Figure 7.2: frequency division duplex (FDD) is where DL and UL subframes occur simultaneously on separate frequencies, and time division duplex (TDD) is where DL and UL subframes occur at different times and usually share the same frequency. SSs can be either full-duplex, that is, they can transmit and receive simultaneously, or half-duplex, that is, they can transmit and receive at nonoverlapping time intervals.

The MAC protocol is connection-oriented: all data communications, for both transport and control, are in the context of a unidirectional connection. At the start of each frame the BS schedules the UL and DL grants to meet the negotiated QoS requirements. Each SS learns the boundaries of its allocation within the current UL subframe by decoding the UL-medium access protocol (MAP) message. On the

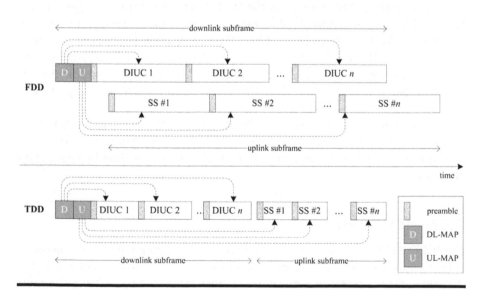

Figure 7.2   Frame structure with frequency and time division duplexes. (From C. Cicconetti, C. Eklund, L. Lenzini, E. Mingozzi, *IEEE Network Magazine*, March 2006. With permission.)

other hand, the DL-MAP message contains the timetable of the DL grants in the forthcoming DL subframe. Both maps are transmitted by the BS at the beginning of each DL subframe, as shown in Figure 7.2.

As the BS controls the access to the medium in the UL direction, bandwidth is granted to SSs on demand. For this purpose, a number of different bandwidth request mechanisms have been specified. With unsolicited granting a fixed amount of bandwidth on a periodic basis is requested during the setup phase of an UL connection. After that phase, bandwidth is never explicitly requested. A unicast poll consists of allocating to a polled UL connection the bandwidth needed to transmit a bandwidth request. If the polled connection has no data awaiting transmission (backlog, for short), or if it has already requested bandwidth for its entire backlog, it will not reply to the unicast poll, which is thus wasted. Instead, broadcast (multicast) polls are issued by the BS to all (multiple) UL connections. The main drawback of this mechanism is that a collision occurs whenever two or more UL connections send a bandwidth request by responding to the same poll. In this case all collided connections* need to resend the bandwidth requests, but a truncated binary exponential backoff algorithm is employed to reduce the chance of colliding again.

Bandwidth requests can also be piggybacked on PDUs. For instance, assume that a bandwidth request is sent by an SS for its connection x. Then, before the entire backlog of connection x is served, more data is received from upper layers. In this case, the SS can notify the BS of the increased bandwidth demands by simply adding a Grant Management subheader to any outgoing PDU of connection x. Finally, while a connection is being served by the BS, the SS can use part of the bandwidth scheduled by the BS for data transmission to send a standalone PDU with no data that updates the amount of backlog notified to the BS. This mechanism is called bandwidth stealing.

It is worth noting that an SS notifies the BS of the amount of bytes awaiting transmission at its connections' buffers, but the BS grants UL bandwidth to the SS as a whole. Due to this hybrid nature of the request/grant mechanism (i.e., requests per connection, grants per SS), an SS also has to implement locally a scheduling algorithm to redistribute the granted capacity to all its connections.

Finally, the BS and SSs can fragment a MAC service data unit (SDU) into multiple PDUs, or they can pack multiple SDUs into a single PDU, so as to reduce the MAC overhead or improve the transmission efficiency. A hybrid analytical simulation study of the impact on the performance of this feature of the MAC layer has been carried out by Hoymann [6]. Results showed that, if the use of fragmentation is enabled, the frame can be filled almost completely, which can significantly increase

---

* Note that bandwidth requests sent in a contention manner are not explicitly acknowledged by the BS, nor it is possible for the SSs to detect collision at the PHY layer because of self-interference effects. Therefore, the IEEE 802.16 specifies that collision is detected implicitly through a timeout.

the frame utilization, depending on the size of SDUs. These optional features have also been exploited in a cross-layer approach between the MAC and application layers, so as to optimize the performance of multimedia streaming [7].

7.2.2 PHY Layer

The IEEE 802.16 standard includes several noninteroperable PHY layer specifications. However, all the profiles envisaged by the WiMAX forum for fixed BWA specify the use of orthogonal frequency division multiplexing (OFDM) with a Fast Fourier Transform (FFT) size of 256, which is thus the primary focus of this study. This PHY layer has been designed to support non-line-of-sight (NLOS) and operates in the 2–11-GHz bands, both licensed and unlicensed. Transmitted data is conveyed through OFDM symbols, which are made up from 200 subcarriers. Part of the OFDM symbol duration, named the cyclic prefix duration, is used to collect multi-path. The interested reader can find a technical introduction to the OFDM system of the IEEE 802.16 in recent survey papers [8,9].

To exploit the location-dependent wireless channel characteristics, the IEEE 802.16 allows multiple burst profiles to coexist within the same network. In fact, SSs that are located near the BS can employ a less robust modulation than those located far from the BS [6]. The combination of parameters that describe the transmission properties, in DL or UL direction, is called a burst profile. Each burst profile is associated with an interval usage code (IUC), which is used as an identifier within the local scope of an IEEE 802.16 network. The set of burst profiles that can be used is periodically advertised by the BS using specific management messages, that is, downlink channel descriptor (DCD) and uplink channel descriptor (UCD). To maintain the quality of the radio frequency communication link between the BS and SSs, the wireless channel is continuously monitored so as to determine the optimal burst profile. The burst profile is thus dynamically adjusted so as to employ the less robust profile such that the link quality does not drop below a given threshold, in terms of the carrier-to-interference-and-noise ratio (CINR) [10]. However, as a side effect of the dynamic tuning of the transmission rate, it is not possible for the stations to compute the transmission time of MAC PDUs *a priori*. Therefore, SSs always issue bandwidth requests in terms of bytes instead of time, without including any overhead due to the MAC and PHY layers.

Although the link quality lies above a given threshold, it is still possible that some data get corrupted. To reduce the amount of data that the receiver is not able to successfully decode, several forward error correction (FEC) techniques are specified, which are employed in conjunction with data randomization, puncturing, and interleaving. Finally, each burst of data is prepended by a short physical preamble (or preamble), which is a well-known sequence of pilot subcarriers that synchronize the receiver. The duration of a preamble is one

OFDM symbol, which can be accounted as PHY layer overhead. In the DL subframe a preamble is prepended to each burst, which can be directed to multiple SSs employing the same burst profile (Fig. 7.2). On the other hand, in the UL subframe, each SS always incurs the overhead of one preamble for each frame where it is served.

7.3 MAC QoS Support

In this section we describe the QoS architecture of the IEEE 802.16 first. Then, we review the mechanisms available at the MAC layer of the IEEE 802.16 for QoS support and discuss the prominent design choices of their implementation.

7.3.1 QoS Architecture

In general, the process of requesting and granting QoS in a network can be logically split in two separate layers: application and network layers. The application layer provides the end-user with a simplified and standardized view of the quality level that will be granted for a given service. This layer is not aware of the technicalities of service requirements (such as bandwidth, delay, or jitter) and it does not depend on the technology-dependent issues related to the actual networks that will be traversed (such as a fiber-optic, wireless, or xDSL). On the other hand, the network layer deals with a set of technical QoS parameters, which it maps on network-specific requirements that have to be fulfilled to provide the end-user with the negotiated quality level. Usually, in wired IP networks the mapping is performed at the network layer. However, such an approach is hardly suitable for wireless networks [11], where there are a number of factors that influence the resource allocation: (*i*) the availability of bandwidth is much more limited with respect to wired networks, (*ii*) there is high variability of the network capacity due, for instance, to environmental conditions, (*iii*) the link quality experienced by different terminals is location-dependent. Therefore, it is often necessary to implement QoS provisioning at the MAC layer, as in IEEE 802.16, so as to gain a better insight of the current technology-dependent network status and to react as soon as possible to changes that might negatively affect QoS.

In IEEE 802.16 the prominent QoS functions of network provisioning and admission control are logically located on the management plane. As already pointed out, the latter is outside the scope of the IEEE 802.16, which only covers the data/control plane, as illustrated in Figure 7.3. Network provisioning refers to the process of approving a given type of service, by means of its network-layer set of QoS parameters that might be activated later. Network provisioning can be either static or dynamic. Specifically, it is said to be static if the full set of services that the BS supports is decided *a priori*. This model is intended for a service provider wishing to specify the full set of services that its subscribers can request, by means of manual or

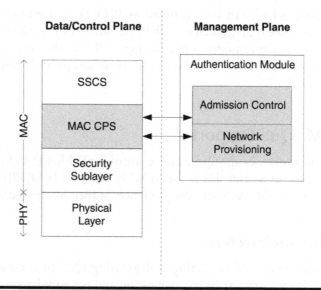

Figure 7.3 Quality-of-service model of the IEEE 802.16.

semiautomatic configuration of the BS's management information base (MIB). On the other hand, with dynamic network provisioning, each request to establish a new service is forwarded to an external policy server (not shown in Fig. 7.3), which decides whether to approve or not. This model allows a higher degree of flexibility, in terms of the types of service that the provider is able to offer to its subscribers, but it requires a signaling protocol between the BS and the policy server, thus incurring additional communication overhead and increased complexity.

Unlike the network provisioning function, which only deals with services that might be activated later, and that are therefore said deferred, the admission control function is responsible for resource allocation. Thus, it will only accept a new service if (*i*) it is possible to provide the full set of QoS guarantees that it has requested, and (*ii*) the QoS level of all the services that have been already admitted would remain above the negotiated threshold. Quite clearly, admission control acts on a time scale smaller than that of network provisioning. This is motivated by the latter being much more complex than the former, as pointed out by a recent study [12] on an integrated end-to-end QoS reservation protocol in a heterogeneous environment, with IEEE 802.16 and IEEE 802.11e [13] devices. Tested results showed that the network provisioning latency of IEEE 802.16 equipments currently available in the market is in the order of several seconds, whereas the activation latency is in the order of milliseconds.

In IEEE 802.16, the set of network layer parameters that entirely defines the QoS of a unidirectional flow of packets resides into a service flow (SF) specification. Each SF can be in one of the following three states: provisioned, admitted, active.

Provisioned SFs are not bound to any specific connection, because they are only intended to serve as an indication of what types of service are available at the BS. Then, when an application on the end-user side starts, the state of the provisioned SF will become admitted, thus booking resources that will be shortly needed to fulfill the application requirements. When the SF state becomes admitted, then it is also assigned a connection identifier (CID) that will be used to classify the SDUs among those belonging to different SFs. However, in this phase, resources are still not completely activated; for instance, the connection is not granted bandwidth yet. This last step is performed during the activation of the SF, which happens just before SDUs from the application starts flowing through the network.

Thus a two-phase model is employed, where resources are booked before the application is started. This is the model employed in traditional telephony applications. At any time it is possible to "put on hold" the application by moving back the state of the SF from active to admitted. When the application stops the SF is set to either provisioned or deleted; in any case, the one-to-one mapping between the service flow identifier (SFID) and the CID is lost, and the CID can be reassigned for other purposes. The SF transition diagram is illustrated in Figure 7.4.

Figure 7.5 shows the blueprint of the functional entities for QoS support, which logically reside within the MAC CPS of the BS and SSs. Each DL connection has a packet queue (or queue, for short) at the BS (represented with solid lines). In accordance with the set of QoS parameters and the status of the queues, the BS DL scheduler selects from the DL queues, on a frame basis, the next SDUs to be transmitted to SSs. On the other hand, UL connection queues (represented in Fig. 7.2 with solid lines) reside at SSs.

Bandwidth requests are used on the BS for estimating the residual backlog of UL connections. In fact, based on the amount of bandwidth requested (and granted) so far, the BS UL scheduler estimates the residual backlog at each UL connection (represented in Fig. 7.5 as a virtual queue, with dashed lines), and allocates future UL grants according to the respective set of QoS parameters and the (virtual) status of the queues. However, as already introduced, although bandwidth requests are

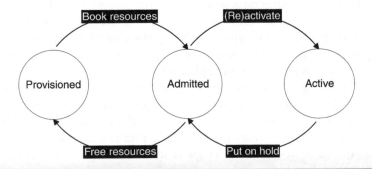

Figure 7.4   Service flow transition diagram.

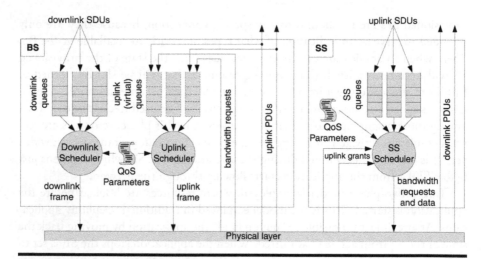

Figure 7.5 Medium Access Control architecture of the Base and Subscriber Stations. (From C. Cicconetti, C. Eklund, L. Lenzini, E. Mingozzi, *IEEE Network Magazine*, March 2006. With permission.)

per connection, the BS nevertheless grants UL capacity to each SS as a whole. Thus, when an SS receives an UL grant, it cannot deduce from the grant which of its connections it was intended for by the BS. Consequently, an SS scheduler must also be implemented within each SS MAC to redistribute the granted capacity to the SS's connections (Fig. 7.5).

7.3.2 Scheduling Services

In IEEE 802.16 each SF is characterized by a set of network-layer QoS parameters that are used by the BS for both network provisioning and admission control, and to serve the UL/DL connections so as to meet the desired level of QoS, while the SF is active. Both traffic requirements and traffic specifications are included, depending on the application type. Specifically, the scheduler of the BS is responsible for guaranteeing the traffic requirements of admitted SFs, provided that their traffic specifications are met by the applications.

Four scheduling services exist in IEEE 802.16, which identify four classes of applications: Unsolicited Grant Service (UGS), Real-Time Polling Service (rtPS), Non-Real-Time Polling Service (nrtPS), and Best Effort (BE). Each scheduling service is characterized by a mandatory set of QoS parameters, reported in Table 7.1, which is tailored to best describe the guarantees required by the applications for which the scheduling service is designed. Furthermore, for UL connections, it also specifies which mechanisms to use to request bandwidth.

Table 7.1   Quality-of-Service (QoS) Parameters of Different Scheduling Services

QoS Parameter		Unsolicited Grant Service	Real-Time Polling Service	Non-Real-Time Polling Service	Best Effort
Minimum reserved traffic rate		✓	✓	✓	
Maximum sustained traffic rate		✓	✓	✓	✓
Maximum latency		✓	✓		
Tolerated jitter		✓			
Traffic priority				✓	✓
Maximum traffic burst				(✓)	
Unsolicited Grant Interval		(✓)			
Unsolicited Polling Interval			(✓)		
Bandwidth Request Mechanisms Allowed (Uplink Only)					
Polling	Unicast	✗	✓	✓	✓
	Broadcast/multicast	✗	✗	✓	✓
Bandwidth stealing		✗	✓	✓	✓
Piggybacking		✗	✓	✓	✓

Note: Parameters with a "✗" symbol cannot be specified by service flows of that scheduling service. A "✓" instead means that the parameter is mandatory according to the IEEE 802.16 standard. Finally, we tagged as "(✓)" the parameters that are specified as optional by the standard, but are required for correct operation in our QoS architecture. The Unsolicited Grant and Polling Interval parameters appeared in the IEEE 802.16e amendment.

Source: IEEE 802.16e-2005, IEEE standard for local and metropolitan area networks—Part 16: air interface for fixed broadband wireless access systems—amendment 2: physical and Medium Access Control layers for combined fixed and mobile operation in licensed bands, 2006.

Traffic requirements are described first. The minimum reserved traffic rate specifies the minimum rate, in b/sec, that must be reserved by the BS for this SF. The maximum latency,* in seconds, upper bounds the interval between the time when

---

* In IEEE 802.16 terminology, the latency is the queueing delay of MAC SDUs. In the packet scheduling literature, the maximum latency is usually known as the delay bound.

an SDU is received by the MAC layer and the time when it is delivered to the PHY layer. The tolerated jitter, in sec, is the maximum delay variation that must be enforced. The traffic priority can be used by the BS to provide differentiated service-to-service flows that have the same QoS requirements. The Unsolicited Grant (Polling) Interval specifies the nominal interval between two consecutive grants (unicast polls) for the SF. The maximum sustained traffic rate and the maximum traffic burst instead are traffic specifications. Thus, if the SF does not comply with these traffic specifications, the BS is not required to provide the negotiated QoS level, in terms of the admitted requirements. Although not requested by the IEEE 802.16 standard, the MAC layer can implement traffic filters, for example, token buckets, so that traffic specifications are never exceeded.

The QoS parameters in Table 7.1 can also be specified by means of service classes, which are collections of prespecified sets of QoS parameters identified by a string, whose scope is local to the BS. Service classes can be used as "macros" by system administrators within the same service provider domain to refer to a complex set of parameters through an easy-to-remember identifier, such as "G.711" or "MPEG4."

7.3.2.1 Unsolicited Grant Service

UGS is designed to support real-time applications with strict delay requirements that generate fixed-size data packets at periodic intervals, such as T1/E1 and VoIP without silence suppression. The guaranteed service, which is illustrated in Figure 7.6, is defined so as to closely follow the packet arrival pattern: grants occur on a periodic basis, with the base period equal to the Unsolicited Grant Interval and the offset upper bounded by the maximum latency. With regard to UL connections, capacity is granted by the BS regardless of the current backlog estimation, thus SSs never send bandwidth requests. In other words, grants are assigned with the same pace as that of packets generation. This can lead to undesirable delays if

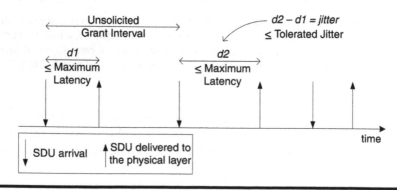

Figure 7.6   Unsolicited Grant Service guarantees.

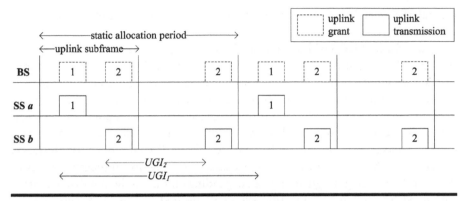

Figure 7.7   Example of Unsolicited Grant Service uplink allocation. The static timetable is reported in Table 7.2. The downlink subframe, including the uplink medium access protocol, is not shown for the ease of readability.

the synchronization between the application and the BS is lost due to clock mismatch [14].

Support of UGS connections can be easily implemented at the BS, due to the periodic nature of the traffic for which UGS has been designed. For instance, grants to the admitted UGS connection can be statically allocated by means of a static timetable updated whenever a new UGS connection is admitted. Specifically, the grant size is the expected size of the SDUs, which is equal to $SDU_x = UGI_x \cdot R_x$, where UGI_x is the Unsolicited Grant Interval and R_x is the minimum reserved traffic rate of connection x. Moreover, the table has to be filled so as to comply with the QoS requirements, in terms of the maximum latency and tolerated jitter. An example of the UGS static allocation is illustrated in Figure 7.7, with two connections 1 and 2, which belong to SS*a* and SS*b*, respectively. The UGI of connection 1 is such that one SDU is generated every two frames, while connection 2 is expected to produce one SDU each frame. The resulting static timetable is reported in Table 7.2. Note that the static allocation is periodic with the period equal to two frames. In

Table 7.2   Example of a Static Timetable to Serve Unsolicited Grant Service Connections (See Fig. 7.7)

Frame Number	Grant Size	Connection Identifier
0	SDU_1	1
0	SDU_2	2
1	SDU_2	2

SDU, service data unit.

general, the timetable span is equal to the least common multiple of the UGI values of all admitted connections.

A drawback of UGS with UL connections is that grants are assigned regardless of the actual backlog at SSs. For instance, UGS is not suitable for VoIP applications with silence suppression. In fact, while packets of fixed size are usually generated at a constant rate during talk-spurt (ON) periods, no packets are produced during silence (OFF) periods. Several modifications to the UGS have been proposed in the literature so as to efficiently support this kind of application by reducing the MAC overhead of unnecessarily assigning UL grants to idle VoIP connections. For instance, Hong et al. [15] proposed two alternative strategies. The first approach consists of dynamically adapting the grant interval, depending on the bandwidth that is actually consumed by the connection. The second is inspired to the UGS with activity detection (UGS/AD) scheduling service of the Data Over Cable Service Interface Specification (DOCSIS) standard [16], whose MAC layer design choices have been broadly reused by the IEEE 802.16 working group. This strategy relies on the SS notifying the BS when it enters/exits an OFF period, respectively. In this way, during the OFF periods, the BS will assign small UL grants instead of full-sized ones, which are used by the SS to notify the BS of the start of the next ON period as soon as the application becomes busy again. This solution has become part of the specifications of the IEEE 802.16e amendment [17] for mobile BWA as Enhanced Real-Time polling service (ertPS). Both these approaches have been shown to perform better than the original UGS, in terms of the MAC overhead. Finally, Lee et al. [18] proposed a modification to the UGS to efficiently support VoIP applications with variable rate, that is, whose packet size changes during the ON periods. More recently they compared all these mechanisms through simulation and showed that their proposal achieves the least MAC overhead with variable rate VoIP applications [19].

In the rest of this chapter we ignore service of UGS connections, as the latter is based on periodic scheduling only. The only effect of having UGS connections is that the amount of capacity available in each subframe varies depending on the allocation of DL/UL grants reserved for them [20].

7.3.2.2 Real-Time Polling Service

The rtPS is designed to support real-time applications with less stringent delay requirements than UGS that generate variable-size data packets at periodic intervals, such as Moving Pictures Expert Group (MPEG) video and VoIP with silence suppression. The key QoS parameters with such connections are the minimum reserved traffic rate, the maximum latency, and the Unsolicited Polling Interval (UPI) (UL only). Because the size of packets with rtPS is not fixed as with UGS-tailored applications, SSs are required to notify the BS of their current bandwidth requirements. However, to grant deterministic access to the medium, the BS periodically sends unicast polls to rtPS connections. For this reason, the latter are refrained from using

Table 7.3   Example of a Static Timetable to Send Unicast Polls to Real-Time Polling Service Connections (See Fig. 7.8)

Frame Number	Connection Identifier
0	1
0	2
1	2

bandwidth request mechanisms on a contention basis. The polling period is equal to the UPI, if specified.

With regard to UL connections, the BS can issue periodic polls by means of a static timetable like that of UGS grants. The only difference is that the grant size is not equal to the expected SDU size, which is not known in advance, but is the number of bytes needed by the SS to transmit a bandwidth request PDU for the polled connection. An example of static timetable of unicast polls is reported in Table 7.3 with reference to Figure 7.8. Note that the bandwidth request mechanism of rtPS connections incurs an additional delay with respect to UGS connections of at least one frame duration. In fact, by responding to a unicast poll an SS notifies the BS of the backlog of one of its connections, but it cannot actually transfer data until the BS reserves an UL grant for it. However, unlike UGS, rtPS connections

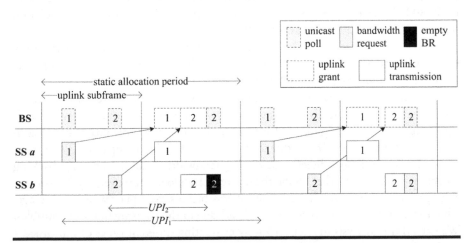

Figure 7.8   Example of a Real-Time Polling Service uplink allocation of unicast polls. The static timetable is reported in Table 7.3. The downlink subframe, including the uplink Medium Access Control, is not shown for the ease of readability. An empty bandwidth request is sent in response to a unicast poll if the connection transmission queue does not contain data-waiting transmission.

can piggyback bandwidth requests on outgoing PDUs, provided that new data arrived before the previous backlog has been entirely served. This way an rtPS connection can anticipate the backlog notification to the BS with respect to the next scheduled unicast poll, thus reducing the aforementioned delay.

Unlike UGS connections, applications served with rtPS have variable bandwidth requirements over time. Therefore, a more sophisticated scheduling algorithm needs to be implemented on the BS, which is left unspecified by the IEEE 802.16 standard. Many scheduling algorithms have been put forward in the literature to support QoS in wired and wireless networks. As a minimum reserved rate is the basic QoS parameter, the class of latency-rate scheduling algorithms [21] is particularly suited for implementing the schedulers in the IEEE 802.16 MAC. Specifically, within this class, we consider the deficit round robin (DRR) [22] to be a good candidate as the DL scheduler to be implemented at the BS, as it combines the ability of providing fair queueing, in the presence of variable-length packets, with the simplicity of implementation. DRR assumes that the size of the head-of-line packet is known at each packet queue, thus it cannot be used by the BS to schedule transmissions in the UL direction. In fact, with regard to the UL direction, the BS is only able to estimate the overall amount of backlog of each connection, but not the size of each backlogged packet. Thus we selected the weighted round robin (WRR) [23] instead, which belongs to the class of rate-latency scheduling algorithms as well. Although there are sophisticated schedulers in the literature that have more advanced theoretical properties than WRR, for example, the weighted fair queueing (WFQ) [24], which is a well-known approximation of the general processor sharing (GPS) ideal scheduler, the improvement in terms of fairness and scheduling latency becomes negligible in IEEE 802.16 due to frame-based transmissions. This has been shown in a simulation study which compared WFQ and WRR [25].

In our previous work [26] we analyzed the impact of different partitioning schemes on the performance of UL rtPS connections, in terms of delay and jitter,* with the BS employing DRR (WRR) as the DL (UL) scheduler. A partitioning scheme can be viewed as the aggregation level of the traffic sources in the IEEE 802.16 network. Specifically, we showed that the performance improves when multiple traffic sources are aggregated into a single connection, or into multiple connections at the same SS. This is due to the statistical multiplexing of traffic sources, which leads to a more efficient use of the bandwidth stealing and piggybacking mechanisms to notify the BS of the current backlog. Moreover, traffic sources aggregation leads to a smaller number of SSs being served per frame, on average, which in turn reduces the physical overhead due to the transmission of preambles. We extend this analysis in Section 7.4 by evaluating different types of multimedia applications.

An alternative approach to rate-latency schedulers is to consider the maximum latency as the main QoS parameter of rtPS instead of the minimum reserved traffic

---

* Delay and jitter are formally defined in Section 7.4.1.

rate. This has been first proposed in the literature by Wongthavarawat et al. [27], who employed the earliest deadline first (EDF) [28] to serve SDUs. Specifically, each rtPS connection plays the role of a task in a real-time operating system, whose deadline is assumed to be equal to its maximum latency. This solution has been shown to provide rtPS connections with bounded delays, provided that the traffic is filtered through a token bucket shaper with rate equal to the minimum reserved traffic rate. More recently, additional simulation studies with the same architecture have been presented [29,30], which confirm the original results obtained.

7.3.2.3 Non-Real-Time Polling Service and Best Effort

Unlike UGS and rtPS scheduling services, nrtPS and BE are designed for applications that do not have any specific delay requirements. The only difference between them is that nrtPS connections are reserved a minimum amount of bandwidth, which can boost performance of bandwidth-intensive applications, such as File Transfer Protocol (FTP) and video on demand (VoD). Both nrtPS and BE UL connections typically use contention-based bandwidth requests; however, the BS should also grant unicast bandwidth request opportunities to nrtPS connections on a large time-scale, so as to enforce the minimum reserved traffic rate commitment even at high network loads. In fact, in these conditions the BS could refrain from assigning slots for broadcast polls to save bandwidth.

When a BE or nrtPS UL connection becomes busy* after it has been idle for a long period, it notifies the BS of its backlog by responding to a multicast/broadcast poll. Therefore, the BS should reserve some capacity as multicast/broadcast polls so as to avoid starvation of idle BE connections. Recall that starvation of nrtPS connections is prevented by the large time-scale unicast polls issued by the BS. For instance, the BS can make the UL capacity that is not scheduled as UL grants available as broadcast polls. In addition, reserving a small number of broadcast polls in each frame has been shown to improve the performance of BE connections, by reducing delays [26]. This is especially suitable for elastic applications, such as Web or FTP, which generate bursts of SDUs at irregular time intervals. One of these bursts may require several frames to be completely dispatched. Should the BS not save some bandwidth to reserve slots for broadcast polls, other connections would not be able to send bandwidth requests during these bursts, which would increase the start-up delay of nrtPS and BE connections. In fact, the corresponding SSs would not be granted any bandwidth until the active burst is over.

Oh et al. [31] performed a hybrid analytical and simulation analysis to derive the optimal number of broadcast polls per frame as a function of the number of SSs.

---

* In the following, we define a connection as busy (or backlogged) when it has one or more buffered SDUs awaiting transmission. Connections that are not busy are said to be idle.

Specifically, they assumed that SSs only send bandwidth requests in a contention manner, that is, piggybacking and bandwidth stealing are not used, and the effect of MAC SDUs queueing is considered to be negligible. Under these simplified assumptions they found that the optimal number of broadcast slots per frame is twice the number of (active) connections. Yan et al. [32] instead found this optimal value to be equal to the number of (active) connections, based on an analytical model validated through Monte Carlo simulation. Basically, they introduced the additional assumptions that there is no binary exponential backoff after a collision, and when the latter occurs one bandwidth request is always successfully decoded by the BS.

In any case, the time needed for an SS to send a bandwidth request using contention is unpredictable, as it depends on the backoff procedure and collision resolution algorithms of all SSs with at least one connection with a pending contention-based bandwidth request. This makes this bandwidth request mechanism much less effective than that employed by rtPS connections, that is, responding to periodic unicast polls. This is quantified in Figure 7.9, which shows the backlog estimation error,* in a typical scenario of multimedia and BE Internet access [33]. As can be seen, the backlog estimation error of rtPS connections is significantly smaller than that of the BE connection, which has an extremely irregular behavior.

We now discuss the issue of selecting the UL and DL schedulers on the BS for nrtPS and BE connections. Wongthavarawat et al. [27] considered WFQ for nrtPS connections. Weights are selected so as to meet the minimum reserved traffic rate

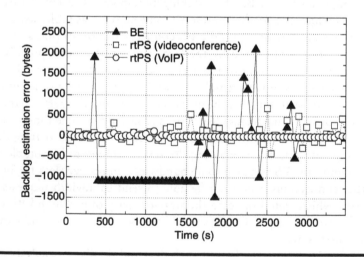

Figure 7.9   Backlog estimation error versus time.

---

* The backlog estimation error is defined as the difference between the BS's estimate of the backlog of a connection (as acquired via bandwidth requests) and the actual backlog of that connection on the SS.

traffic specification of all admitted connections. BE connections are served in a simple round-robin manner. Finally, a strict priority discipline is enforced among connections with different scheduling service, so that a different scheduler can be employed in a hierarchical manner within each service class. This strict priority discipline between scheduling services can lead to starvation of BE connections when the entire capacity available for data transmission is consumed by rtPS and nrtPS connections. On the other hand, we argue [26,33] that a single instance of the DRR/WRR scheduler can be used for sharing the DL/UL bandwidth among all the connections. As this approach requires a minimum rate to be reserved for each connection being scheduled, although not required by the 802.16 standard, BE connections should also be guaranteed a minimum rate. This opportunity can be taken to avoid BE traffic starvation in overloaded scenarios.

We also proposed DRR as the scheduler to be implemented on the SSs on receiving grants from the BS. This way the UL capacity is shared fairly by the connections of each SS proportionally to their minimum reserved rates, as shown through simulation in the following. To this aim we set up a scenario* with a variable number (N) of SSs, each with one BE and one nrtPS connection, both carried with a Web source with a mean rate of 147 kb/sec. The reserved rate of nrtPS connections is equal to the mean rate of the Web traffic, whereas that of BE connections is set to a nominal value of 14.7 kb/sec. Figure 7.10 shows the average access delay, when the number of SSs increases from 20 to 80. As expected, the higher the

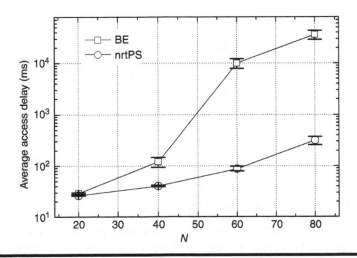

Figure 7.10 Service differentiation with deficit round robin as the Subscriber Station scheduler. Best Effort (BE) versus Non-Real-Time Polling Service (nrtPS), with the minimum reserved rate of BE set to one-tenth of that of nrtPS connections.

---

* The simulation environment is described in Section 7.4.1.

offered load, the better is the service experienced by the nrtPS connections. In fact, when the network is underloaded (i.e., $N \le 40$) the available bandwidth is enough to serve both BE and nrtPS connections completely, thus yielding comparable delays. On the other hand, when the network becomes overloaded (i.e., $N > 40$) the BE curve increases steeply due to the 1:10 proportional share of the UL grants scheduled by the BS.

An intermediate solution between hierarchical scheduling and complete sharing of bandwidth among all scheduling services has been proposed by Settembre et al. [34], who employed a two-stage priority scheduler. First, rtPS and nrtPS are served by a WRR scheduler, with weights proportional to their minimum reserved traffic rates; then, if there is still capacity available, BE connections are served in a round-robin manner. While this approach can lead to starvation of BE connections, it is more efficient than solutions with deadline-based schedulers [27,29,30], due to the reduced implementation complexity of WRR.

Finally, with regard to the DL scheduler on the BS only, Xergias et al. [35] proposed an innovative algorithm that takes into account the inherent properties of the IEEE 802.16 MAC. Specifically, the incoming SDUs are arranged in a complex data structure, based on the connection, scheduling service, arrival time, MCS. The schedule of the next frame is obtained in a simple manner by accessing this data structure through sequential reading. The main advantage of this solution is that the scheduling complexity is spread over time, as SDUs' arrivals are not synchronized with the MAC frame.

7.4   Performance Evaluation

In this section we extend our previous analysis [26] of the IEEE 802.16 with the DRR/ WRR schedulers on the BS, and the DRR scheduler on SSs, as described in Section 7.3.2.2. Specifically, we now evaluate the impact on the performance of different types of multimedia applications, served under the rtPS and nrtPS scheduling services. We first describe in Section 7.4.1 the simulation environment, including the description of the traffic models and metrics employed. Results are reported in Section 7.4.2.

7.4.1   Simulation Environment

The simulations were carried out by means of a prototypical simulator of the IEEE 802.16 protocol. The simulator is event-driven and was developed using C++. The MAC layer of SSs and the BS are implemented, including all procedures and functions for UL/DL data transmission, and UL bandwidth request/grant. The simulator provides an accurate abstraction of the IEEE 802.16 MAC layer and it has been extensively used for evaluating the performance of the 802.16 MAC layer in our recent works [26,33]. The system parameters are those specified in the standard and referred to as system profile "profP3_7," which is designed for use in a 7-MHz

Table 7.4 System Parameters

Parameter	Value(s)
Duplex mode	Time division duplex
Orthogonal frequency division multiplexing (OFDM) symbol duration	34 μsec
Channel bandwidth	7.0 MHz
Frame duration	5 msec, 10 msec, 20 msec
Data modulation	QPSK, 16-QAM, 64-QAM
Medium access protocol modulation	BPSK
Forward error correction code	RS-CC
SSTTG/SSRTG duration	One OFDM symbol

licensed band, and are reported in Table 7.4. The DL and UL subframes have the same duration and are duplexed in a TDD manner.

Although accurately modeling channel conditions is a key aspect of simulation for network provisioning and resource management, this study only focuses on the functions and mechanisms available at the MAC layer to provide QoS, thus we assumed ideal channel conditions, that is, no packet corruption due to the wireless channel. In addition, we analyzed the system while in a steady-state, where the set of admitted connections does not change over time.* The simulation scenarios consisted in several SSs located at various distances from the BS. In realistic conditions, the nearer the SS to the BS, the more robust is the PHY profile. We assumed that the SSs employ the following modulations: QPSK 3/4, 16-QAM 3/4, 64-QAM 3/4, which are evenly partitioned among SSs.

The simulation analysis has been carried out using the method of independent replications [36]. Specifically, we ran ten independent replications of 300 sec each, with a 30 sec initial warm-up period. In all the simulation runs, we estimate the 95 percent confidence interval of the measured throughput. Confidence intervals are not reported whenever negligible.

In our analysis we consider three traffic models: VoIP, videoconference (VC), and VoD. We model VoIP traffic as an ON/OFF source with voice activity detection (VAD). Packets of 66 bytes are generated only during the ON periods, at fixed intervals of 20 msec, so that a GSM adaptive multirate encoder at 3.3 KB/sec is simulated [37]. The duration of the ON and OFF periods is distributed exponentially [38].

---

* We do not assess the performance of the signaling protocol between the BS and SSs for establishing new connections and the admission control procedures at the BS.

Table 7.5   Parameters of Pre-Encoded Trace Files

Traffic	Mean Rate	Peak Rate	Frame Interval
Videoconference—MPEG4	58 kb/sec	690 kb/sec	40 msec
Videoconference—H.263	64 kb/sec	400 kb/sec	40 msec
Video on demand	1 Mb/sec	6.98 Mb/sec	33 msec

VC traffic is based on pre-encoded MPEG4 and H.263 traces, from a real-life lecture [39]. Both MPEG4 and H.263 encoders produce video frames of variable size, depending on the frame type and on the input. However, the former generates packets at a fixed rate, whereas the latter generates packets at intervals that are multiple of a base period. VoD traffic is based on a pre-encoded MPEG4 trace [39], from the movie "Jurassik Park." The trace parameters are reported in Table 7.5.

VoIP and VC connections employ the rtPS scheduling service, whereas VoD connections use nrtPS, due to the less stringent delay requirements of the latter. Moreover, the minimum reserved traffic rate of VC and VoIP connections is equal to the peak rate of their respective traffic source, and the UPI is equal to the minimum packet interarrival time (i.e., 20 msec for VoIP and 40 msec for VC). On the other hand, the minimum reserved traffic rate of VoD is equal to the mean rate of the MPEG4 trace file.

Because our analysis is focused on QoS traffic, with strict delay requirements, the most relevant metric is the delay variation (or jitter), which is defined as follows. The delay is defined as the time between the arrival of the packet at the MAC transmit buffer of the source node (SS/BS) and the time that this packet is completely delivered to the upper protocol layer of the destination node (BS/SS). The jitter is then defined as the difference between the 99th percentile of the delay and the packet transmission delay, that is, the time it takes for a packet of minimum length to be transmitted over the air from the source to the destination.

7.4.2   Simulation Analysis

We now evaluate the impact on the performance of VoIP, VC, and VoD traffic, in terms of the jitter, with respect to the number of SSs (N) and the frame duration. We assume that each SS has only one connection, which carries exactly one traffic source of the specified type (i.e., one of VoIP, VC, or VoD).

We start assessing the performance of QoS traffic by setting up a scenario with a variable number of SSs ranging from 5 to 43, with a frame duration of 10 msec. Only one SS is provisioned with a VoD connection, whereas the remaining is partitioned between VoIP and VC traffic evenly. We repeated the scenario with both the MPEG4 and the H.263 trace files for VC traffic. The jitter of DL connections is

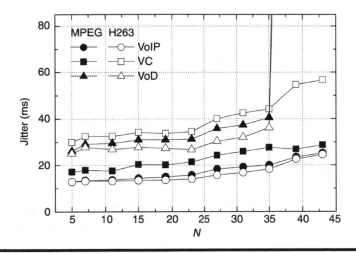

Figure 7.11 Jitter of downlink connections versus number of Subscriber Stations, with different videoconference codecs.

reported in Figure 7.11. As can be seen, when the network is underloaded (i.e., $N \leq 27$) the jitter is always smaller than the interarrival time of packets of each traffic source. As the number of SSs further increases, the VoD curves increases steeply. In fact, with a high offered load, the VoD traffic performance degrades because the rate provisioned for the VoD connection is equal to the mean rate of the application. This results in the performance of VoIP and VC connections being isolated from that of VoD traffic. Note that this has been achieved without enforcing a strict priority between rtPS and nrtPS connections, which are served by the same instance of the DRR schedulers.

Additionally, the VC codec significantly impacts on the performance. In particular, with $N < 35$, in the case of H.263, the jitter results slightly higher than that of the VoD source whereas the MPEG case exhibits a lower jitter. This can be explained as follows. H.263 codecs produce frames at variable time intervals. Therefore, the VC connection queue can potentially become idle due to inactivity periods. As soon as the queue becomes backlogged again, it is re-inserted at the tail of the DRR list of connections waiting to be served. Therefore, the new arrived packet has to wait until all the other connections have been served, which increases the jitter. Such a situation is less likely to happen with the MPEG4 codec because video frames are generated at fixed time intervals, with no inactivity periods.

With regard to the UL connections, results are reported in Figure 7.12. As expected, the jitter is higher than that in the DL case, because UL connections experience the additional delay of notifying the BS of their bandwidth requests. However, the curves are almost constant when the offered load increases (except for the H.263 VC case), because the BS schedules unicast polls on a periodic basis,

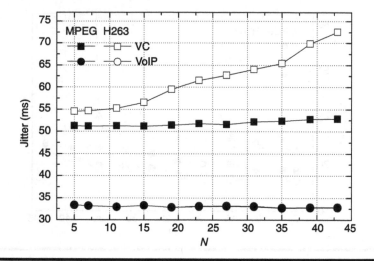

Figure 7.12 Jitter of downlink connections versus number of Subscriber Stations, with different videoconference codecs.

with the period equal to the interarrival time of SDUs of each connection. The anomaly of the H.263 curve with respect to VoIP and VC-MPEG4 is due to the variable interarrival of video frames. In fact the transmission queue of a H.263 connection can potentially become idle when polled from the BS. Hence, a connection that misses the poll then needs to wait an entire polling interval before it will have a subsequent chance to send a bandwidth request.

Finally, we setup a scenario with VoIP and VC-MPEG4 traffic only, where N ranges between 5 and 45, and with variable frame duration. The jitter of UL connections is reported in Figure 7.13. As can be seen, the longer the frame duration, the higher are the curves. This can be explained as follows. As scheduling is performed at the beginning of each frame, the higher the frame duration, the longer (on average) an SS has to wait before using its grant. In other words, with longer frames the BS is less responsive to the SSs' bandwidth requests. Similar considerations also hold for the DL case, where an SDU received by the BS has to wait at least until the next frame (i.e., the transmission of the next DL-MAP) before it can be served.

7.5 Conclusions

The IEEE 802.16 standard specifies support for QoS at the MAC layer. Four scheduling services are defined: UGS, rtPS, nrtPS, and BE. While UGS is well suited for constant bit-rate traffic only, where fixed-size packets are generated a constant rate, rtPS and nrtPS are much more flexible classes and can be used

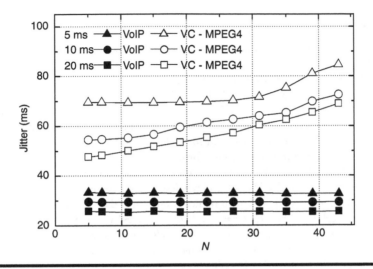

Figure 7.13   Jitter of uplink connections, both Voice over IP and videoconference, versus number of Subscriber Stations, with variable frame duration.

efficiently with current multimedia applications. In particular rtPS is tailored to interactive applications, which have stringent delay requirements. Finally, elastic traffic with no specific QoS requirements is expected to be served with the BE scheduling service.

In order for the QoS guarantees to be met, the BS and SSs have to implement appropriate scheduling algorithms. Several approaches have been put forward in the literature, whose advantages and disadvantages have been highlighted, with respect to our proposal of employing the DRR and WRR as the DL and UL schedulers on the BS, respectively. This design choice has been analyzed in our previous works, whose results have been summarized here. We have then evaluated the performance of varied multimedia traffic, that is, VoIP, videoconference, and VoD, under different scenarios. The results have shown that the delay variation, which is a key QoS metric, is stable with varying offered load. In addition, small-to-medium variations have been observed depending on the frame duration and the multimedia traffic type. This is especially true for UL connections, whereas the performance of DL connections is more resilient to these changes.

Acknowledgments

The author would like to thank Prof Luciano Lenzini, who provided guidance and detailed suggestions, Prof Enzo Mingozzi for his insightful discussions on IEEE 802.16, and his colleague, Alessandro Erta, for his invaluable help in obtaining the numerical results.

References

1. International Telecommunication Union, ITU Internet reports: birth of broadband, 2003.
2. WiMAX forum. Business case models for fixed broadband wireless access based on WiMAX technology and the 802.16 standard, 2004.
3. D. I. Axiotis, T. Al-Gizawi, K. Peppas, E. N. Protonotarios, F. I. Lazarakis, C. Papadias, and P. I. Philippopoulos, Services in interworking 3G and WLAN environments, *IEEE Wireless Communications*, 11 (5) (2004) 14–20.
4. WiMAX forum, http://www.wimaxforum.org/.
5. IEEE 802.16-2004, IEEE standard for local and metropolitan area networks—Part 16: air interface for fixed broadband wireless access systems, 2004.
6. C. Hoymann, Analysis and performance evaluation of the OFDM-based metropolitan area network IEEE 802.16, *Computer Networks* (Elsevier), 49 (3) (2005) 341–363.
7. S. Sengupta, M. Chatterjee, S. Ganguly, and R. Izmailov, Exploiting MAC flexibility in WiMAX for media streaming, *Proceedings of WoWMoM 2005*, pp. 338–343, Taormina (Italy), June 13–16, 2005.
8. A. Ghosh, D. R. Wolter, J. G. Andrews, and R. Chen, Broadband wireless access with WiMAX/802.16: current performance benchmarks and future potential, *IEEE Communications*, 43 (2) (2005) 129–136.
9. I. Koffman and V. Roman, Broadband wireless access solutions based on OFDM access in IEEE 802.16, *IEEE Communications*, 40 (4) (2004) 96–103.
10. C. Eklund, R. B. Marks, K. L. Stanwood, and S. Wang, IEEE standard 802.16: a technical overview of the WirelessMAN air interface for broadband wireless access, *IEEE Communications*, 40 (6) (2002) 98–107.
11. Y. Cao and O. K. Li, Scheduling algorithms in broadband wireless networks, *Proceedings of the IEEE*, 89 (1) (2001) 76–87.
12. P. Neves, S. Sargento, and R. L. Aguiar, Support of real-time services over integrated 802.16 metropolitan and local area networks, *Proceedings of IEEE ISCC 2006*, pp. 15–22, Cagliari (Italy), June 26–29, 2006.
13. IEEE Std 802.11e Wireless LAN Medium Access Control and physical layer specifications, Medium Access Control quality of service enhancements, 2005.
14. Y. Yao and J. Sun, Study of UGS grant synchronization for 802.16, *Proceedings of ISCE 2005*, pp. 105–110, Macau (Hong Kong), June 14–16, 2005.
15. S.-E. Hong and O.-H. Kwon, Considerations for VoIP Services in IEEE 802.16 Broadband Wireless Access Systems, *Proceedings of VTC 2006*, pp. 1226–1230, Melbourne (Australia), May 7–10, 2006.
16. CM-SP-RFIv2.0-108-050408, Data-over-cable service interface specifications DOCSIS 2.0: radio frequency interface specification, Cable Television Laboratories, April 2005.
17. IEEE 802.16e-2005, IEEE standard for local and metropolitan area networks—Part 16: air interface for fixed broadband wireless access systems—amendment 2: physical and Medium Access Control layers for combined fixed and mobile operation in licensed bands, 2006.
18. H. Lee, T. Kwon, and D.-H. Cho, An efficient uplink scheduling algorithm for VoIP services in IEEE 802.16 BWA systems, *Proceedings of IEEE VTC 2004*, pp. 3070–3074, Los Angeles (USA), September 26–29, 2004.

19. H. Lee, T. Kwon, and D.-H. Cho, Extended-rtPS Algorithm for VoIP Services in IEEE 802.16 systems, *Proceedings of IEEE ICC 2006*, pp. 2060–2065, Istanbul (Turkey), June 11–15, 2006.

20. D.-H. Cho, J.-H. Song, M.-S. Kim, and K.-J. Han, Performance analysis of the IEEE 802.16 wireless Metropolitan Area Network, *Proceedings of the DFMA 2005*, pp. 130–137, Besançon (France), February 6–9, 2005.

21. D. Stiliadis and A. Varma, Latency-rate servers: a general model for analysis of traffic scheduling algorithms, *IEEE/ACM Transactions on Networking*, 6 (11) (1998) 611–624.

22. M. Shreedhar and G. Varghese, Efficient fair queueing using deficit round robin, *IEEE/ACM Transactions on Networking*, 4 (3) (1996) 375–385.

23. M. Katevenis, S. Sidiropoulos, and C. Courcoubetis, Weighted round-robin cell multi-plexing in a general-purpose ATM switch chip, *IEEE JSAC*, 9 (8) (1991) 1265–1279.

24. A. Parekh and R. Gallager, A generalized processor sharing approach to flow control in integrated services networks: the single node case, *IEEE/ACM Transactions on Networking*, 1 (3) (1993) 344–357.

25. O. Gusak, N. Oliver, and K. Sohraby, Performance evaluation of the 802.16 Medium Access Control layer, *LCNS* (Springer-Verlag), 3280 (2004) 228–237.

26. C. Cicconetti, A. Erta, L. Lenzini, and E. Mingozzi, Performance evaluation of the IEEE 802.16 MAC for QoS support, *IEEE Transactions on Mobile Computing*, 6 (1) (2007) 26–38.

27. K. Wongthavarawat and A. Ganz, Packet scheduling for QoS support in IEEE 802.16 broadband wireless access systems, *International Journal of Communication Systems*, 16 (1) (2000) 81–96.

28. C. L. Liu and J. W. Layland, Scheduling algorithms for multiprogramming in a hard-real-time environment, *Journal of ACM*, 20 (1) (1973) 46–61.

29. J. Chen, W. Jiao, and H. Wang, A service flow management strategy for IEEE 802.16 broadband wireless access systems in TDD mode, *Proceedings of the ICC 2005*, pp. 3422–3426, Seoul (Korea), May 16–20, 2005.

30. K. Wongthavarawat and A. Ganz, IEEE 802.16 based last mile broadband wireless military networks with quality of service support, *Proceedings of IEEE MILCOM 2003*, pp. 779–784, Boston (USA), October 13–16, 2003.

31. S.-M. Oh and H.-H. Kim, The analysis of the optimal contention period for broad-band wireless access network, *Proceedings of PerCom 2005*, pp. 215–219, Kauai Island (USA), March 8–12, 2005.

32. J. Yan and G.-S. Kuo, Cross-layer design of optimal contention period for IEEE 802.16 BWA systems, *Proceedings of ICC 2006*, pp. 1807–1812, Istanbul (Turkey), June 11–15, 2006.

33. C. Cicconetti, C. Eklund, L. Lenzini, and E. Mingozzi, Quality of service sup-port in IEEE 802.16 networks, *IEEE Network*, 20 (2) (2006) 50–55.

34. M. Settembre, M. Puleri, S. Garritano, P. Testa, R. Albanese, M. Mancini, and V. Lo Curto, Performance analysis of an efficient packet-based IEEE 802.16 MAC support-ing adaptive modulation and coding, *Proceedings of IEEE ISCN 2006*, pp. 11–16, Istanbul (Turkey), June 16–18, 2006.

35. S. A. Xergias, N. Passas, and L. Merakos, Flexible resource allocation in IEEE 802.16 wireless metropolitan area networks, *Proceedings of IEEE LANMAN 2005*, Chania (Greece), September 18–21, 2005.

36. A. M. Law and W. D. Kelton, Simulation modeling and analysis, 3rd edn, McGraw-Hill, 2000.
37. Cisco Press, *Traffic analysis for Voice over IP*, 2001.
38. P. T. Brady, A model for generating on-off speech patterns in two-way conversation, *Bell System Technical Journal*, 48 (1969) 2445–2472.
39. F. H. P. Fitzek and M. Reisslein, MPEG4 and H.263 video traces for network performance evaluation, *IEEE Network*, 15 (6) (2001) 40–54.

Chapter 8

Configuration Issues and Best Effort Performance Investigation of IEEE 802.16e

Alessandro Bazzi, Giacomo Leonardi,
Gianni Pasolini, and Oreste Andrisano

The IEEE 802.16 standard family enables the convergence of fixed and mobile broadband networks through a common wide area broadband access technology and a flexible network architecture. In this paper we provide an overview of the recent IEEE 802.16e-2005 amendment, aimed at supporting user mobility, and we discuss several issues related to a proper choice of system parameters and strategies to be adopted to fully exploit the technology potential. Moreover, we propose a scheduling strategy aimed at providing a fair service to best effort users. The effectiveness of this proposal is shown via simulation.*

---

* Portions reprinted, with permission, from Proceedings of IEEE International Conference on Communications 2007 (ICC 2007). © 2007 IEEE.

8.1   Introduction

The IEEE 802.16-2004 Air Interface Standard,[1] which is the basis of the WiMAX technology, is the most promising solution for the provision of fixed broadband wireless services in a wide geographical scale and proved to be a really effective solution for the establishment of wireless metropolitan area networks (WirelessMAN). On February 2006, the IEEE 802.16e-2005 amendment[2] to the IEEE 802.16-2004 standard has been released, which introduced a number of features aimed at supporting also users mobility, thus originating the so-called Mobile-WiMAX profile.

The result of the IEEE 802.16 standardization activity is a complete standard family that specifies the air interface for both fixed and mobile broadband wireless access systems, thus enabling the convergence of fixed and mobile networks through a common wide area radio access technology.

What is really remarkable, with regard to the Mobile-WiMAX profile, is the high number of degrees of freedom that are left to the manufacturers. Among all possibilities foreseen by the specifications, the final decision on a lot of very basic and crucial aspects, such as, just to cite few of them, the bandwidth, the frame duration, the duplexing scheme, and the kind of Automatic Repeat reQuest (ARQ) strategy, are left to implementers.

Moreover, besides the aforementioned basic choices, a number of strategies not defined by the specifications, whose design is traditionally left to manufacturers, have also to be conceived and implemented to fully exploit the Mobile-WiMAX potential and fulfill users' requirements: scheduling algorithms are an example.

As a consequence, the set up of a Mobile-WiMAX network requires a tricky "parameters and strategies" decision and implementation phase, that should be carried out carefully considering both Mobile-WiMAX peculiarities (in the design of the scheduling procedure, for instance) and interactions among different possible strategies to be adopted (at Transport level and Data Link level, for instance), which could affect, and in some cases greatly reduce, the perceived performance.

In the following we discuss some of these issues, with reference to the mobility enhancements provided by the latest IEEE 802.16e amendment, by means of a simulation approach; hereafter, in particular, we provide some insight on the performance level achievable with this technology, we discuss a fair resource scheduling strategy for best effort services and we highlight possible detrimental interactions among the transport control protocol (TCP) and the ARQ strategy adopted at the Data-Link level.

8.2   IEEE 802.16 Overview

The IEEE 802.16 standard family supports four physical (PHY) modes:

■ WirelessMAN-SC, which has been mainly developed for back-hauling in line-of-sight (LOS) conditions and operates in the 10–66 GHz frequency range adopting a single-carrier (SC) modulation scheme.

- WirelessMAN-SCa, which has the same characteristics of WirelessMAN-SC but operates in frequency bands below 11 GHz.
- WirelessMAN-OFDM, which has been developed for fixed wireless access in non-LOS conditions and adopts the orthogonal frequency division multiplexing (OFDM) modulation scheme in frequency bands below 11 GHz.
- WirelessMAN-OFDMA, which has been conceived for mobile access and adopts the orthogonal frequency division multiple access (OFDMA) scheme in the 2–6 GHz frequency range.

Because we are interested in the mobility enhancement provided by the latest IEEE 802.16e amendment, we focus our attention on WirelessMAN-OFDMA in the following.

8.2.1 *WirelessMAN-OFDMA*

WirelessMAN-OFDMA PHY mode is based on the OFDMA multiple-access/multiplexing technique which is, on its turn, based on an N_{FFT} subcarriers OFDM modulation scheme[3,4] with N_{FFT} equal to 128, 512, 1024, or 2048.

The N_{FFT} subcarriers form an OFDM symbol, which can be further divided into three main groups:

- Data subcarriers, used for data transmission.
- Pilot subcarriers, used for estimation and synchronization purposes.
- Null subcarriers, transmitting no signal: DC subcarrier and guard subcarriers.

Considering sequences of OFDM symbols, it is easy to understand that transmission resources, in an OFDMA system, are available both in the time domain, by means of groups of OFDM symbols, and in the frequency domain, by means of groups of subcarriers (subchannels).

With reference to WirelessMAN-OFDMA, in particular, the OFDMA multiplexing/multiple-access technique provides OFDM-based multiplexing of data streams from multiple users onto downlink (DL) subchannels as well as uplink multiple-access by means of uplink (UL) subchannels.

8.2.2 *Subchannelization*

An OFDM symbol can be divided into several subchannels by grouping its subcarrier. WirelessMAN-OFDMA allows two different subcarrier grouping methods to realize the subchannelization: distributed and adjacent permutation.

Distributed permutation, which can be implemented as DL-FUSC (downlink full usage of subchannels), DL-PUSC (downlink partial usage of subchannels), UL-PUSC (uplink partial usage of subchannels), or UL-TUSC (uplink tile usage of subchannels) on the basis of pilot subcarriers amount and positions, pseudo-randomly draws a given amount of data subcarriers, from the entire subcarriers set of a symbol, to form a subchannel. It provides frequency diversity and interference averaging.

With adjacent permutation, on the contrary, a subchannel is formed grouping a given amount of contiguous data subcarriers. In general, distributed permutation performs well in mobile applications whereas adjacent permutation is well suited for fixed, portable, or low-mobility environments.[5]

With reference to the distributed permutation scheme, which is of interest for the mobility case considered here, the minimum OFDMA frequency-time resource that can be allocated is one OFDMA-Slot, which correspond to 48 data subcarriers that can be accommodated in one, two, or three OFDMA symbols, depending on which kind of permutation scheme (DL-FUSC, DL-PUSC, UL-PUSC, and so on) is adopted, as explained hereafter in the case of OFDM symbols formed by $N_{FFT} = 2048$ subcarriers:

- *DL-FUSC.* It can be adopted in the DL phase. All 48 subcarriers of one OFDMA-Slot are mapped into a single OFDM symbol. Because DL-FUSC provides 32 subchannels within a single OFDM symbol, it follows that each OFDM symbol contains 1536 (i.e., 32×48) data subcarriers. The 166 pilot subcarriers foreseen by DL-FUSC are divided into two main groups: fixed (whose position is the same in every symbol) and variable (whose position is different between even and odd symbols). The remaining 346 null subcarriers are distributed as follows: 173 left guard subcarriers, 172 right guard subcarriers, and 1 DC subcarrier.

- *DL-PUSC.* In this case, which is allowed in the DL phase, the 48 subcarriers of an OFDMA-Slot are equally divided in two OFDM symbols (24 + 24). Because DL-PUSC provides 60 subchannels in a single OFDM symbol, it follows that each OFDM symbol carries 1440 (i.e., 60×24) data subcarriers. The 240 pilot subcarriers foreseen in this case in each OFDM symbol occupy different positions in even and odd symbols. The remaining 368 null subcarriers are distributed as follows: 184 left guard subcarriers, 183 right guard subcarriers, and 1 DC subcarrier.

- *UL-PUSC.* With UL-PUSC, which is allowed in the UL phase, the 48 subcarriers of an OFDMA slot cover three consecutive OFDM symbols, with the following, unequal, division: 12-24-12. Because the amount of subchannels provided by UL-PUSC in each OFDM symbol is still fixed and equal to 70, it follows that the number of pilot subcarrier is not the same in all symbols. In particular, symbols 1 and 3 of each OFDMA-Slot carry 840 (70×12) data subcarriers and 840 pilot subcarriers, whereas symbol 2 carries 1680 (70×24) data subcarriers and no pilot subcarrier. The remaining 368 null subcarriers are distributed as follows: 184 left guard subcarriers, 183 right guard subcarriers, and 1 DC subcarrier.

- *DL-OPTIONAL-FUSC.* It can be adopted in the DL phase. Similar to DL-FUSC all 48 subcarriers of one OFDMA-Slot are mapped into a single OFDM symbol. It provides 32 subchannels and 192 pilot subcarriers within a single OFDM symbol. The 1728 used subcarriers (1536 + 192) are divided into nine contiguous subcarriers in which one pilot carrier is allocated. The position of the pilot carrier in nine contiguous subcarriers varies according to the index of the OFDMA symbol which contains the subcarriers. The remaining 320 null subcarriers are distributed as follows: 160 left guard subcarriers, 159 right guard subcarriers, and 1 DC subcarrier.
- *UL-OPTIONAL-PUSC.* With UL-OPTIONAL-PUSC, which can be adopted in the UL phase, the 48 subcarriers of an OFDMA-Slot cover three consecutive OFDM symbols, with the following, unequal, division: 18-12-18. Because the amount of subchannel provided by UL-OPTIONAL-PUSC in each OFDM symbol is still fixed and equal to 96, it follows that the number of pilot subcarriers is not the same in all symbols. In particular, symbols 1 and 3 of each OFDMA-Slot carry 1728 (96×18) data subcarriers and no pilot subcarriers, whereas symbol 2 carries 1152 (96×12) data subcarriers and 576 (96×6) pilot subcarriers. The remaining 320 null subcarriers are distributed as follows: 160 left guard subcarriers, 159 right guard subcarriers, and 1 DC subcarrier.
- *DL-TUSC1 and DL-TUSC2.* DL-TUSC1 corresponds in structure to UL-PUSC, whereas DL-TUSC2 corresponds in structure to UL-OPTIONAL-PUSC; thus in both cases an OFDMA-Slot covers three OFDM symbols. Both DL-TUSC1 and DL-TUSC2 permutations shall only be used within an adaptive antenna system (AAS) zone.

8.2.3 Channel Coding and Modulation

Data to be transmitted are subjected to a channel-coding process which includes randomization, forward error correction (FEC) encoding, and bit interleaving.

The mandatory channel coding is performed by means of a tail-biting punctured convolutional code with variable coding rate R_c, and provides seven fixed combinations of modulation and coding rate, hereafter denoted as Modes (Table 8.1). Block Turbo Codes, Convolutional Turbo Codes, and Low-Density Parity Check codes are optional.

Each mode allows to transmit differently sized data payloads in a single OFDMA-Slot. Concatenation of a number of OFDMA-Slots up to a mode-dependent limit (fourth column of Table 8.1) can be performed to have larger codewords (fifth column of Table 8.1).

For instance, adopting the QPSK modulation scheme with $R_c = \frac{1}{2}$ we can transmit a 6-byte useful data payload in a single OFDMA-Slot and we can concatenate up to six OFDMA-Slots, for a maximum of 36 byte useful data payload to be encoded.

Table 8.1 Mode Characteristics

Mode ID	Modulation (R_c)	Useful Data Per Slot (bytes)	Maximum Number of Concatenated Slots	Maximum Data Payload (bytes)
0	QPSK $\left(\frac{1}{2}\right)$	6	6	36
1	QPSK $\left(\frac{3}{4}\right)$	9	4	36
2	16QAM $\left(\frac{1}{2}\right)$	12	3	36
3	16QAM $\left(\frac{3}{4}\right)$	18	2	36
4	64QAM $\left(\frac{1}{2}\right)$	18	2	36
5	64QAM $\left(\frac{2}{3}\right)$	24	1	24
6	64QAM $\left(\frac{3}{4}\right)$	27	1	27

Source: O. Andrisano, A. Bazzi, G. Leonardi, G. Pasolini. *IEEE 802.16e Best Effort Performance Investigation. ICC Proceedings 2007*. With permission.

8.2.4 Frame Structure

IEEE 802.16e-2005[2] supports time division duplex (TDD), frequency division duplex (FDD), and half duplex FDD operations, but the initial release of Mobile WiMAX certification profiles includes only TDD, as it enables adjustment of the DL/UL ratio to support asymmetric DL/UL traffic. Furthermore, TDD receiver implementation is less complex and expensive.

Both FDD and TDD duplexing schemes are based on a time-frequency frame structure to accommodate DL/UL data flows.

In Figure 8.1, the TDD frame structure is depicted. Each TDD frame is divided into DL and UL subframes, separated by transmit/receive and receive/transmit transition gaps (TTG and RTG).

The OFDMA frame may include "multiple zones"; this means that the permutation method can be changed within the same frame, thus moving, for instance, from DL-PUSC to DL-FUSC (Fig. 8.1).

The basic elements of a TDD frame, depicted in Figure 8.1, are hereafter detailed.

- *Preamble.* Each frame begins with a preamble OFDM symbol, used for synchronization.
- *Frame control header (FCH).* Subchannels 0 to 3 in the first two OFDM symbols following the preamble convey the FCH field. FCH is transmitted using QPSK, $R_c = \frac{1}{2}$, DL-PUSC and contains frame configuration information.
- *DL-MAP and UL-MAP.* They convey subchannel allocations, providing pointers to bursts (see in the following) allocated to users, and other control information for the respective subframes.

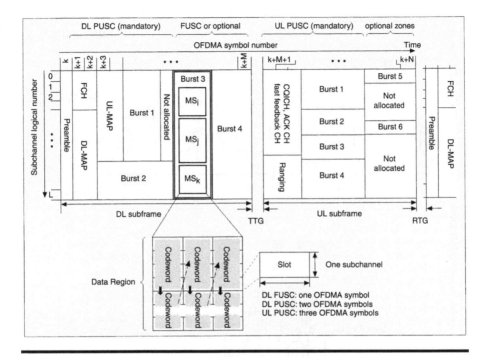

Figure 8.1 Time division duplex frame structure. (From O. Andrisano, A. Bazzi, G. Leonardi, G. Pasolini. *IEEE 802.16e Best Effort Performance Investigation. ICC Proceedings 2007.* **With permission.)**

- *UL ranging.* The UL ranging field is allocated for mobile stations to perform closed-loop time, frequency and power adjustments as well as bandwidth requests.
- *UL CQICH.* The UL CQICH field is allocated to convey feedback channel-state information.
- *UL ACK-CH.* The UL ACK-CH is allocated to convey fast feedback for DL Hybrid-ARQ.
- *Burst.* A burst is a portion of the frame which carries the coded data adopting the same set of parameters (such as mode and permutation method) assigned to a Base Station ↔ Mobile Station (BS ↔ MS) communication in a given DL/UL direction.

8.2.5 Data Mapping

In Figure 8.2, the process of Data Mapping, starting from the Internet Protocol (IP) packet down to the physical level allocation, is illustrated when the ARQ mechanism is active.

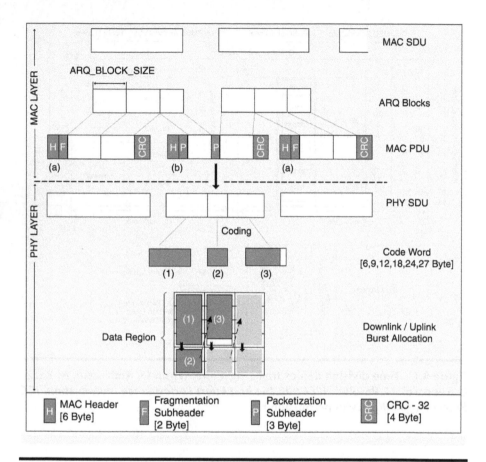

Figure 8.2 Packets processing. *Abbreviations:* MAC, Medium Access Control; PHY, Physical; PDU, Protocol data units; SDU, Service data units; ARQ, Automatic Repeat reQuest. (From O. Andrisano, A. Bazzi, G. Leonardi, G. Pasolini, *IEEE 802.16e Best Effort Performance Investigation. ICC Proceedings 2007.* With permission.)

At the data-link level each received IP packet corresponds to one service data unit (SDU). SDUs are divided into ARQ blocks of *ARQ_BLOCK_SIZE* bytes, which is a value negotiated during connection. The last ARQ block of each SDU will generally be smaller than *ARQ_BLOCK_SIZE*.

Following the available PHY level resources, at the Medium Access Control (MAC) level one or more ARQ blocks will be inserted into a protocol data unit (PDU), with some overhead: in particular, a MAC header will be added, plus either (a) one fragmentation subheader if all ARQ blocks are contiguous and related to the same SDU or (b) a packetization subheader per each group of contiguous ARQ blocks belonging to the same SDU; a CRC (Cyclic Redundancy Check) tail of 32 bits will be added at the end of the PDU, to check the integrity of received data.

At the physical level, MAC PDUs will be divided into groups of bytes, that will be coded and modulated. In particular, coded data of a burst, processed as described in Section 8.2.3, are mapped into an OFDMA data region, which is a two-dimensional allocation of a group of contiguous subchannels (i.e., with contiguous identifying numbers) in a group of contiguous OFDM symbols. A data region is commonly depicted as a rectangle, as shown in the lower parts of both Figures 8.1 and 8.2. Please note that allocations refer to logical subchannels, represented by their identifying numbers, which can be, in general, constituted by sparse subcarriers.

Because, as recalled in Section 8.2.2, the minimum OFDMA frequency-time resource that can be allocated is the OFDMA-Slot, whose extension in terms of OFDM symbols depends on the adopted distributed permutation scheme (DL-PUSC, DL-FUSC, UL-PUSC, and so on), it follows that data region dimensions must be such to accommodate a finite number of OFDMA-Slots.

8.3 Resource Request and Scheduling Policies

IEEE 802.16e-2005 divides all possible data services into five classes: Unsolicited Grant Service (UGS), Real-Time Polling Service (rtPS), Extended Real-Time polling Service (ErtPS), Non-Real-Time Polling Service (nrtPS), and Best Effort (BE). Each service is associated with a set of quality of services (QoS) parameters that quantify aspects of its characteristics: (a) maximum sustained rate, (b) minimum reserved rate, (c) maximum latency tolerance, (d) jitter tolerance, and (e) traffic priority.

The aforementioned parameters are the basic inputs for the service scheduler placed in the BS, whose design and implementation are left to the manufacturers, which is aimed at fulfilling service-specific QoS requirements.

Within IEEE 802.16, in particular, the scheduler task is to define both uplink and downlink resource allocation maps (UL-MAP and DL-MAP) on the basis of the users' needs. With reference to BE services, an insight on the bandwidth request mechanism is provided in the following subsections and a fair and efficient scheduling strategy is proposed.

8.3.1 BE Resource Requests

In order to support BE services (FTP, web browsing, and so on), a resource request mechanism is needed to make the scheduler aware of MSs bandwidth requirements in both directions.

As for the DL direction, however, it is immediate to understand that the scheduler has a perfect knowledge of MSs needs, because they coincide with the amount of data waiting to be transmitted in the respective BS transmission queues.

As far as the UL is concerned, on the contrary, a request mechanism is introduced by the standard, which allows MSs to make use of (a) contention request opportunities, (b) unicast request opportunities, and (c) unsolicited data grant burst types. In the first case a bandwidth request is transmitted during the appropriately shared UL allocation, whereas in the other two cases each MS is given a reserved UL resource to convey its request.

All requests for bandwidth are made in terms of number of bytes needed to carry MAC PDUs.

Once an MS is given the UL resource, further bandwidth requests may come as a piggyback request, thus avoiding to resort again to one of the three bandwidth request mechanisms introduced before (a, b, and c).

Bandwidth requests may be incremental or aggregate; in the former case the BS adds the amount of bandwidth requested to its current perception of the bandwidth needs of the connection, whereas in the latter case it replaces its perception of the bandwidth needs of the connection with the amount of bandwidth requested. The type field in the bandwidth request header indicates whether the request is incremental or aggregate. Because piggyback bandwidth requests do not have a type field, piggyback bandwidth requests shall always be incremental.

The mechanism of piggyback incremental requests can be conveniently exploited to reduce the time wastage due to the complete bandwidth request mechanism: it is reasonable to operate in such a way that the first time an MS performs a bandwidth request (adopting one of the three previously introduced procedures), it notifies the BS the dimension of the entire amount of data waiting in its transmission queue. This way the BS has an exact knowledge of each MS need and can update it (decreasing) each time PHY-level resources are assigned to that connections and the related transmissions are correctly acknowledged.

Possible further data incoming in the MS queue while transmissions are still ongoing (i.e., before the MS queue gets empty) can be notified to the BS scheduler through the incremental piggyback bandwidth requests, which will determine a variation of the perception of MS bandwidth needs.

8.3.2 BE Scheduling

In this section we still focus on BE data services and we show how to provide, at the same time, a fair and efficient resource sharing, carefully considering IEEE 802.16e-2005 specific characteristics.

As far as the fairness issue is concerned, here we considered a round robin (RR) scheduling policy among all BE services. Although this choice seems the most suited to this kind of service, it has to be pointed out that its implementation, as well as the implementation of any other scheduling policy, requires some preliminary consideration on the nature of IEEE 802.16e-2005 radio resource (the OFDMA-Slot).

Because the different modes that can be adopted by users convey, in a single OFDMA-Slot, different amounts of data, it follows that, to provide a really fair scheduling among BE users, a different amount of slots has to be allocated to each of them.

To meet as much as possible the aforedescribed fairness requirement, hence avoiding that users with huge amount of data to be transmitted gather the most of resources, it is convenient to define an elementary resource unit, hereafter called Virtual Resource Unit (VRU), consisting in a fixed amount of data bytes, which is the basic element assigned by the scheduler in each RR cycle to BE users needing resources.

To better understand the scheduler behavior, let us focus our attention, for instance, on the DL subframe: the scheduler task is, in this case, to define the DL-MAP, assigning PHY-level resources to the different BS → MS links requiring DL resources.

As long as there is room (in terms of available slots) in the subframe and there are pending data that can be allocated in it, the scheduler moves from an user to the next, performing the following actions: (*i*) it assigns a VRU to the user, adding it to its Virtual Resource Budget (VRB); then, (*ii*) it virtually generates the biggest PDU that can be allocated, and (*iii*) correspondingly makes a slot booking; this means that although the scheduler does not effectively allocate the PDU at this time, that amount of slots is definitively reserved to that user; at the next round a bigger PDU or more PDUs may be allocated following the increased VRB, thus increasing the amount of booked slots. Please note that the scheduler would not do any slot reservation until VRB is not enough to allocate a PDU containing a single ARQ block. When the RR cycle ends (i.e., no more room or no more data to be allocated), the PDUs are effectively generated and the coding-mapping processes described in Section 8.2.3 can take place.

The separation between the virtual and physical allocations guarantees a flexible management: as an example, doubling the VRB (during the second round) may allow, for instance, to generate a PDU that includes two ARQ blocks and this could be preferred than generating and transmitting two separate PDUs with a single ARQ block each.

At the end of the allocation procedure users' VRBs are not reset, to allow stations with larger ARQ blocks to gain the same long-term priority as the others. Furthermore, VRB will be increased at most up to the amount of pending data.

The same procedure is performed also for the definition of the UL-MAP, which is carried out on the basis of UL resource request made by users.

To preserve fairness at most, in the next subframe of the same kind (UL or DL) the cyclic scheduling procedure will start from the user that follows, in the cyclic order, the last user served in the current frame.

8.4 Simulations Settings

To evaluate the performance of the proposed scheduling algorithm, we realized an apposite IEEE 802.16e simulator, which has been developed within the framework of a general simulation platform.[6] It is important to underline that such simulator reproduces all aspects related to each single level of the protocol pillar, from the physical to the application level.

All the aforementioned peculiarities of IEEE 802.16e, such as subchannelization using different permutation methods, channel coding and OFDMA-Slots concatenation, data mapping, frame management, and so on, have been carefully implemented. As for the transport level protocol, we implemented the New Reno version of TCP.

Finally, to evaluate the saturation throughput perceived at application level, we considered DL FTP sessions with infinite duration.

Propagation Model: A realistic channel model has been adopted reproducing the behavior of the wireless channel, including time and frequency correlation. In particular, we considered the path-loss law related to terrain type *C* (sub-urban) and the channel model SUI1 (at 3 Km/h), both reported in reference.[7]

Simulations Settings: As for the most important system parameters adopted in our simulations, please refer to Table 8.2.

With reference to the TDD version of IEEE 802.16e, we decided to adopt an *x* to 1 DL/UL asymmetry, where *x* is as near to 3 as possible. Three symbols are assumed to be reserved in the DL subframe for preamble, FCH, and DL/UL MAPs and three symbols in the UL subframe for UL ACK-CH, ranging, and so on. Thus,

Table 8.2   Simulation Parameters

System Parameters	
Carrier frequency	3.5 GHz
BS antenna height	15 m
MS antenna height	2 m
BS antenna gain	15 dBi (120°)
MS antenna gain	5 dBi
BS transmission power	30 dBm
MS transmission power	22 dBm
BS noise figure	4 dB
MS noise figure	5 dB
OFDMA Parameters	
Channel bandwidth	7 MHz
N_{FFT}	2048
Useful symbol time	256 μs
Guard time	8 μs
Symbol time	264 μs

BS, base station; MS, mobile station.
Source: O. Andrisano, A. Bazzi, G. Leonardi, G. Pasolini. *IEEE 802.16e Best Effort Performance Investigation. ICC Proceedings 2007*. With permission.

adopting a 10-ms frame duration, corresponding to 31 symbols left for data (over a total of 37), 22 were left for DL (11 couples of OFDMA symbols, adopting DL PUSC) and 9 for UL (three triples of OFDMA symbols, adopting UL PUSC). Recalling that we have 60 subchannels in DL and 70 in UL, it follows that nominal data rates range between 3.168 Mbps (mode 0, see Table 8.1) and 14.256 Mbps (mode 6) in DL and between 1.008 Mbps and 4.536 Mbps in UL. The exact DL/UL asymmetry is thus $x_{10ms} = (11 \times 60)/(3 \times 70) = 3.14$.

Similarly, adopting a 20-ms frame duration, corresponding to 69 data symbols (over a total of 75), 48 were left for DL (24 couples of OFDMA symbols) and 21 for UL (seven triples of OFDMA symbols). Nominal data rates range between 3.456 Mbps and 15.552 Mbps in DL and between 1.176 Mbps and 5.292 Mbps in UL. The exact DL/UL asymmetry is thus $x_{20ms} = (24 \times 60)/(7 \times 70) = 2.94$.

Obviously, adopting a 20-ms frame duration we have an increase of the achievable data rate, due to the reduction of the impact of the six symbols reserved for DL and UL management; please note that in the DL phase this benefit is slightly reduced, with respect to the previous case, by the lower asymmetry rate. It has to be observed, furthermore, that the previously introduced data rate is referred to data link-level data bits, which exclude, on one hand, preamble, FCH, DL/UL MAPs, UL ACK-CH, and so on, but include, on the other hand, the overhead introduced by the data link level and by the upper protocol levels. It follows that the application level throughput, considered in the numerical results section, will be slightly lower than data link level data rate, owing to the presence of all overheads.

Another important aspect to be considered is the definition of *ARQ_BLOCK_ SIZE* and *VRU*. First of all, we decided to never allocate more than one ARQ Block into a single PDU. Please note that, although including more ARQ blocks into a single PDU allows to reduce MAC overhead, this increases the overall overhead due to retransmissions: CRC calculation is done, in fact, on a PDU basis and the uncorrect reception of one PDU may lead to the retransmission of all the ARQ blocks it includes. Future works with different choices may reveal better performance, but a detailed analysis of this aspect is out of the scope of the present discussion.

Following this choice and trying to allocate an integer number of OFDMA-Slots to each PDU, whichever the adopted mode, the PDU should be allocated into the least common multiplier of all useful data per slot (see the second column of Table 8.1), which results in 216 bytes. Thus, *ARQ_BLOCK_SIZE* = 204 bytes and *VRU* = 216 bytes (*ARQ_BLOCK_SIZE* + MAC header + fragmentation Subheader + CRC) were chosen.

8.5 Data Link and TCP Issues

Few words are to be spent on data link and TCP interactions; it must be remarked, in fact, that a data link configuration without considering TCP behavior may incur in heavy performance degradation.

A problem that deserves particular attention, for instance, is that a significant number of TCP packets can be delivered to an user in a single subframe. If one or more of these incur in transmission errors, while the following are correctly received and passed to the upper protocol layers of the receiver side, multiple duplicate TCP acknowledgments are issued, which could cause an unwanted and useless congestion window reduction at the sender TCP side, thus reducing the experienced throughput.

To solve this problem, the optional *ARQ_DELIVER_IN_ORDER* functionality must be activated at data link level for each TCP session. When *ARQ_DELIVER_IN_ORDER* is enabled, a MAC SDU is handed to the upper layers as soon as all the ARQ blocks of the MAC SDU have been correctly received within the defined time-out values and all blocks with sequence numbers smaller than those of the completed message have either been discarded due to time-out violation or delivered to the upper layers.

For this reason, where not specified, in this paper the ARQ mechanism is adopted following IEEE 802.16e specifications and the *ARQ_DELIVER_IN_ORDER* option is activated. A numerical proof on what is asserted here will be given in Section 8.6.

8.6 Numerical Results

The following numerical results characterize the WirelessMAN-OFDMA performance in terms of DL application-level throughput in saturation conditions and fairness, adopting the parameters detailed in Section 8.4. To better understand the numerical results, the reader should have in mind that application-level throughput is a performance metric strictly related to the users perception of the service quality as it takes into account all aspects affecting communications (with particular reference to the physical, data link, and transport levels).

As far as the fairness in concerned, here we assessed it by means of the parameter F, whose definition is reported in Appendix A, which is an indicator of the equity in resource allocation and ranges between 0 (total unfairness) and 1 (total fairness).

In Figure 8.3, the DL application-level saturation throughput (hereafter, throughput) perceived by a single user is reported as a function of the distance, for each one of the seven possible modes (modulations and FEC rates). Here a frame duration $T_f = 10$ ms is considered. As can be observed, the throughputs experienced in optimal conditions (nearby the BS) are slightly lower with respect to the nominal data rates reported in Section 8.4; as already observed, this difference is mainly due to the protocol overheads that reduce the experienced application level throughput.

To assess the upper bound to the achievable performance here we do not consider any link adaptation and power control strategy.

It can be noted that not all the possible modes would be needed in the considered scenario. In particular, mode 5 never outperforms mode 6 within a reasonable

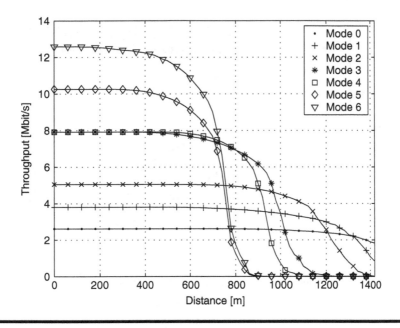

Figure 8.3 Throughput perceived by a single user as a function of the distance for all possible modes; 10-ms frame duration.

range of achievable throughput. Moreover, modes 3 and 4 provide the same maximum throughput, as they both carry 18 bytes per slot (see Table 8.1); thus, the more reliable mode 3 can be always preferred.

It must be noted that an efficient link adaptation algorithm will let the connection be always on the envelope of curves depicted in Figure 8.3, or theoretically even better on a varying channel: in such a case, in fact, different modes may gain higher throughputs in different time intervals. However, as the analysis of an efficient link adaptation technique is out of the scope of the present chapter, the envelope of throughputs perceived adopting the various modes will be considered in the following.

In Figure 8.4, the throughput perceived by a single user is reported as a function of the distance in both cases of *ARQ_DELIVER_IN_ORDER* active or not. This figure clearly shows the great impact of interactions between transport (here the New Reno TCP version is considered) and data link levels. Here, in particular, the performance degradation due to unnecessary reductions of the TCP congestion window when the *ARQ_DELIVER_IN_ORDER* option is not active can be clearly appreciated (see Section 8.5).

In Figure 8.5, the impact of different frame durations T_f on the average throughput perceived by users is shown as a function of the distance. Here we assumed $T_f = 10$ ms and $T_f = 20$ ms; $N = 5$, 10, and 20 active users at the same distance from

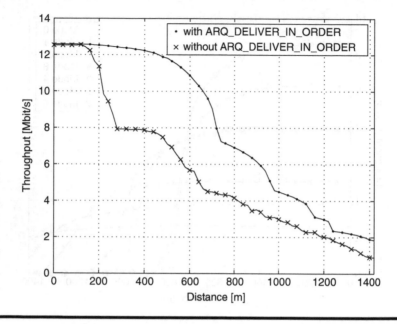

Figure 8.4 Throughput perceived by a single user as a function of the distance, with and without the adoption of the *ARQ_DELIVER_IN_ORDER* option; 10-ms frame duration.

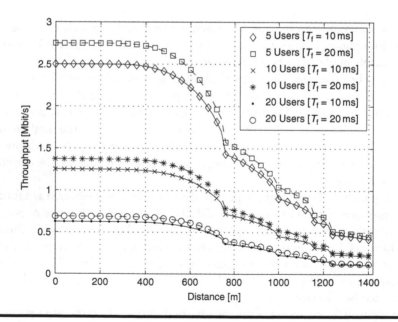

Figure 8.5 Average throughput of 5, 10, or 20 users at the same distance from the Base Station, considering 10-ms and 20-ms frame durations.

the BS are considered. As can be observed, the average throughputs multiplied by N are almost constant, whereas the difference between those referring to a 10-ms frame and a 20-ms frame in optimal conditions (nearby the BS) is almost identical to that between nominal data rates evaluated in Section 8.4. These considerations show the effectiveness of the proposed scheduling algorithm, which fully exploits system resources. Let us observe, moreover, that the throughput increase due to a longer frame is not so relevant to justify the impact it would have on possible active real-time sessions (20-ms frame duration may be too high).

Finally, in Figure 8.6 the system performance is analyzed considering MSs at different distances from the BS, that is, experiencing very different path losses. Here, in particular, $N = 5$ MSs are equally spaced from 200 m and d_{max}, where d_{max} ranges from 600 m to 1400 m.

Figure 8.6 shows both the average perceived throughput per user (curves) and the achieved fairness level F (labels) as a function of d_{max} in the two cases $T_f = 10$ ms and $T_f = 20$ ms. Because, as can be observed, F is almost equal to one in all cases, it follows that the average perceived throughput almost coincides with the single user perceived throughput. The scheduler effectiveness in pursuing the fairness objective, notwithstanding the very different channel conditions experienced by users, is therefore proved.

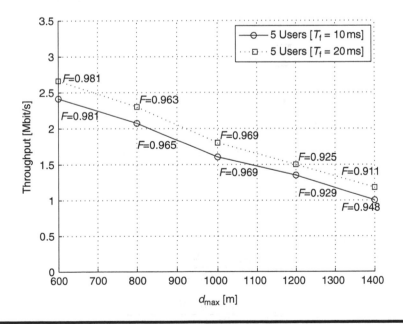

Figure 8.6 Average throughput of five equally spaced users with distances in the interval between 200 m and d_{max} from the Base Station. Labels refer to the related achieved fairness F.

Appendix A   Fairness Parameter Definition

Given N users, we define the fairness parameter F of the network as follows:

$$F = 1 - \frac{AD}{AT} \cdot \frac{N}{2(N-1)}, \tag{A.1}$$

$$AD = \frac{\sum_{i=1}^{N} |T_i - AT|}{N}, \tag{A.2}$$

where AT is the average throughput, AD is the average distance from the average throughput, and T_i is the throughput perceived by user i ($i = 1, 2, \ldots, N$). It must be noted that F ranges from 1 to 0, where 1 is achieved when all users perceive the same throughput ($AD = 0$), while 0 is achieved when one user perceives a given throughput whereas all other users perceive no throughput.

Proof. Suppose that there are N active users, among which one user (user 1) perceives a given throughput $T_1 = T$ whereas all other users perceive $T_i = 0$, where $i = 2, 3, \ldots, N$. Then, F can be calculated through the following steps:

$$AT = \sum_{i=1}^{N} T_i / N = T / N,$$

$$\begin{aligned}
AD &= \frac{\sum_{i=1}^{N} |T_i - AT|}{N} \\
&= \frac{(T - T/N) + (N-1)T/N}{N} = \frac{2(N-1)T}{N^2};
\end{aligned}$$

hence

$$\begin{aligned}
F &= 1 - \frac{AD}{AT} \cdot \frac{N}{2(N-1)} \\
&= 1 - \frac{2(N-1)T/N^2}{T/N} \cdot \frac{N}{2(N-1)} \\
&= 1 - 1 = 0.
\end{aligned}$$

In the opposite case, when all users perceive the same application-level through-put T, it is immediate to verify that

$$AT = \sum_{i=1}^{N} T_i/N = \frac{NT}{N} = T,$$

$$AD = \frac{\sum_{i=1}^{N} |T_i - AT|}{N} = \frac{\sum_{i=1}^{N} |T_i - T|}{N} = 0;$$

hence

$$F = 1 - \frac{AD}{AT} \cdot \frac{N}{2(N-1)} = 1.$$

References

1. *IEEE std 802.16-2004 IEEE Standard for Local and Metropolitan Area Networks. Part 16: Air Interface for Fixed Broadband Wireless Access Systems* (IEEE, Piscataway, NJ, USA, 2004).
2. *IEEE std 802.16e-2005 and IEE std 802.16-2004/Cor1-2005 IEEE Standard for Local and Metropolitan Area Networks. Part 16: Air Interface for Fixed and Mobile Broadband Wireless Access Systems Amendment 2: Physical and Medium Access Control Layers for Combined Fixed and Mobile Operation in Licensed Bands and Corrigendum 1* (IEEE, Piscataway, NJ, USA, 2005).
3. L. J. Cimini, Analysis and simulation of digital mobile channel using orthogonal frequency division multiplexing, *IEEE Transaction on Communications*, 33(7), 665–675 (1985).
4. R. V. Nee and R. Prasad, *OFDM for Wireless Multimedia Communications* (Artech House, Norwood, MA, USA, 2000).
5. Mobile wimax part I: A technical overview and performance evaluation, *White paper* (August 2006). http://www.wimaxforum.org/news/downloads/.
6. A. Bazzi, C. Gambetti, and G. Pasolini. Shine: simulation platform for heterogeneous interworking networks. In *Proc. IEEE International Conference on Communications 2006 (ICC 2006)*, Istanbul, Turkey (2006).
7. V. Erceg and Alii. Channel model for fixed wireless applications. In *IEEE 802.16 Broadband Wireless Access Working Group*. http://www.ieee802.org/16/tga/docs/80216a-03_01.pdf.

Chapter 9

QoS Architecture for Efficient Bandwidth Management in the IEEE 802.16 Mesh Mode

Parag S. Mogre, Matthias Hollick, Ralf Steinmetz, and Christian Schwingenschlögl

The IEEE 802.16 standard series represents the state-of-the-art in technology for metropolitan area broadband wireless access networks. The point-to-multipoint (PMP) mode of IEEE 802.16 has been designed to enable quality of service (QoS) in operator-controlled networks and, thus, is foreseen to complement existing third-generation cellular networks. In contrast, the optional mesh (MESH) mode of operation in IEEE 802.16 enables the setup of self-organizing wireless multi-hop mesh networks. A distinguishing characteristic of the IEEE 802.16 standard series is its support for QoS at the Medium Access Control (MAC) layer. However, the QoS specifications and mechanisms for the PMP and the MESH mode are not consistent. This article presents a novel QoS architecture as a key enhancement to the IEEE 802.16 MESH mode of operation. The architecture is based on the QoS mechanisms outlined for the PMP mode and, thus, enables a seamless coexistence of the PMP and the MESH mode. In particular, we look at the various options the standard provides and the trade offs involved when implementing QoS support in the 802.16 MESH mode, with a focus on the efficient management of the available bandwidth resources. This article is meant to provide researchers and implementers crucial anchor points for further research.

9.1   Introduction and Motivation

The demand for ubiquitous connectivity is the driving force for innovation in the field of wireless networks. To satisfy the differing demands of users a huge variety of wireless network platforms has developed over the years. Figure 9.1a shows some of the contemporary wireless access technologies.

From the figure, one can see that the IEEE 802.16 standard as published in [1] intends to support metropolitan area networks, rural networks, or enterprisewide networks. Initially these networks are expected to support only static nodes (subscriber stations). The standard IEEE 802.16-2004 [1] specifies two modes of

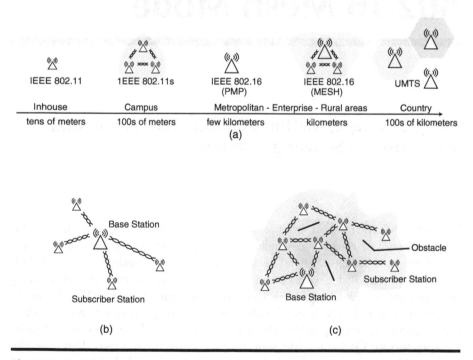

Figure 9.1   (a) Overview of contemporary wireless access network technologies showing the geographical scale of the wireless technologies. (b) Point-to-multipoint (PMP) mode of operation in 802.16. (c) Mesh mode (MESH) of operation in 802.16.

operation as shown in Figure 9.1b, c. In the point-to-multipoint mode (PMP) all the Subscriber Stations (SSs) are required to be in direct range of the Base Station (BS). The SSs can directly communicate only with the BS. Direct communication between two SSs is not supported in the PMP mode. On the other hand, when operating in the MESH mode, the SSs are allowed to establish communication links with neighboring nodes and are able to communicate with each other directly. In addition, they are also able to send traffic to and receive traffic from the BS (a MESH BS is a SS, which provides backhaul services to the mesh network). The MESH mode of 802.16 allows for flexible growth in the coverage of the mesh network and also increases the robustness of the network due to the provision of multiple alternate paths for communication between nodes. An overview of the 802.16 standard is provided in [2]. In addition, current efforts in the 802.16e task group have led to the publication of the IEEE 802.16e specification [3]. The latter mainly offers enhancements to the IEEE 802.16-2004 to support mobility.

A distinguishing feature of the IEEE 802.16 standard is the extensive support for QoS at the MAC layer. In addition, the IEEE 802.16 standard outlines a set of physical layer (PHY) specifications which can be used with a common MAC layer. This flexibility allows the network to operate in different frequency bands based on the users' needs and the corresponding regulations. The QoS support and flexibility at the PHY layer in the 802.16 standard make it an optimal base to support multi-service networks. Thus, 802.16 networks are expected to play a significant role in next-generation broadband wireless access (BWA) networks. Such networks cater to the demand for the so-called "Triple Play" networks, that is, a single network supporting broadband Internet access, telephony, and television services. They are thus expected to replace conventional Digital Subscriber Line (DSL)-based access networks. The needs of each of these application categories are however varying. For example, applications such as interactive video conferencing, telephony, etc. require predictable response time and a static amount of bandwidth continuously available for the life-time of the connection. On the other hand, traffic like variable rate compressed video streams (e.g., to support television services) relies on accurate timing between the traffic source and destination but does not require a static amount of bandwidth over the duration of the connection. Some other applications such as data transfer using File Transfer Protocol (FTP) have no inherent reliance on time synchronization between the traffic source and destination. However, these applications benefit when the network attempts to provide a guaranteed bandwidth or latency. Some other services may not be very important from the providers' point of view, and traffic belonging to this class may be serviced on space-available basis. This type of traffic has usually no reliance on time synchronization between the traffic source and destination. An example of application generating the latter type of traffic is web surfing. The 802.16 standard defines different data scheduling services to support these types of traffic; thus, providing tools for network operators to support multi-service networks.

The optional MESH mode of operation specified by the standard allows for organic growth in coverage of the network, with low initial investment in infrastructure. In addition, a mesh inherently provides a robust network due to the possibility of multiple paths for communication between nodes. Thereby, a mesh can help to route data around obstacles or provide coverage to areas which may not be covered using the PMP setup with a similar position for the BS as in the MESH mode. A comprehensive description ofthe MESH mode can be found in Ref. [4]. A mesh also enables the support of local community networks as well as enterprise-wide wireless backbone networks. This application scenario makes the MESH mode very attractive to network providers, companies, and user communities. This article focuses on efficient management of bandwidth for realization of QoS in the MESH mode. In particular we look shortly at the nuts and bolts involved and the options provided by the standard. Although the IEEE 802.16-2004 standard specifies an extensive set of messages and mechanisms to realize QoS, the algorithms to realize QoS and manage bandwidth are left open to foster innovation and provide scope for vendor optimization. This article provides the readers with an overview of the scope for innovation and some critical challenges that need to be addressed to obtain a robust and efficient implementation of the IEEE 802.16 MESH mode.

In the next section we provide an overview of the QoS specification for the MESH mode as specified by the 802.16 standard. In particular, we introduce the mechanisms available in the PMP mode and the MESH mode. In future, IEEE 802.16-based networks are expected to support seamless interworking of nodes operating in the PMP and the MESH modes. The QoS specifications for the two modes however are not consistent. We provide an overview of our proposed QoS architecture for management of the bandwidth in the MESH mode to enable support of the data scheduling services similar to those available in the PMP mode.

Finally, we provide an insight into the benefits and effects of deploying our proposed QoS architecture which we have been able to observe via an intensive simulation study. The simulation study was carried out using a MESH mode simulator we built into the JiST/SWANS [5] environment. We will here also highlight some promising areas for further investigation and outline areas for research which can build up on and extend the QoS architecture described by us.

9.2 Realizing QoS Using the 802.16 Standard

In this section we first provide an overview of the QoS support mechanisms specified in the standard for the PMP mode followed by an overview of those for the MESH mode. The focus of this article is on the bandwidth management mechanisms required to efficiently support the different classes of traffic. The admission control as well as queueing and priority mechanisms needed to support hard QoS requirements of individual connections are not in the focus of this article.

9.2.1 QoS Support in the 802.16 PMP Mode

The 802.16 MAC is connection-oriented. QoS is provisioned in the PMP mode on a per-connection basis. All data, either from the SS to the BS or vice versa is transmitted within the context of a connection, identified by the connection identifier (CID) specified in the MAC protocol data unit (PDU). The CID is a 16-bit value that identifies a connection to equivalent peers in the MAC at both the BSs as well as the SSs. It also provides a mapping to a service flow identifier (SFID). The SFID defines the QoS parameters which are associated with a given connection (CID). The SFID is a 32-bit value and is one of the core concepts of the MAC protocol. It provides a mapping to the QoS parameters for a particular data entity.

Figure 9.2 shows the core objects involved in the QoS architecture as specified in the standard for the PMP mode. As is seen from Figure 9.2, each MAC PDU is transmitted using a particular CID, which is in turn associated with a single service flow identified by a SFID. Thus, many PDUs may be transmitted within the context of the same service flow but a single MAC PDU is associated with exactly one service flow. Figure 9.2 also shows that there are different sets of QoS parameters associated with a given service flow. These are the "*ProvisionedQoSParamSet,*" "*AdmittedQoSParamSet,*" and "*ActiveQoSParamSet.*" The provisioned parameter set is a set of parameters provisioned using means outside the scope of the 802.16 standard, such as with the help of a network management system. The admitted parameter set is a set of QoS parameters for which resources (bandwidth, memory, and so on) are being reserved by the BS (SS). The active parameter set is the set of QoS parameters defining the service actually being provided to the active flow. For example, the BS transmits uplink and downlink maps specifying bandwidth allocation for the service flow's active parameter set. Only an active service flow is allowed to transmit packets. To enable the dynamic setup and configuration of service flows, the standard specifies a set of MAC management messages, the so-called

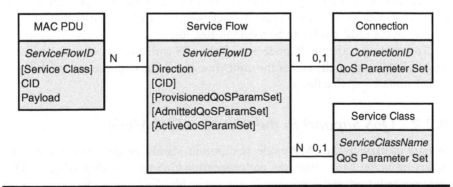

Figure 9.2 Quality-of-service (QoS) object model [1] for IEEE 802.16-2004 point-to-multipoint mode.

dynamic service messages (DSx messages). These are the dynamic service addition (DSA), dynamic service change (DSC), and the dynamic service deletion (DSD) messages. The various QoS parameters associated with a service flow are negotiated using these messages.

Typical service parameters associated with a service flow are traffic priority, minimum reserved rate, tolerated jitter, maximum sustained rate, maximum traffic burst, maximum latency, and scheduling service. The BS may optionally create a service class as shown in Figure 9.2. A service class is a name given to a particular set of QoS parameters, and can be considered as a macro for specifying a set of QoS parameters typically used. The value for the scheduling service parameter in the QoS parameter set specifies the data scheduling service associated with a service flow. The 802.16 standard currently defines the following data scheduling services: Unsolicited Grant Service (UGS), Real-Time Polling Service (rtPS), Non-Real-Time Polling Service (nrtPS), and Best Effort (BE). The UGS is meant to support real-time data streams consisting of fixed-size data packets issued periodically. The rtPS is meant to support data streams having variable-sized data packets issued at periodic intervals. The nrtPS is designed to support delay-tolerant streams of variable-sized data packets for which a minimum data rate is expected. The BE traffic is serviced on a space-available basis. For service flow associated with the scheduling service UGS, the BS allocates a static amount of bandwidth to the SS in every frame. The amount of bandwidth granted by the BS for this type of scheduling service depends on the maximum sustained traffic rate of the service flow. For rtPS service flows, the BS offers real-time, periodic, unicast request opportunities meeting the flow's requirements and allowing the SS to request a grant of the desired size. For nrtPS the BS, similar to the case of a rtPS service flow, offers periodic request opportunities. However, these request opportunities are not real-time, and the SS can also use contention-based request opportunities in addition to the unicast request opportunities for a nrtPS service flow as well as the unsolicited data grant types. For a BE service flow no periodic polling opportunities are granted. The SS uses contention request opportunities, unicast request opportunities, and unsolicited data grant burst types. A brief overview and evaluation of the QoS support in the PMP mode can be found in Ref. [6].

To summarize, the PMP mode provides the BS with efficient means to manage the bandwidth optimally and at the same time satisfy the requirements of the individual admitted service flows.

9.2.2   QoS Support in the 802.16 MESH Mode

In stark contrast to the PMP mode, the QoS in MESH mode is provisioned on a packet-by-packet basis. Thus, the per-connection QoS provisioning using the DSx messages as introduced previously is not applicable. This design decision helps to reduce the complexity of implementing the MESH mode considerably. However, the MESH mode even with this simplification is quite complex.

The CID in the MESH mode is shown in Figure 9.3. The mesh CID is used to differentiate the forwarding service a PDU should get at each individual node. As can be seen from Figure 9.3 it is possible to assign a priority to each MAC PDU. Based on the priority the transmission scheduler at a node can decide if a particular PDU should be transmitted before another. The field reliability specifies the number of retransmissions for the particular MAC PDU (if needed). The drop precedence specifies the dropping likelihood for a PDU during congestion. Messages with a higher drop precedence are more likely to be dropped. In effect, QoS specification for the MESH mode is limited to specifying the priority of a MAC PDU, the reliability, and its drop precedence. Given the same reliability and drop precedence and MAC PDU type (see Fig. 9.3), the MAC will attempt to provide a lower delay to PDUs with higher priority. This QoS mechanism, however, does not allow the node to estimate the optimal bandwidth requirement for transmissions on a particular link. This is because (just based on the previous interpretation as presented in the 802.16 standard), the node is not able to identify the expected arrival characteristics of the traffic and classify it into the different categories as traffic requiring UGS, rtPS, nrtPS, or BE service.

To summarize, QoS mechanisms in the MESH mode are not consistent with those provided for the PMP mode. In addition, the per-packet QoS specification for the MESH mode does not allow a node to optimally estimate the amount of bandwidth required for transmission on a link, as no information about the data scheduling service required for the traffic is included explicitly in the QoS specification in the mesh CID.

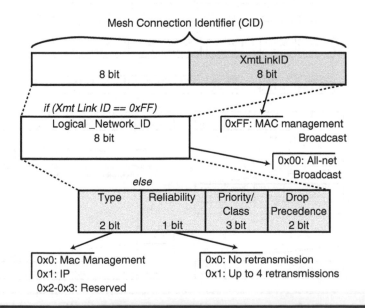

Figure 9.3 MESH connection identifier (CID).

We next give an overview of the existing bandwidth request and grant mechanisms specified for the MESH mode of 802.16. This is followed by a description of our proposed QoS architecture, which enables efficient bandwidth management in the MESH mode and allows support of the data scheduling services consistent with those outlined for the PMP mode.

9.3 Frame Structure and Bandwidth Management in the MESH Mode

The 802.16 network supports only time division duplex (TDD) in the MESH mode [1]. Figure 9.4 shows the corresponding frame structure. The time axis is divided into frames of a specified length decided by the mesh BS. Each frame is in turn composed of a control subframe and a data subframe. There are two types of control subframes, namely the network control subframe and the schedule control subframe. Network control subframes are used to transmit network configuration information as well as to allow new nodes to register and join the network. The schedule control subframe is used by nodes to transmit scheduling information, and to request and grant bandwidth for transmission. All data transmissions take place in the data subframe using slots previously reserved by the node for

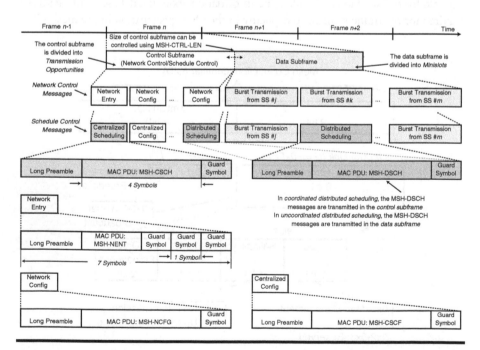

Figure 9.4   MESH frame structure. *Abbreviations:* **MAC, Medium Access Control; PDU, protocol data units.**

transmission. The control subframe is divided into a number of transmission opportunities and the data subframe is divided into a number of minislots. The length of the control subframe depends on the mesh configuration in use. This decides the number of transmission opportunities in the control subframe and the number of minislots in the data subframe. The MESH mode supports coordinated centralized scheduling, and coordinated as well as uncoordinated distributed scheduling for allocating bandwidth for transmission on individual links in the MESH mode of operation. The mesh configuration specifies a maximum percentage of minislots in the data subframe allocated to centralized scheduling. The remainder of the data subframe as well as any minislots not occupied by the current centralized schedule can be used for distributed scheduling.

In centralized scheduling, the bandwidth is managed in a more centralized manner than when using distributed scheduling. Thus, although the computation of the actual transmission schedule is done by the individual nodes independently (in a distributed manner), the grants for each individual node are controlled centrally by the BS in coordinated centralized scheduling (also called centralized scheduling). The BS uses centralized scheduling to manage and allocate bandwidth for transmissions up and down the routing tree (scheduling tree, see Fig. 9.5 for an example) from the BS to the SSs up to a specified maximum hop limit. The routing

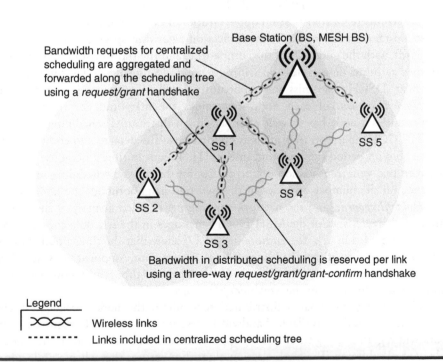

Figure 9.5 Overview of scheduling in the MESH mode.

tree is advertised by the BS periodically using MSH-CSCF messages. The BS in the mesh network gathers resource requests from individual SSs within the maximum hop range. Each SS in the scheduling tree accumulates the requests from its children and adds to it its own requirement for uplink bandwidth before forwarding the request upwards along the scheduling tree (uplink here implies transmission along a link in the scheduling tree from a SS to another SS that is closer to the BS; downlink will be considered to be a transmission down the tree in the opposite direction). The BS collects all the requests and transmits the grants to its children. The grants for each individual SS are then propagated down the scheduling tree hop-by-hop. Nodes use MSH-CSCH messages to propagate requests and grants for centralized scheduling.

The grants propagated to the SSs in the scheduling tree do not contain the actual schedule. Each SS computes the schedule using a predetermined algorithm and the parameters obtained from the grant. Using centralized scheduling, transmissions can be scheduled only along the links in the scheduling tree. To reserve bandwidth for transmission on links not in the scheduling tree, distributed scheduling has to be used.

Distributed scheduling is used by a node to reserve bandwidth for transmission on a link to any other neighboring node (also for links included in the centralized scheduling tree). Nodes use distributed scheduling to coordinate their transmissions in their two-hop neighborhood. The nodes use a distributed election algorithm to compete for transmission opportunities in the schedule control subframe. A pseudo-random function (the mesh election algorithm specified in the 802.16 standard), with the node IDs of the competitors and the transmission opportunity number as input determines the winning node. The losing nodes compete for the next DSCH transmission opportunity until they win. The parameter *XmtHoldoffExponent* of each node determines the magnitude of transmission opportunities a node has to wait after sending a distributed scheduling message (MSH-DSCH) in a won transmission opportunity. The details as to computation of the hold off period can be found in Ref. [1]. The mean time a node has to wait between two won transmission opportunities for distributed scheduling messages depends on the number of nodes in the two-hop neighborhood, the node's own *XmtHoldoffExponent*, and the network topology. A detailed analysis of the transmission characteristics of the MSH-DSCH messages in the schedule control subframe is provided in [7]. The authors in Ref. [7] show that the time a node has to wait between two distributed scheduling transmission opportunities it wins increases with an increase in the number of two-hop neighbors and moreover with an increase in the value of the *XmtHoldoffExponent*.

When using coordinated distributed scheduling, the nodes broadcast their individual schedules (available bandwidth resources, bandwidth requests, and bandwidth grants) using transmission opportunities won by the node in the schedule control subframe. The mesh election algorithm ensures that when a node wins a transmission opportunity in the schedule control subframe for transmission, no other node in its two-hop neighborhood will simultaneously transmit. Thus, it is ensured that the scheduling information transmitted by a node in the schedule control subframe can be received by all of the nodes' neighbors. To enable a conflict-free schedule to be negotiated each node maintains the status of all individual minislots in the frame. A minislot at any point in time may be either in status available (node can receive or transmit data in minislot), receive available (node can only receive data in minislot), transmit available (node can only transmit data in the minislot), or unavailable (node may not transmit or receive data in the minislot).

The schedule negotiated using coordinated distributed scheduling is such that it does not lead to conflict with any of the existing data transmission schedules in the two-hop neighborhood of the transmitter. On the other hand, nodes can also establish their transmission schedule by directed uncoordinated requests and grants between two nodes. In contrast to coordinated distributed scheduling requests and grants which are sent in the schedule control subframe, the uncoordinated requests and grants are sent in the data subframe. The latter scheduling mechanism is called uncoordinated scheduling. When a node *SS3* wants to reserve slots for transmission to a neighbor node *SS4*, they exchange scheduling information using slots in the data subframe reserved for transmissions between the two nodes (see Fig. 9.5). Nodes individually need to ensure that their scheduled transmissions do not cause collisions with the data as well as with control traffic scheduled by any other node in their two-hop neighborhood. Transmissions in the data subframe using slots reserved for transmission to a particular neighbor may not be received by all the other neighbors due to other simultaneous transmissions. Thus, the schedule negotiated using the data subframe (uncoordinated scheduling) may not be known to all the neighbors of the nodes involved in the uncoordinated schedule. The neighbors of these nodes may then schedule conflicting transmissions due to lack of the previous uncoordinated schedule information. Hence, uncoordinated scheduling may lead to collisions and is not suitable for long-term bandwidth reservations. Nodes use MSH-DSCH messages to transmit the bandwidth requests grants and negotiate schedules when using distributed scheduling (both coordinated as well as uncoordinated distributed scheduling).

In contrast, centralized scheduling allows the setup of a transmission schedule for transmissions only along links in the scheduling tree, and hence, is not very suitable for enabling a wireless mesh network in the traditional sense [8]. We next outline our novel proposed QoS architecture for bandwidth management in the MESH mode. Without loss of generality and to avoid confusion in the following discussion we assume that the nodes in the mesh network use only distributed scheduling.

The proposed QoS architecture using distributed scheduling is easily extensible and can be adapted for use in centralized scheduling, too. The proposed architecture uses a combination of coordinated distributed scheduling and uncoordinated distributed scheduling to efficiently manage the bandwidth in the network.

9.4 Proposed QoS Architecture for the 802.16 MESH Mode

Figure 9.6 shows our proposed QoS architecture for efficient management of bandwidth in the MESH mode. For the current discussion we assume the Internet Protocol (IP) as the network layer protocol. The module packet classifer shown in the figure provides the functionality of the service-specific convergence sublayer (see scope of the IEEE 802.16 standard [1]). Figure 9.7 shows the mapping we used to classify traffic from the network layer using the IP type of service (TOS) field and the corresponding values assigned to fields of the mesh CID by our classifier. Based on the values for the fields priority, drop precedence, and reliability we use the mapping shown in Figure 9.7 to identify the scheduling service (UGS, rtPS, nrtPS, or BE) to be provided for the data packet. A similar mapping function may be implemented for other network protocols.

After classification of data received from the upper layers, the packets are sent to the data management module as shown in Figure 9.6. The data management module enqueues the arriving packets in the corresponding queue. Based on the

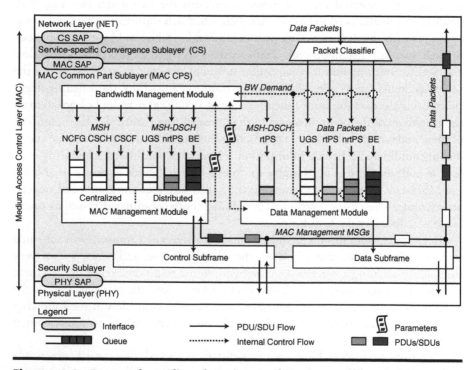

Figure 9.6 Proposed quality-of-service architecture. *Abbreviations*: UGS, Unsolicited Grant Service; rtPS, Real-Time Polling Service; nrtPS, Non-Real-Time Polling Service; BE, Best Effort.

Network Layer Priority (e.g., IP Type of Service)	802.16 Service Class	802.16 MSH CID Priority/Class	802.16 MSH CID Drop Precedence	802.16 MSH CID Reliability
0	Best Effort (BE)	0	3	0
1	BE	1	3	1
2	non-real-time Polling Service (nrtPS)	2	2	0
3	nrtPS	3	2	1
4	real-time Polling Service (rtPS)	4	1	0
5	rtPS	5	1	1
6	Unsolicited Grant Service (UGS)	6	0	0
7	UGS	7	0	1

Figure 9.7 Table showing mapping from the Internet Protocol (IP) type of service (TOS) to the appropriate mesh connection identifier (CID) and data scheduling service.

congestion situation, it can also decide which packets may be dropped. Besides handling the data received for transmission from the upper layers, the module also manages the MSH-DSCH messages to be transmitted in the data subframe (uncoordinated distributed scheduling). The data management module keeps an account of the minislots reserved for transmission for each link to a neighbor at a node. It then sends the appropriate data packet from its queues for transmission on the wireless medium to the lower layer in a minislot reserved for transmission. The data management module can deploy sophisticated queueing and scheduling algorithms internally to meet the QoS requirements of the different types of traffic in its queues. For the proof-of-concept evaluation of our QoS architecture we used a simple weighted fair queueing (WFQ) scheduler. Our simple scheduler services the MSH-DSCH queue (MAC management messages) with a higher priority than the data queues. Within the data queues the WFQ scheduler serves the UGS, rtPS, nrtPS, and BE queues with weights in decreasing order. As previously mentioned, the focus of this article is to provide insights into tools for efficient bandwidth management in the MESH mode and not to verify the satisfaction of hard QoS requirements for each kind of traffic. The data management module can use an admission control policy and a QoS scheduling scheme similar to the one outlined in [9] to meet hard per-hop QoS requirements for each kind of traffic. Thus, the data management module is responsible for handling all transmissions during the data subframe. In addition this module keeps a running estimate of the incoming data rate in each queue and, based on the policy to be implemented, notifies the bandwidth management module of the current bandwidth requirements for each class of traffic.

The MAC management module shown in Figure 9.6 is responsible for handling all kinds of MAC management messages. It handles MAC management messages received from the lower layer. If the MAC management message corresponds to a bandwidth request or a grant or grant-confirmation, this module updates the respective internal tables and extracts the relevant parameters (information elements, IEs, contained in the message). These parameters are then sent to the bandwidth management module for further processing when required. In addition, it is also responsible for processing MAC management messages received during the network control subframe. This module maintains information about the schedules of the neighbors, the node identifiers of the neighbors, details about the physical two-hop neighborhood, the link IDs assigned for transmission to and reception from each neighboring node. The MAC management module is responsible for executing the mesh election algorithm specified in the standard to decide if management messages may be transmitted in a given transmission opportunity in the control subframe. We, for our QoS architecture, introduce the concept of traffic classified as belonging to various data scheduling services. We also provide similar means to allow nodes to distinguish the MSH-DSCH and find out the service class to which the requests contained in the MSH-DSCH message correspond. This enables the bandwidth management module at the node receiving the MSH-DSCH request to give an appropriate grant based on the expected traffic behavior. For example, when the requested bandwidth is to serve traffic of class UGS (constant bit rate traffic with time synchronization requirements between sender and receiver), it is better to grant a fixed number of minislots for a longer period of time as the data traffic can be expected to be sent at a constant bit rate for a longer period. The existing MSH-DSCH message structure is shown in Figure 9.8. To enable a receiver of a MSH-DSCH message to find out which scheduling service the MSH-DSCH corresponds to, we propose to use the two reserved bits (see Fig. 9.8) in the MSH-DSCH message to map the MSH-DSCH message to one of the four data scheduling services.

The bandwidth management module shown in Figure 9.6 is responsible for generating bandwidth requests when more bandwidth is required, or generating cancel requests to free bandwidth when it is no longer required. It is also responsible

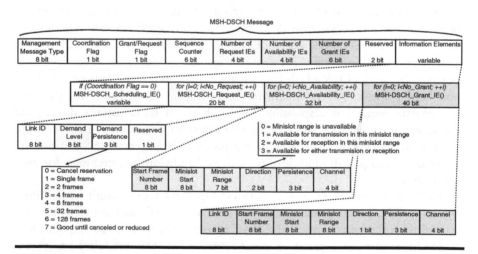

Figure 9.8 **Structure of the MSH-DSCH message and information elements contained in the MSH-DSCH message.**

for processing bandwidth requests received from the neighboring nodes and taking appropriate action when a grant or grant-confirmation is received. All these requests, grants, and grant-confirmations are sent as information elements within a MSH-DSCH message as shown in Figure 9.8. The bandwidth management module receives information about instantaneous bandwidth demand from the data management module. The bandwidth management module maintains internally a set of *MSH-DSCH_Availability_IEs* (see Fig. 9.8). The complete set of *MSH-DSCH_Availability_IEs* describes the local status of individual minislots over all frames in the future. When generating a MSH-DSCH message to request bandwidth for transmission, the bandwidth management module creates a *MSH-DSCH_Request_IE* (see Fig. 9.8) describing the amount of minislots required (specified by the demand level field in the *MSH-DSCH_Request_IE*) in a frame and the number of frames over which the bandwidth is required (denoted by the demand persistence field in the *MSH-DSCH_Request_IE*). Due to the classification of traffic into the different scheduling services for the MESH mode as proposed by us, the bandwidth management module is able to estimate the arrival characteristics of traffic and make an intelligent choice for the persistence value to be sent with the request. As an example, in our proof of concept implementation, the bandwidth management module requests minislots with persistence 7 (good until canceled or reduced, see Fig. 9.8) only when the data scheduling service associated with the traffic is UGS. This maps the UGS service provided in the PMP mode where a node receives a constant amount of bandwidth for the lifetime of the connection.

In the PMP mode the rtPS scheduling service is meant to support real-time data streams consisting of variable-sized data packets arriving periodically. To support such a service in the MESH mode one requires opportunities for requesting bandwidth in real-time. Using coordinated distributed scheduling a node, however, has to compete with other nodes in its two-hop neighborhood for transmission opportunities in which a bandwidth request can be sent. Nodes using distributed scheduling need to complete the three-way request/grant/grant-confirm handshake procedure before data can be transmitted using the reserved bandwidth. It is thus not possible to complete the handshake in real-time if we use only coordinated distributed scheduling and the topology is highly connected. To ensure an upper bound on the handshake delay, our QoS architecture proposes to reserve at least a single slot on each link to a neighbor with persistence 7 (i.e., the slot is available for transmission all the time). This slot can then be used for transmitting MSH-DSCH messages containing requests and grants for the rtPS service class. This ensures that the handshake completes in the next few frames irrespective of the topology or the value of *XmtHoldoffExponent* (in the best possible case within four frames). More details about the dependence of the handshake duration on the topology and the *XmtHoldoffExponent* parameter at the node can be found in [7]. Hence, as can be seen from Figure 9.6 the bandwidth management module sends all MSH-DSCH messages for the rtPS to the data management module for transmission. In addition, internally, to ensure a minimum delay, the traffic from the rtPS class can borrow (be transmitted in) bandwidth reserved for UGS traffic. UGS traffic can then borrow bandwidth back from the reserved bandwidth for the rtPS class as soon as the uncoordinated scheduling handshake is over. A characteristic of rtPS is that it has a variable bit rate. Thus, it is highly inefficient to request a fixed amount of slots for transmission for rtPS with persistence 7. This may lead to many of these slots being unused in many frames. As a solution, in our proof-of-concept implementation, we used an estimation of the number of slots required per frame to send the arriving rtPS data, and request those slots with a persistence 5 (reservation is valid for 32 frames). Using uncoordinated scheduling to reserve bandwidth for a long term is not recommended as it may lead to collisions as explained earlier in this article.

For the nrtPS class we require periodic request opportunities, which need not be in real-time. nrtPS traffic is moreover delay-tolerant. Thus, we can use an estimator to find out the amount of minislots required per frame and send requests with a persistence smaller than 7. As a result, we can periodically (using transmission opportunities in the schedule control subframe) reserve the exact amount of bandwidths required for transmitting nrtPS data. The BE service is very similar to the nrtPS service with the difference that it is served on a space-available basis. Thus, for BE the estimated number of minislots is reserved with a *persistence* less than 7. The difference to nrtPS is that traffic belonging to UGS and rtPS are allowed to borrow band width reserved for BE traffic.

Every request has to be accompanied by a set of *MSH-DSCH_Availability_IEs* as shown in Figure 9.8. A maximum of 16 *MSH-DSCH_Availability_IEs* may be transmitted with the request. This set of *MSH-DSCH_Availability_IEs* notifies the receiver of the request of the minislot range within which the bandwidth is to be granted. Thus, a poor choice of the set of *MSH-DSCH_Availability_IEs* to transmit with the request will lead to a failure of the request. In our proof-of-concept implementation outlined in this article we first select a subset of *MSH-DSCH_Availability_IEs* at the node which are just able to satisfy the request. Then a set of 16 of the above *MSH-DSCH_Availability_IEs* is selected randomly to be sent with the request. To understand what we mean by a *MSH-DSCH_Availability_IE* just satisfying the request consider the following example. Let us assume that we need a single slot for all future frames, then all availability information elements with *persistence* less than 7 are not able to satisfy this request. Now consider *MSH-DSCH_Availability_IEs* (see Fig. 9.8), all having one minislot and persistence 7, however a different value for the direction field (see Fig. 9.8). It should be clear that transmission is not possible in minislots with direction 0 (unavailable) or 2 (available for reception only). Thus, *MSH-DSCH_Availability_IEs* having direction 0 or 2 will not be able to satisfy the request at the sender and should not be sent along with the request. The *MSH-DSCH_Availability_IEs* with directions 1 and 3 will be able to satisfy the request and may be sent along with the request. A poor choice may not only lead to a failure of the handshake but also result in less slots with status 3 (available for both transmission and reception) and 1 (transmit available) remaining at the nodes in the network.

On receiving a request, the bandwidth management module is also responsible for processing the request to find a mutually suitable set of slots for a grant which is able to satisfy the request. The internal structure of a grant information element (MSH-DSCH_Grant_IE) can be seen in Figure 9.8. A poor choice for the grant would be for example a grant starting at a frame before the three-way handshake can be completed, this means that the slots in that range will remain unused (data transmission using the granted slots may not start till the three-way handshake is complete as required by the standard). On the other hand, if the grant starts from a frame much in the future after completion of the three-way handshake it leads to additional delay before transmission can start. We, in our proof-of-concept implementation selected grants which would start at least four frames in the future after reception of the request.

A three-way handshake (request/grant/grant-confirmation) may fail after the grant has been sent. Nodes in the neighborhood of the node sending the grant update the status of the minislot range being granted as being in use. Thus, these slots are no longer available for transmission at the nodes receiving the grant. If the grant was sent with persistence 7 (good until canceled) these slots will not be available for transmission for all frames in the future at the nodes which received the grant. When the handshake now fails, the grant-confirmation will not be sent, and hence the slots will never be used for data transmission. Despite the fact that the slots will not be used, the IEEE 802.16 standard currently lacks a mechanism to indicate that the grant sent previously has become invalid (due to failure of the handshake). Thus, these slots are "lost" forever. To avoid this phenomenon one can either use a soft-state reservation mechanism or introduce an explicit revoke of the grant. We, for our architecture propose, modify the *MSH-DSCH_Grant_IE* to include a revoke bit. When a grant-confirmation (for a grant with persistence 7) is not received within a specified time-out, the node which sent the grant sends a copy of the grant with the revoke bit set (we call it the grant-revoke message). This enables the bandwidth management module at nodes receiving the grant-revoke to take appropriate action and update the status of *MSH-DSCH_Availability_IEs* stored locally. No grant-revoke confirmation is sent as the grant-confirmation was not sent either.

The bandwidth management module is also responsible for maintaining an up-to-date status of the *MSH-DSCH_Availability_IEs* stored locally at a node. This involves updating the status when receiving or transmitting either a grant or grant-confirmation. The exact details about each of these algorithms are out of scope of this chapter and hence have not been presented in favor of keeping the chapter easily accessible and understandable.

Thus, as seen from Figure 9.6, the bandwidth management module, data management module, and the MAC management module comprise the MAC common part sublayer (see Ref. [1]) in our QoS architecture. In Figure 9.6, arrows passing through the boxes labeled "control subframe" and "data subframe" represent transmissions/receptions in the control and data subframes, respectively.

9.5 Conclusion and Directions for Future Research

To test our proposed QoS architecture we implemented a standard-conform version of the distributed scheduler of the IEEE 802.16-2004 MESH mode using the JiST/SWANS [5] simulation environment and integrated our QoS architecture in this simulation environment. In the current section we will highlight the key findings of the extensive simulation study we carried out using the implemented 802.16 MESH simulator. One of the key features of our QoS architecture is that it adapts the bandwidth requests and grants keeping in mind the traffic class for the bandwidth requests. In addition to the per-hop differentiated handling (QoS) that can be provided to each packet (as specified in the standard for the MESH mode), our QoS architecture allows the network to tailor the bandwidth available at a node per QoS class and link. We expected that this would lead to an optimized usage of bandwidth. At the same time, the QoS model requests bandwidth sufficient to satisfy the QoS requirements (throughput and delay requirements) of the different traffic service classes supported. The QoS model enables the network to support scheduling services similar to those outlined for the PMP mode, namely, UGS, rtPS, nrtPS, and BE.

Through our simulation study we observed the following advantages of the proposed QoS architecture. Bandwidth for UGS flows (mainly constant bit rate type of traffic) is reserved with persistence 7 (good until canceled). This reservation profile is highly suitable for CBR type of traffic which maintains a constant throughput over a period of time. The reservation with persistence 7 avoids the need for periodic requests (grants and grant-confirmations as well) for the same constant amount of bandwidth. This leads to more free bandwidth in the scheduling control subframe which can then be used for other purposes. When sufficient bandwidth has been requested for UGS (with persistence 7), the QoS architecture is able to guarantee steady delay and jitter characteristics for UGS traffic over each hop. For rtPS and nrtPS the bandwidth requests are expected to be highly varying over time, so the proposed QoS architecture avoids reserving the estimated bandwidth required for traffic flows belonging to these scheduling services for a longer duration (i.e., with higher persistence). Thus, the proposed QoS architecture sends bandwidth requests for these traffic classes and also grants and grant-confirmations with persistence less than 7. In our study we used persistence 5 for these traffic classes. This helps to optimize the bandwidth usage as compared with the case when the bandwidth would be reserved only with persistence 7. In addition, an important parameter for rtPS traffic is the delay (for both the transmission of data, as well as that for completing the three-way handshake for bandwidth arbitration). The results we obtained through our simulations tally with the analysis for the distributed scheduling handshake carried out by the authors in [7]. The time needed for completion of a three-way handshake increases with an increase in the number of competing nodes and with the holdoff time per node. This is criticial in the case of the rtPS handshake. A delay in reserving bandwidth for the rtPS traffic means that it may no longer be possible to satisfy the QoS requirements for that traffic class (in absence of long-term, persistence 7, reservations which in turn waste bandwidth). To overcome this problem, the QoS architecture uses uncoordinated distributed scheduling to setup bandwidth for rtPS flows. Here, unlike coordinated distributed scheduling, the messages for the three-way handshake are transmitted in the data subframe. For reserving rtPS bandwidth for a link the MAC management messages (request, grant, and grant-confirm) are transmitted in minislots already reserved for transmission of data on the links between the two neighboring nodes connected by this link. This leads to a guaranteed maximum delay for the three-way handshake when the MAC management messages in the data subframe have a higher priority as compared with the data messages. The short duration of validity of the reservation setup using uncoordinated distributed scheduling ensures that a very small (in most cases a negligible fraction) amount of rtPS messages could not be correctly received (due to a parallel transmission setup in the receiver's neighborhood via coordinated/uncoordinated distributed scheduling). The bandwidth savings hold for the case of nrtPS data too. For nrtPS data the throughput is important and the handshake delay plays a relatively insignificant role. Hence, our QoS architecture uses the control subframe (coordinated distributed scheduling) for the nrtPS three-way handshake. For BE traffic, our architecture tries to use the remaining unused bandwidth reserved for the other three scheduling services. It also additionally requests a minimal possible number of minislots per frame for the BE traffic with a persistence less than 7. We observed via our simulations that this led to a starvation of the BE traffic when a strict priority mechanism was used for the three-way handshake and the scheduling of data. We therefore used a weighted fair queueing approach for scheduling the BE requests and data transmissions.

The additional grant-revoke mechanism implemented by us helps to recover bandwidth when the three-way handshake fails. We observed a small amount of revokes being sent as compared with the total amount of grants. However, a single revoke message leads to the bandwidth being recovered at all nodes in the neighborhood of the node transmitting the revoke. This in turn translates into significant bandwidth savings. This also helped to prove that the revoke mechanism functions as expected.

To summarize, good bandwidth management algorithms are crucial to the robust and efficient working of the MESH mode of IEEE 802.16. We presented a novel scheme for managing bandwidth in the 802.16 MESH mode of operation with an aim to support the data scheduling services similar to those currently supported by the PMP mode. In addition, we presented and introduced a bandwidth revocation mechanism which allows the recovery of bandwidth in case the three-way handshake fails. We also provided detailed insights into the working of the IEEE 802.16 MESH mode. The insights obtained should help researchers and implementors tackle the various challenges mentioned by us in this chapter. The presented QoS architecture provides a solid and extendable foundation for future work. In particular, areas such as fair bandwidth distribution and fragmentation of bandwidth need to be looked into.

References

1. IEEE Computer Society and IEEE Microwave Theory and Techniques Society, 802.16 IEEE Standard for Local and metropolitan area networks, Part 16: Air Interface for Fixed Broadband Wireless Access Systems. IEEE Std. 802.16-2004 (October, 2004).
2. C. Eklund, R. Marks, K. Stanwood, and S. Wang, IEEE standard 802.16: a technical overview of the WirelessMAN air interface for broadband wireless access. *IEEE Communications Magazine*, 40(6), 98–107 (June, 2002).
3. IEEE Computer Society and IEEE Microwave Theory and Techniques Society, 802.16 IEEE Standard for Local and metropolitan area networks, Part 16: Air Interface for fixed Broadband Wireless Access Systems, Amendment 2: Physical and Medium Access Control Layers for Combined Fixed and Mobile Operation in Licensed Bands and Corrigendum 1. IEEE Std 802.16e-2005 (February, 2006).
4. P. S. Mogre and M. Hollick, The IEEE 802.16-2004 MESH Mode Explained. Technical Report KOM-TR-2006-08, Multimedia Communications Lab, Department of Electrical Engineering and Information Technology, Technische Universität Darmstadt, Germany (2006). ftp://ftp.kom.tu-darmstadt.de/pub/TR/KOM-TR-2006-08.pdf.
5. R. Barr, JiST/SWANS user guides. http://jist.ece.cornell.edu/docs.html (2004).
6. C. Cicconetti, L. Lenzini, E. Mingozzi, and C. Eklund, Quality of service support in IEEE 802.16 networks. *IEEE Network*, 20(2), 50–55 (March/April, 2006).
7. M. Cao, W. Ma, Q. Zhang, X. Wang, and W. Zhu, Modelling and performance analysis of the distributed scheduler in IEEE 802.16 mesh mode. In *MobiHoc '05*, pp. 78–89 (2005).
8. I. F. Akyildiz, X. Wang, and W. Wang, Wireless mesh networks: a survey. *Computer Networks*, 47(4), 445–487 (2005).
9. K. Wongthavarawat and A. Ganz, Packet scheduling for QoS support in IEEE 802.16 boadband wireless access systems. *International Journal of Communication Systems*, 16, 81–96 (2003).

Chapter 10

Quality-of-Service Scheduling for WiMAX Networks

Nicola Scalabrino, D. Miorandi,
Francesco De Pellegrini, R. Riggio,
Imrich Chlamtac, and E. Gregori

The broadband wireless world is moving toward the adoption of WiMAX (the commercial name of the IEEE 802.16 standard) as the standard for broadband wireless Internet access. This will open up a very large market for industry and operators, with a major impact on the way Internet access is conceived today. On the other hand, the emergence of innovative multimedia broadband services is going to impose severe quality of service (QoS) constraints on underlying network technologies. In this work, after a brief review of the IEEE 802.16 standard, we intend to present an in-depth discussion of its QoS support features. We point out the scheduling algorithm as the critical point in QoS provisioning over such networks, and discuss architectural and algorithmic solutions for an efficient support of multimedia flows. Performance measurements obtained from an experimental test-bed are also presented. The chapter concludes with a description of the key research challenges in the area, and provides a roadmap for the research in the field.

217

10.1 Introduction

The IEEE 802.16 standard [1], promoted by the WiMAX (Worldwide Interoperability for Microwave Access) forum [2], will be the leading technology for the wireless provisioning of broadband services in wide area networks. Such technology is going to have a deep impact on the way Internet access is conceived, by providing an effective wireless solution for the last mile problem.

The market for conventional last mile solutions (e.g., cable, fiber, and so on) presents indeed high entrance barriers, and it is thus difficult for new operators to make their way into the field. This is due to the extremely high impact of labor-intensive tasks (i.e., digging up the streets, stringing cables, and so on) that are required to put the necessary infrastructure into place. On the other hand, the market is experiencing an increasing demand for broadband multimedia services [3], pushing toward the adoption of broadband access technologies. In such a situation, broadband wireless access (BWA) represents an economically viable solution to provide Internet access to a large number of clients, thanks to its infrastructure-light architecture, which makes it easy to deploy services where and when it is needed. Furthermore, the adoption of *ad hoc* features, such as self-configuration capabilities in the Subscriber Stations (SSs) would make it possible to install customer premises equipment without the intervention of a specialized technician, so boosting the economical attractiveness of WiMAX-based solutions. In this context, WiMAX is expected to be the key technology for enabling the delivery of high-speed services to the end users.

Typical BWA deployments will rely on a point-to-multipoint (PMP) architecture, as depicted in Figure 10.1a, consisting of a single Base Station (BS) wirelessly interconnecting several SSs to an Internet gateway. The standard also supports, at least in principle, mesh-based architectures, like the one plotted in Figure 10.1b. While WiMAX-based mesh deployments could play a relevant role in the success of such technology, the current standard [1] is far from offering a real support to such architecture. Therefore, we intend to restrict the scope of our work to the PMP architecture only.

In terms of raw performance, WiMAX technology is able to achieve data rates up to 75 Mb/s with a 20 MHz channel in ideal propagation conditions [4]. But regulators will often allow only smaller channels (10 MHz or less) reducing the maximum bandwidth. Although 50 km distance is achievable under optimal conditions and with a reduced data rate (a few Mb/sec), the typical coverage will be around 5 km in non-line-of-sight conditions and around 15 km with an external antenna in a line-of-sight situation. Moreover, such a wide coverage makes it possible, and economically viable to provide broadband connectivity in rural and remote areas, a market which is usually not covered by traditional service providers.

The fundamental requirements for WiMAX to define itself as a possible winning technology are data reliability and the ability to deliver multimedia contents. Indeed, the provision of QoS guarantees will be a pressing need in the next

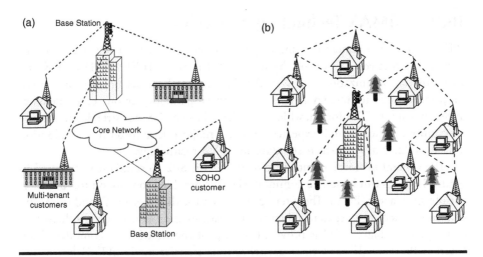

Figure 10.1   Typical WiMAX system configuration: (a) point-to-multipoint; (b) mesh.

generation of Internet, to enable the introduction of novel broadband multimedia applications. Users are actually getting more and more interested in broadband applications (e.g., video streaming, video conferencing, online gaming, and so on) that require assurances in terms of throughput, packet delay and jitter, to perform well. This applies also to WiMAX networks, which have also to face all the problems related to the hostile wireless environment, where time-varying channels and power emission mask constraints make it difficult to provide hard QoS guarantees. This entails the definition of a Medium Access Control (MAC) protocol which is able to effectively support such multimedia applications, while on the other hand, it efficiently exploits the available radio resources. The IEEE 802.16 standard encompasses four classes of services, with different QoS requirements and provides the basic signaling between the BS and the SS to support service requests/grants. However, the scheduling algorithms to be employed in the BS and the SS are not specified and are left open for the manufacturers to compete.

In this paper, after a brief review of the standard fundamentals, we will provide an in-depth overview and discussion on the QoS support provided by WiMAX technology. Particular attention will be devoted to scheduling algorithms for WiMAX networks. We will survey the existing literature, and point out some common issues involved in well-known technologies (e.g., wireless ATM), from which a system designer can draw to design an efficient scheduler without starting from scratch. Performance measurements obtained from an experimental test-bed are also presented. This chapter concludes with an overview of the actual research challenges, pointing out and detailing the most promising directions to pursue for research in this field.

10.2 WiMAX Technology Overview

WiMAX is the commercial name of products compliant with the IEEE 802.16 standard. Effectively replicating the successful history of IEEE 802.11 and Wi-Fi, an industrial organization, the WiMAX Forum has been set up to promote the adoption of such technology and to ensure interoperability among equipments of different vendors. This forum, which includes all the major industrial leaders in the telecommunication field, is expected to play a major role in fostering the adoption of IEEE 802.16 as the *de facto* standard for BWA technology.

The general protocol architecture of the IEEE 802.16 standard is depicted in Figure 10.2. As can be seen, a common MAC is provided to work on top of different physical layers (PHY). The interface between the different PHYs and the MAC is accomodated as a separate sublayer, the transmission convergence sublayer. A convergence sublayer (CS) is provided on top of the MAC, to accomodate both Internet Protocol (IP) as well as asynchronous transfer mode (ATM)-based network technologies. A basic privacy support is provided at the MAC layer.

In its first release in 2001, the 802.16 standard addressed applications for a static scenario in licensed frequency bands in the range between 10 and 66 GHz, where the use of directional antennas are mandatory to obtain satisfactory performance figures. In a metropolitan sub-area, however, line-of-sight operations cannot be ensured due to the presence of obstacles, buildings, foliage, and so on. Hence, subsequent amendments to the standard (802.16a and 802.16-2004) have extended the 802.16 air interface to non-line-of-sight applications in licensed and unlicensed bands in the 2–11-GHz frequency band. With the revision of IEEE standard document 802.16e, also some mobility support will be provided. Revision 802.16f is

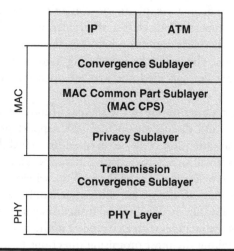

Figure 10.2   IEEE 802.16 protocol architecture.

intended to improve multi-hop functionality, and 802.16g is supposed to deal with efficient handover and improved QoS.

WiMAX technology can reach a theoretical 50 km coverage radius and achieve data rates up to 75 Mb/s [4], although actual IEEE 802.16 equipments are still far from these performance figures. As an example, in Ref. [5] the authors report the outcomes of some bit-level numerical simulations performed assuming a channel width of 5 MHz and a Multiple-Input Multiple-Output (MIMO) 2 × 2 system (which reflects the most common actual equipment), showing that, under ideal channel conditions, data rates up to 18 Mb/s can be attained.

Duplexing is provided by means of either time division duplexing (TDD) or frequency division duplexing (FDD). In TDD, the frame is divided into two subframes, devoted to downlink and uplink, respectively. A time division multiple access (TDMA) technique is used in the uplink subframe, the BS being in charge of assigning bandwidth to the SSs, whereas a time division multiplexing (TDM) mechanism is employed in the downlink subframe. In FDD, the uplink and downlink subframes are concurrent in time, but are transmitted on separate carrier frequencies. Support for half-duplex FDD SSs is also provided, at the expense of some additional complexity. Each subframe is divided into physical slots. Each TDM/TDMA burst carries MAC protocol data units (PDUs) containing data toward SS or BS, respectively.

The transmission convergence sublayer operates on top of the PHY and provides the necessary interface with the MAC. This layer is specifically responsible for the transformation of variable-length MAC PDUs into fixed-length PHY blocks [6].

The necessity to provide secure data transmissions has led to the native inclusion of a privacy sublayer, at the MAC level. Such protocol is responsible for encryption/decryption of the packet payload, according to the rules defined in the standard [1].

As IEEE 802.16 uses a wireless medium for communications, the main target of the MAC layer is to manage the resources of the radio interface in an efficient way, while ensuring that the QoS levels negotiated in the connection setup phase are fulfilled. The 802.16 MAC protocol is connection-oriented and is based on a centralized architecture. All traffic, including inherently connectionless traffic, is mapped into a connection which is uniquely identified by a 16-bit address.

The common part sublayer is responsible for the segmentation and the reassembly of MAC service data units (SDUs), the scheduling and the retransmission of MAC PDUs. As such, it provides the basic MAC rules and signaling mechanisms for system access, bandwidth allocation, and connection maintenance. The core of the protocol is bandwidth requests/grants management. An SS may request bandwidth, by means of a MAC message, to indicate to the BS that it needs (additional) upstream bandwidth. Bandwidth is always requested on a per-connection basis to allow the BS uplink scheduling algorithm (which is not specified in the standard) to consider QoS-related issues in the bandwidth assignment process.

As depicted in Figure 10.2, the MAC includes a convergence sublayer which provides three main functionalities:

1. *Classification.* The CS associates the traffic coming from upper layer with an appropriate service flow (SF) and connection identifier (CID).
2. *Payload header suppression (PHS).* The CS may provide PHS at the sending entity and reconstruction at the receiving entity.
3. Delivery of the resulting CS PDU to the MAC common part sublayer in conformity with the negotiated QoS levels.

The standard defines two different CSs for mapping services to and from IEEE 802.16 MAC protocol. The ATM convergence sublayer is defined for ATM traffic, whereas the packet CS is specific for mapping packet-oriented protocol suites, such as IPv4, IPv6, Ethernet, and virtual LAN. As regards IP, the packets are classified and assigned to the MAC layer connections based on a set of matching criteria, including the IP source and the destination addresses, the IP protocol field, the type of service (TOS) or DiffServ Code Points (DSCP) fields for IPv4, and the traffic class field for IPv6. However, these sets of matching criteria are not in the standard and their implementation is left open to vendors.

10.2.1   QoS Architecture

As described before, the data packets entering the IEEE 802.16 network are mapped into a connection and an SF is based on a set of matching criteria. These classified data packets are then associated with a particular QoS level, based on the QoS parameters of the SF they belong to. The QoS may be guaranteed by shaping, policing, or prioritizing the data packets at both the SS and BS ends. The BS allocates upstream bandwidth for a particular upstream service flow based on the parameters and service specifications of the corresponding service scheduling class negotiated during connection setup. The IEEE 802.16 standard defines four QoS service classes: Unsolicited Grant Service (UGS), Real-Time Polling Service (rtPS), Non-Real-Time Polling Service (nrtPS), and Best Effort (BE) [6,7]. These four classes are characterized as follows.

■ The UGS service is defined to support constant bit rate (CBR) traffic, such as audio streaming without silence suppression. Unsolicited grants allow SSs to transmit their PDUs without requesting bandwidth for each frame. The BS provides fixed-size data grants at periodic intervals to the UGS flows. Because the bandwidth is allocated without request contention, the UGS provides hard guarantees in terms of both bandwidth and access delay. The QoS parameters defined for this service class are the size of the grant to be allocated, the nominal interval length between successive grants and the tolerated grant jitter, defined as the maximum tolerated variance of packet access delay.

- In the case of variable bit rate (VBR) video traffic, such as MPEG streams, the bandwidth requirements for the UGS grant interval cannot be determined at connection setup time. As a result, peak stream bit rate-based CBR allocation would lead to severe network underutilization, whereas the average bit rate CBR allocation can result in unacceptable packet delay and jitter. The rtPS service has been introduced to accomodate such flows. For this service, indeed, the BS provides periodic transmission opportunities by means of a basic polling mechanism. The SS can exploit these opportunities to ask for bandwidth grants, so that the bandwidth request can be ensured to arrive at the BS within a given guaranteed interval. The QoS parameters relevant to this class of services are the nominal polling interval between successive transmission opportunities and the tolerated poll jitter.

- The nrtPS is similar in nature to rtPS but it differs in that the polling interval is not guaranteed but may depend on the network traffic load. This fits bandwidth-demanding non-real-time SFs with a variable packet size, such as large file transfers. In comparison with rtPS, the nrtPS flows have fewer polling opportunities during network congestion, whereas the rtPS flows are polled at regular intervals, regardless of the network load. In heavy traffic conditions, the BS cannot guarantee periodic unicast requests to nrtPS flows, so that the SS would also need to use contention and piggybacking to send requests to the BS uplink scheduler.

- For BE traffic, no periodic unicast requests are scheduled by the BS. Hence, no guarantees in terms of throughput or packet delay can be given. The BE class has been introduced to provide an efficient resource utilization for low-priority elastic traffic, such as telnet or HTTP.

While these services provide the basics for supporting QoS guarantees, the "real" core, that is, traffic scheduling, policing, shaping, and admission control mechanisms, is not specified by the standard. In the next section, we will present and review some possible QoS architectures for WiMAX-based PMP networks.

10.3 QoS Scheduling in WiMAX Networks

To offer an efficient QoS support to the end user, a WiMAX equipment vendor needs to design and implement a set of protocol components that are left open by the standard. These include traffic policing, traffic shaping, connection admission control (CAC), and packet scheduling.

Due to the highly variable nature of multimedia flows, traffic shaping and traffic policing are required by the SS, to ensure an efficient and fair utilization of network resources. At connection setup, the application requests network resources according to its characteristics and to the required level of service guarantees. A traffic shaper is necessary to ensure that the traffic generated actually conforms to the pre-negotiated traffic specification. However, traffic shaping may not guarantee such conformance between the influx traffic and service requirements. This is dealt with by a traffic policer, which compares the conformance of the user data traffic with the QoS attributes of the corresponding service and takes corresponding actions, for example, it rejects or penalizes nonconformance flows.

QoS profiles for SS are usually detailed in terms of Committed Information Rate (CIR) and Maximum Information Rate (MIR) for the various QoS classes [8,9]. The CIR (defined for nrtPS and rtPS traffic) is equal to the information transfer rate that the WiMAX system is committed to carry out under normal conditions. The MIR (defined for nrtPS and BE QoS types) is the maximum information rate that the system will allow for the connection. Both these QoS parameters are averaged over a given interval time.

To guarantee that the newly admitted traffic does not result in network overload or service degradation for existing traffic, a (centralized) CAC scheme also has to be provided.

Although all the aforementioned components are necessary to provide an efficient level of QoS support, the core of such a task resides in the scheduling algorithm. An efficient scheduling algorithm is the essential *conditio sine qua non* for the provision of QoS guarantees, and it plays an essential role in determining the network performance. Besides, a traffic shaper, policer, and CAC mechanisms are tightly coupled with the scheduler employed. Therefore, the rest of this section is devoted to such an issue.

Although the scheduling is not specified in the standard, system designers can exploit the existing rich literature about scheduling in wireless ATM [10], from which WiMAX has inherited many features. If this allows one not to start from scratch, existing schemes need to be adapted to match the peculiar features (e.g., traffic classes, frame structure) of the IEEE 802.16 standard.

As an example, the IEEE 802.16 scheduling mode can be seen as an outcome of the research carried out on hierarchical scheduling [11]. This is rooted in the necessity of limiting the MAC exchange overhead by letting the BS handle all connections of each SS as an aggregated flow. As explained in the previous section, according to the standard, the SSs request bandwidth on per-connection basis; however, the BS grants bandwidth to each individual SS, so that the resources are allocated to the aggregation of active flows at each SS. Each SS is then in charge of allocating the granted bandwidth to the active flows, which can be done in an efficient way because the SS has complete knowledge of its queues status. This, however, requires the introduction of a scheduler at each SS, enhancing the complexity (and consequently the cost) of the SS equipment. A detailed operational scheme is depicted in Figure 10.3, outlining the role played by each component and the requests/grants mechanism at the basis of WiMAX QoS support.

Schedulers work on multiple connections to ensure the negotiated throughputs, delay bounds, and loss rates. The target of a scheduling algorithm is to select which

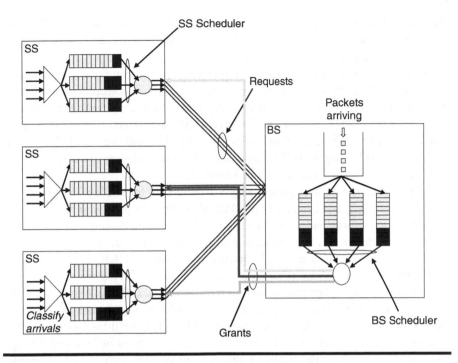

Figure 10.3   Graphic representation of hierarchical scheduling.

connection has to be served next. This selection process is based on the QoS require-ments of each connection. An efficient scheduling algorithm at the BS must be pro-vided guarantee proper performance. To better explain the scheduler's role, let us first assume that the BS performs the scheduling functions on a per-connection basis.* To schedule packets correctly, information such as the number of pending connections, their reserved throughputs, and the statues of session queues is needed. While this information is easily accessible as concerns downlink connections, the SSs need to send their bandwidth requests and queue status to the BS for the uplink. This has a twofold effect. On the one hand, it increases the signalling overhead, while, on the other hand, it provides the BS with information that may be not up-to-date (e.g., due to contention delays, and so on). In downlink, the scheduler has complete knowledge of the queue status, and, thus, may use some classical scheduling schemes, such as weighted round robin (WRR), weighted fair queueing (WFQ), etc. [10]. Priority-oriented fairness features are also important in providing differentiated services in WiMAX networks. Through priority, different traffic flows can be treated almost as isolated when sharing the same radio resource. However, due to the nature of WiMAX TDD systems, the BS scheduler is non-work-conserving, as the output link can be

---

* This was "grant per connection" in the original 2001 standard.

idle even if there are packets waiting in some queues. Indeed, after downlink flows are served in their devoted subframe, no additional downlink flows can be served till the end of the subsequent uplink subframe.

Scheduling uplink flows is more complex because the input queues are located in the SSs and are hence separated from the BS. The UL connections work on a request/grant basis. Using bandwidth requests, the uplink packet scheduling may retrieve the status of the queues and the bandwidth parameters. The literature is not rich in terms of QoS scheduling schemes specifically designed for WiMAX networks. In the following, we will briefly describe the most relevant works that address such a topic, to the best of the authors' knowledge.

In Ref. [12], the authors present a QoS architecture for IEEE 802.16 based on priority scheduling and dynamic bandwidth allocation. In particular, they propose a scheduling process divided into two parts. The first one, executed by the uplink scheduler inside the BS, is performed to grant resources to the SSs in response to bandwidth requests. This is done by means of a classical WRR [13]. At each SS, bandwidth assignments are computed by starting from the highest priority class (i.e., UGS flows) and then going down to rtPS, nrtPS, and BE. In this way, a strict priority among service classes is guaranteed. The scheduling schemes employed for the various classes are different. A classical WFQ [14] is used for UGS and rtPS, whereas a simpler WRR is used for nrtPS service class. BE traffic is served through a simple FIFO policy. By means of this prioritized approach (which resembles somehow multiclass priority fair queueing [11]), the proposed architecture is able to guarantee a good performance level to UGS and rtPS classes, to the detriment of lower priority traffic (i.e., nrtPS and BE flows).

Finally, in Ref. [7] the authors have assesed, via simulation, the performance of an IEEE 802.16 system using the class of latency-rate [15] scheduling algorithms where a minimum reserved rate is the basic QoS parameter negotiated by a connection within a scheduling service. Specifically, within this class, they selected defict round robin (DRR) as the downlink scheduler to be implemented in the BS, as it combines the ability to provide fair queueing in the presence of variable length packets with the simplicity of implementation. In particular, DRR requires a minimum rate to be reserved for each packet flow being scheduled. Therefore, although not required by the IEEE 802.16 standard, BE connections should be guaranteed a minimum rate. This fact can be exploited to both avoid BE traffic starvation in overloaded scenarios, and let BE traffic take advantage of the excess bandwidth which is not reserved for the other scheduling services. On the other hand, DRR assumes that the size of the head-of-line packet is known at each packet queue; thus, it cannot be used by the BS to schedule transmissions in the uplink direction. In fact, with regard to the uplink direction, the BS is only able to estimate the overall amount of backlog of each connection, but not the size of each backlogged packet. Therefore, the authors selected WRR as the uplink scheduler. Like DRR, WRR belongs to the class of rate-latency scheduling algorithms. At last, DRR is implemented in the SS scheduler, because the SS knows the sizes of the head-of-line packets of its queues.

10.4 Case Study: Voice-Over-IP Support in WiMAX Networks

In this section we present some preliminary results, obtained from an experimental test-bed, on the ability of WiMAX systems to support Voice-over-IP (VoIP) applications. The measurements reported next, assess WiMAX capability to support VoIP flows. In particular, the voice quality was evaluated through the E-Model [16] by using the R-factor [17].

10.4.1 Test-Bed Configuration

Our test-bed is based on Alvarion equipment operating in the (licensed) 3.5-GHz-based frequency band and compliant with the IEEE 802.16d specifications. The experimental data has been collected exploiting a four-node wireless test-bed deployed in a rural environment, located in northern Italy, implementing a PMP architecture. The BS is equipped with a sectorial antennas with a gain of 14 dBi covering all the three SSs. The default maximum output power at antenna port is 36 dBm for both the BS and the SS. The distance between the BS and SS1, SS2, and SS3 is 8.4 km, 8.5 km, and 13.7 km, respectively. The average Signal-to-Noise (SNR) ratio is above 30 dB, thus enabling the higher modulation, that is, 64 QAM, for each connection. The SSs work in line-of-sight conditions under FDD half-duplex. All nodes run a Linux distribution based on a 2.4.31 kernel. The measurements are performed exploiting an Alvarion BreezeMAX platform [18] operating in the 3.5-GHz licensed band and using a 3.5-MHz wide channel in FDD mode. Each node is attached through an Ethernet connection to the WiMAX equipment.

10.4.2 Parameters Setting

MIR and CIR are specified for each SS according to the negotiated service level agreement (SLA); the compliance to the negotiated SLA is assessed over a reference window, called committed time (CT). In what follows we assume that n SSs make MIR and CIR requests to the BS. We let R_{max} the maximum traffic rate available at the WiMAX Downlink Air Interface, and denote CIR_k and MIR_k the request of the k-th SS,* where $0 \leq CIR_k \leq MIR_k \leq R_{max}$.

The BS dynamically allocates the BE Service Rate R_{BE} (bit/s) and the Real-Time (RT) Service Rate R_{RT} (bit/s) with a cumulative upper bound of R_{max}, making sure that the RT service traffic has a higher priority than the BE service traffic: $R_{RT} + R_{BE} \leq R_{max}$. The residual capacity is allocated as R_{BE}. Let N_{tot} be the total number of

---

* The Alvarion BreezeMAX device does not allow to set the MIR parameter for real-time traffic.

downstream service flows consisting of N_{VoIP} VoIP flows and N_{TCP} Transport Control Protocol (TCP) persistent connection, so that $N_{\text{tot}} = N_{\text{TCP}} + N_{\text{VoIP}}$.

Let $R_{\text{TCP}}(m)$ be the service rate that the BS can provide to the m-th TCP SF, the aggregated BE service rate is $R_{\text{BE}} = \sum_{m=1}^{N_{\text{TCP}}} R_{\text{TCP}}(m)$; similarly, if $R_{\text{VoIP}}(m)$ is the service rate that the BS provides to the m-th VoIP SF, the aggregated RT service rate becomes: $R_{\text{RT}} = \sum_{m=1}^{N_{\text{VoIP}}} R_{\text{VoIP}}(m)$. The Alvarion equipment used in the test-bed provides resource allocation mechanisms corresponding to three cases.

In the first case, the downlink bandwidth is over-provisioned, meaning that the aggregated traffic service rate for the WiMAX network is deterministically lower than R_{max}, that is, $\sum_{m=1}^{N_{\text{TCP}}} \text{MIR}(m) + \sum_{n=1}^{N_{\text{VoIP}}} \text{MIR}(n) \leq R_{\text{max}}$ and no congestion occurs: the allocation in this case is fairly simple and the BS sets $R_{\text{VoIP}}(n) = \text{MIR}(n)$ and $R_{\text{TCP}}(m) = \text{MIR}(m)$.

The opposite case occurs when the aggregate of the CIR requested by VoIP subscribers exceeds R_{max}, that is $\sum_{n=1}^{N_{\text{VoIP}}} \text{CIR}(n) > R_{\text{max}}$; then the BS sets $R_{\text{VoIP}}(n) = R_{\text{max}}/N_{\text{VoIP}}$, and $R_{\text{TCP}}(m) = 0$ for every SS $n = 1, 2, \ldots, n$.

The remaining case is such that:

$$\sum_{m=1}^{N_{TCP}} \text{MIR}(m) + \sum_{n=1}^{N_{VoIP}} \text{MIR}(n) > R_{\text{max}};$$

$$\sum_{n=1}^{N_{VoIP}} \text{CIR}(n) \leq R_{\text{max}}. \tag{10.1}$$

This is the case when the BS guarantees the minimum service rate for the VoIP traffic and can reallocate the remaining bandwidth to the BE services, namely

$$R_{\text{VoIP}}(n) = \text{CIR}(n);$$

$$R_{\text{TCP}}(m) = \frac{(R_{\text{max}} - R_{\text{RT}})}{N_{\text{TCP}}}. \tag{10.2}$$

This is also the case that was considered for our measurement, as it is the probing case when QoS guarantees must be provided in spite of concurrent data traffic.

Notice that the actual implementation of the resource allocation depends on the scheduling implemented at the BS and vendors usually do not disclose such a critical detail to customers. Nevertheless, with appropriate probing, we could get some insight into the system behavior (see Section 10.4.3). Finally, the IP's DSCP [19] field is exploited to enforce a certain QoS class service. Traffic flows belonging to different service categories are tagged using the `iptables` software [20]. During our measurements, all SSs share the same QoS, as summarized in Table 10.1.

Table 10.1   Mapping Rules of Alvarion BreezeMAX

Traffic Class	DSCP	CIR (kbps)	MIR (kbps)
BE	1	n.a.	12,000
nrtPS	2–31	3000	12,000
rtPS	32–63	300	n.a.

Abbreviations: BE, Best Effort; nrtPS, Non-Real-Time Polling Service; rtPS, Real-Time Polling Service; DSCP, Diffserv Code Point; CIR, Committed Information Rate; MIR, Maximum Information Rate.

10.4.3   Performance Measurements

Data flows and VoIP flows were generated via the Distributed Internet Traffic Generator (D-ITG), a freely available software tool [21]. VoIP codecs feed RTP packet flows and two commonly used codecs have been considered, that is, G.729.2 and G.723.1. VoIP connections are mapped into the rtPS class, whereas TCP-controlled traffic is mapped into the BE class. Mapping of CBR sources into the rtPS class made much easier to trace the behavior of the system, as the actual scheduling policies were unknown on our side. To collect reliable measure of delays, before each experiment we synchronized all nodes using NTP [22]. All SSs sustain the same traffic, consisting of a given number of VoIP session plus one persistent TCP connection, modeling background traffic. Measurements were performed over 5-min intervals and averaged over ten runs.

In the first set of measurements, we determined the voice capacity, that is, the maximum number of sustained VoIP calls with high quality $(70 < R < 80)$ and related parameters. Here, we report only the downlink results, as it was found to be the bottleneck.

Figures 10.4 and 10.5 depict the measurement results we collected for the delay and the packet loss, respectively. Particularly, the delay saturates at 300 ms, whereas, after the saturation point, packet loss increases almost linearly. The G.723.1 codec outperforms clearly G.729.2; such a difference is due to the higher G.729.2 packet generation rate, coupled to the large overhead of packet headers of the RTP/ UDP/ IP/MAC protocol stack ($\simeq$ 45 percent for the G.729.2). Such effect is well known in VoIP over WLANs: in practice, it is convenient to employ larger speech trunks per packet and consequently larger interpacket generation intervals [23,24]. Finally, Figure 10.6 provides a comprehensive picture in terms of the R-factor. There exist roughly three regions: in the leftmost region, G.729.2 provides a fairly good quality, but after ten calls, G.723.1 obtains much better performance. In the end, for the given CIR, the system under exam supports up to 17 G.723.1 VoIP calls, and 10 G.729.2 calls per SS.

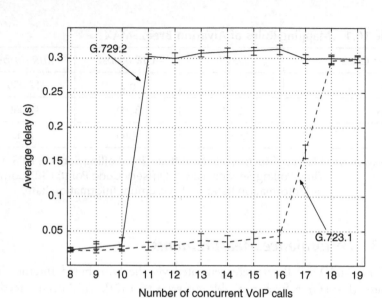

Figure 10.4 Average delay versus number of Subscriber Station voice-over-IP (VoIP) flows; minimum value and maximum value delimiters superimposed.

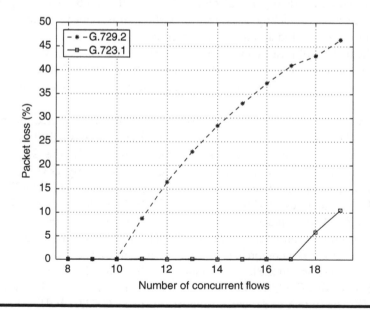

Figure 10.5 Packet loss rate of voice-over-IP (VoIP) flows per Subscriber Station using different codes.

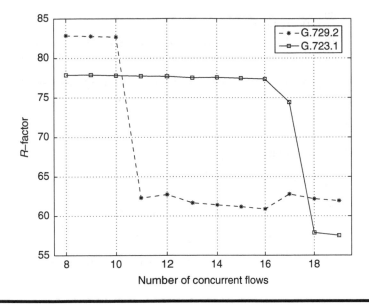

Figure 10.6   Average *R*-factor versus number of Subscriber Station voice-over-IP (VoIP) flows.

To determine the voice capacity, we restricted to the downlink, claiming that it is the bottleneck. As reported in Figure 10.7, in fact, the *R*-factor is higher for the uplink, irrespective of the index of the SS VoIP flow and of the code. Furthermore, we sampled the cumulative density function (cdf) of the packet delay around the voice capacity. Figure 10.8 represents the delay cdf for downlink VoIP flows using a G.723.1 codec. Although the scheduling policy is undisclosed, we can infer that it is not simply the average delay to degrade, at the increase of the offered VoIP traffic, but, the whole delay distribution is shifted around higher delay values. The BS operates a strict threshold control policy: in case a SS exceeds a certain threshold above the CIR, all the flows of the violating SS are penalized. Only for 17 G.723.1 VoIP calls the excess above the CIR appears in a delay spreading as clearly shown in Figure 10.8. At the SS side, this strict BS policy calls for admission control of VoIP flows, to prevent service degradation. We repeated the same measurements for the uplink and the results were similar. As emerged from the *R*-factor measurements, the uplink performs better than the downlink. This contradicts the simulation results obtained in Ref. [7], where larger uplink delays, compared with the downlink, were ascribed to the bandwidth request mechanism and to the PHY overhead. In the case at hand, the uplink delay due to bandwidth request did not prove significant; we ascribe this fact to the activation of the piggybacking mechanisms for bandwidth reservation provided by WiMAX.

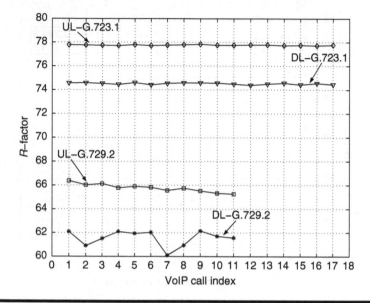

Figure 10.7   Uplink and downlink *R*-factor versus Subscriber Station voice-over-IP (VoIP) session index, using 11 and 17 concurrent calls with the G.729.2 and G.723.1 codecs, respectively.

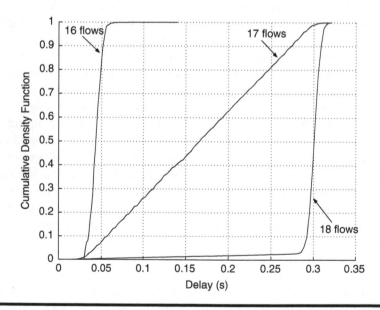

Figure 10.8   Packet delay cumulative density function, G.723.1 codec, downlink direction.

10.5   Research Challenges

Though WiMAX is the most promising technology for enabling BWA systems to be widely deployed, many issues need to be addressed to make it effectively support the requirements and constraints of end-users' multimedia flows. To do so, according to the discussion mentioned previously, an efficient QoS-enabled scheduling algorithm has to be designed and implemented. In this section, we point out and briefly describe the most promising, as well as challenging, directions in such a field, by outlining a research roadmap for QoS provisioning in WiMAX networks. As we considered the scheduling algorithm in isolation in the last section, we shall now present cross-layer approaches, in which performance improvements are obtained by making an appropriate use of information which comes from the lower or upper layers.

■ *Multiantenna architectures for WiMAX networks.* In recent years, intensive research efforts have led to the development of spectrally efficient multi-user transmission schemes for wireless communications based on the use of multiple antenna systems. The use of multiple antennas in combination with appropriate signal processing and coding is indeed a promising direction which aims to provide a high-data rate and a high-quality wireless communications in the access link. In this sense, multiantenna systems can be seen as a way to enhance the cell capacity while offering a better and more stable link quality at the same time. On the other hand, antenna arrays can be used also to achieve beam-forming capabilities, with a remarkable improvement in terms of network performance. Adaptive Antenna Systems (AAS) are encompassed by the IEEE 802.16 standard to improve the PHY-layer characteristics. However, AAS can also act as enablers of spatial division multiple access (SDMA) schemes. In this way, multiple SSs, separated in space, can simultaneously trasmit or receive on the same subchannel. This, obviously, demands the realization of a scheduling algorithm able to effectively exploit the presence of such beam-forming capabilities. In this way, through a cross-layer approach, striking results can be obtained in terms of QoS support. An AAS-aware scheduling could indeed profit from the additional degree of freedom (i.e., the spatial dimension [25]) provided by the underlying PHY techniques. Although this may lead to better performance, it also leads to an increase in the complexity of the scheduler itself. Nonetheless, we believe that the use of this and other related multiantenna techniques (e.g., space-time codes) represent a research direction with big potential in terms of throughput optimization. To fully take advantage of the power provided by multiple antenna systems, innovative QoS-enabled scheduling algorithms, able to work in both space and time dimensions, need to be designed and engineered.

■ *Opportunistic scheduling.* In wireless networks, channel conditions may vary over time because of user mobility or propagation phenomena. These effects are usually referred to as shadowing and fading, depending on their typical time-scales. They have been traditionally considered as harmful features of the radio interface due to their potentially negative impact on the quality of communication. However, recent research has shown that the time-varying nature of the radio channel can be used for enhancing the performance of data communications in a multi-user environment. Indeed, time-varying channels in multi-user environments provide a form of diversity, usually referred to as multi-user diversity, that can be exploited by an "opportunistic" scheduler, that is, a scheduler that selects the next user to be served according to the actual channel status [26]. This approach may also be applied, at the cost of some additional complexity and signaling between PHY and MAC, to WiMAX networks. Opportunistic scheduling schemes do not usually apply to flows that require QoS guarantees, due to the unpredictable delays that may come from the channel dynamics. However, their use may actually lead to an enhanced QoS support. For example, improving the effect of non-real-time traffic (i.e., nrtPS and BE traffic) would free some additional resources to higher priority traffic. In this way, opportunistic scheduling schemes may actually help to increase the QoS capabilities of WiMAX networks. Moreover in this case, novel scheduling schemes are required to exploit multi-user diversity while providing QoS guarantees to the active traffic flows at the same time. It may be interesting to note that multiple antenna systems can actually be used to build up multi-user diversity by means of random beam-forming mechanisms (usually referred to in the literature as "dumb" antennas [27]). Although this direction is somehow orthogonal in nature to the one (based on "smart antennas") outlined before, it could be worth investigating whether these two techniques may be implemented to coexist (e.g., in a time-sharing fashion) to obtain the advantages of both approaches.

■ *QoS support in mesh-based architectures.* The techniques we have presented earlier as research challenges are aimed at providing a better QoS support in PMP architecture. However, they are still subject to the limits imposed by such an architectural choice in terms of service coverage, network capacity, and system scalability. One possible solution to overcome such problems could be the adoption of a mesh-based architecture [28]. In mesh topologies, direct communication among neighboring SSs is allowed, so enhancing the network coverage and possibly enabling the deployment of a fully wireless backbone connecting to an Internet gateway. While mesh-based architectures offer interesting possibilities thanks to its inherent flexibility, they also present many research challenges to be addressed in terms of MAC and packet routing. This is even more challenging in the case of QoS support for multimedia flows, where reliable levels of services have to be ensured by means of distributed algorithms. In this framework, a "double cross-layer" approach (where information is shared among PHY, MAC, and NET layers) may lead to potentially dramatic performance improvements compared with conventional layered solutions. This clearly entails the definition of radically innovative scheduling protocols, which are able to work in a distributed and collaborative way, so cooperating with the routing algorithms to provide QoS guarantees to SFs based on some PHY information. For example, the integration of scheduling and routing protocols can be based on the actual channel conditions, as well as on the level of interference in the network.* The application of these concepts to WiMAX networks is not straightforward, as it would imply some major modifications to the actual standard, in terms of both signaling (necessary for pursuing cross-layer optimization) as well as definition of basic functionalities and interfaces of the routing protocol to be employed.

References

1. IEEE Standard for Local and Metropolitan Area Networks Part 16: Air Interface for Fixed Broadband Wireless Access Systems (2004).
2. Wimax forum (2001). URL http: //www.wimaxforum.org.
3. Nokia. Broadband Media Services (2001). URL http://www.nokia.com/downloads/solutions/operators/broadband_media_services_tutorial_net.pdf. White Paper.
4. Alcatel. WiMAX, making ubiquitous high-speed data services a reality (2004). URL http://www.alcatel.com/com/en/appcontent/apl/s0406-WiMAX-EN_tcm172-4479 1635.pdf.Strategy White Paper.
5. A. Ghosh, D. R. Wolter, J. G. Andrews, and R. Chen, Broadband wireless access with WiMax/802.16: Current performance benchmarks and future potential. *IEEE Communications Magazine*, 43(2), 129–136 (February 2005).
6. C. Eklund, R. Marks, K. Standwood, and S. Wang, IEEE standard 802.16: A technical overview of the WirelessMAN air interface for boadband wireless access. *IEEE Communications Magazine*, 40(6), 98–107 (June 2002).
7. C. Cicconetti, L. Lenzini, E. Mingozzi, and C. Eklund, Quality of service support in IEEE 802.16 networks. *IEEE Network Magazine*, 20(2), 50–55 (March 2006).
8. N. J. Muller and R. P. Davidson, *The Guide to Frame Relay & Fast Packet Networking* (1991).
9. P. Tzerefos, V. Sdralia, C. Smythe, and S. Cvetkovic, Delivery of low bit rate isochronous streams over the DOCSIS 1.0 cable televison protocol. *IEEE Transactions on Broadcasting*, 45(2), 206–214 (June 1999).
10. Y. Cao and V. O. Li, Scheduling algorithms in broad-band wireless networks. *Proceedings of the IEEE*, 89(1), 76–87 (January 2001).
11. J. Moorman and J. Lockwood, Multiclass priority fair queuing for hybrid wired/wireless quality of service support. In *Proceedings of Second ACM International Workshop on Wireless Mobile Multimedia, WOWMOM* (August 1999).
12. G. Chu, D. Wang, and S. Mei, A QoS architecture for the MAC protocol of IEEE 802.16 BWA system. *IEEE Conference on Communications, Circuits and Systems and West Sino Expositions*, 1, 435–439 (June 2002).

---

* Note that this requires the introduction of novel metrics for path selection in routing algorithms [28].

13. K. Mezger, D.W. Petr, and T.G. Kelley, Weighted fair queuing vs. weighted round robin: a comparative analysis. In *Proceedings of IEEE Wichita Conference on Communications, Networks and Signal Processing* (1994).

14. H. Fattah and C. Leung, An overview of scheduling algorithms in wireless multi-media networks. *IEEE Wireless Communications*, 9(5), 76–83 (October 2002).

15. D. Stiliadis and A. Varma, Latency-rate servers: a general model for analysis of traffic scheduling algorithms. *IEEE/ACM Transactions on Networking*, 6(2), 611–624 (October 1998).

16. N. O. Johannesson, The ETSI computation model: a tool for transmission planning of telephone networks. *IEEE Communications Magazine*, 35(1), 70–79 (January 1997).

17. ITU. The E-model, a computational model for use in transmission planning. Recommendation G.107, Telecommunication Standaridization Sector of ITU, Geneva, Switzerland (December 1998).

18. Alvarion. URL http://www.alvarion.com/upload/contents/291/alv_BreezeMAX%20PRO_CPE.pdf.

19. IETF. Definition of the differentiated services field (ds field) in the ipv4 and ipv6 headers. URL http://www.ietf.org/rfc/rfc2474.txt.

20. The netfilter.org project. URL http://www.netfilter.org/.

21. D-ITG, Distributed Internet Traffic Generator. URL http://www.grid.unina.it/software/ITG.

22. Simple Network Time Protocol (SNTP) Version 4. URL http: //www.apps.ietf.org/rfc/rfc2030.html.

23. D. P. Hole and F. A. Tobagi. Capacity of an IEEE 802.11b wireless LAN supporting VoIP. In *Proceedings of IEEE ICC*, Paris (20–24 June 2004).

24. C. Hoene, H. Karl, and A. Wolisz, A perceptual quality model intended for adaptive VoIP applications: Research articles. *International Journal of Communication Systems*, 19(3), 299–316 (2006). ISSN 1074-5351. doi: http://dx.doi.org/10. 1002/dac.v19:3.

25. B. H. Walke, N. Esseling, J. Habetha, A. Hettich, A. Kadelka, S. Mangold, J. Peetz, and U. Vornefeld, IP over wireless mobile ATM-guaranteed wireless QoS by Hiper-LAN/2. *Proceedings of the IEEE*, 89(1), 21–40 (January 2001).

26. X. Liu, E. Chong, and N. Shroff, Opportunistic transmission scheduling with resource-sharing constraints in wireless networks. *IEEE Journal on Selected Areas in Communications*, 19(10), 2053–2064 (October 2001).

27. P. Viswanath, D. Tse, and R. Laroia, Opportunistic beamforming using dumb antennas. *IEEE Transactions on Information Theory*, 48(6), 1277–1294 (June 2002).

28. R. Bruno, M. Conti, and E Gregori, Mesh networks: commodity multihop ad hoc networks. *IEEE Communications Magazine*, 43(3), 123–131 (March 2005).

Chapter 11

Handoff Management with Network Architecture Evolution in IEEE Broadband Wireless Mobility Networks

Rose Qingyang Hu, David Paranchych,
Mo-Han Fong, and Geng Wu

Mobility management is an essential element of wireless communications. It is the key to enabling always-on wireless connectivity to mobile users. With the advance of the next generation of wireless technologies and different network architecture evolution options, the capability to achieve fast and seamless handoff becomes a critical factor in evaluating the success of a new technology and the efficiency of the network architecture. This chapter discusses the possible handoff solutions to new broadband wireless technologies such as WiMAX and 802.20 under different network architectures. The key handoff issues, challenges, and architectural options faced in designing the next-generation wireless broadband mobility system are addressed.

11.1   Introduction

Until recently, wireless users have had to choose between broadband connectivity and mobility. The infrastructure of existing cellular networks is optimized for high mobility users. However, the cost per bit for data delivery over the air to mobility users is relatively high, in particular when compared with 802.11-based Wi-Fi systems. The next generation of wireless networks is expected to offer a broadband experience to high mobility users at a significantly reduced cost. To achieve this goal, the wireless industry and the research community have worked extensively to advance a number of key technology areas, including state-of-the-art orthogonal frequency division multiplexing (OFDM)-based multiple access technology, advanced Multiple Input Multiple Output (MIMO) antenna technology for broadband access, more efficient handoff management for robust high-speed mobility support at reduced overhead cost, and innovative and flexible network architecture to minimize the cost of network ownership.

There are many important publications on OFDM and MIMO technologies [1,2], which formed the foundation of the next-generation wireless broadband air interface. OFDM-MIMO dramatically boosted the peak data rate to beyond 100 Mbps and increased the system capacity by several times. As the key capabilities of the next-generation wireless networks significantly exceed that of existing 3G networks, the wireless industry is looking into new network architectural options to fully leverage the increased capability in a wide range of deployment environments.

Traditional 3G network architectures have been highly hierarchical, designed to accommodate mobility users in outdoor macrocellular environments. But it also has high deployment cost and high network latency. Wi-Fi radio access points (APs) based on IEEE 802.11 standards employ a flat network architecture, which has no intermediate network above the AP to connect to the Internet Protocol (IP) network and thus has a low deployment cost. Nevertheless, high-speed user mobility is not well supported in Wi-Fi networks. In evolving the network architecture, we not only want to keep all the cost benefits provided by flat architectures, but also want to keep the good mobility performance from hierarchical architectures. This may require new architecture considerations, together with innovative handoff mechanisms, to achieve that goal.

The rest of the chapter is organized as follows. A general description of the next-generation broadband wireless technologies is given in Section 11.2. The handoff processes are illustrated in Section 11.3, with particular attentions to the components of handoff in OFDM-MIMO-based WiMAX and 802.20 networks. Section 11.4 highlights the architecture impacts on mobility management. The chapter concludes with Section 11.5.

11.2 Next-Generation IEEE 802 Broadband Wireless Technologies

The IEEE 802.16 standard [3,4] has emerged as an important technology for delivering packet data service in a wide area cellular network. IEEE 802.16 was initially targeted for fixed wireless deployments, but the 802.16e-2005 amendment [4], published in early 2006, introduced significant enhancements to enable the support of mobility for the OFDMA physical (PHY) layer. To promote interoperability among products based on the IEEE 802.16 standard and manage certification of vendor equipments, the WiMAX Forum was formed. The forum has the important role of defining interoperability tests for a subset of the 802.16 features, and making arrangements for third party certification labs to test vendor equipments. The forum also defines standard interfaces for upper protocol layers, as the scope of the IEEE 802.16 standard only encompasses the Medium Access Control (MAC) and PHY layers.

Unlike WiMAX, which was incubated inside IEEE 802.16 family and evolved from earlier 802.16 technologies, 802.20 [5] or Mobile-Fi was designed from ground up as a technology to support high-mobility services. It aims to support mobility as high as 250 km/h and a peak rate of up to 260 Mbps in the licensed spectrum below 3.5 GHz. The enabling technologies are also OFDM, MIMO, and beam-forming. The draft standard is still under the IEEE standardization process.

Both WiMAX and 802.20 use OFDM-MIMO, which is emerging as the main technology for future cellular packet data networks, including 3GPP long-term evolution and 3GPP2 air interface evolution as well. The introduction of OFDM-MIMO carries with it a number of considerations for handoff management that may be different from those in the existing 3G CDMA systems. We briefly consider those factors here.

Resource control in OFDM-MIMO networks is done at the lowest level in the network hierarchy, at the Base Station (BS), with frame-by-frame scheduling. The implications to handoff of frame-by-frame scheduling on both uplink and downlink mainly concern soft handoff (SHO), where multiple APs are simultaneously either transmitting to or receiving from a given mobile. If SHO is supported, tight coordination between the APs in SHO is required for the benefits of dynamic scheduling to be maintained. On the forward link, the SHO transmissions may be either in the same resource block or tone-symbols on multiple APs, which allows combining at the radio frequency (RF) level (as long as the differential delay between sites is within the cyclic prefix duration of the OFDM symbol), or in different resource blocks, in which case the mobile listens to both allocations and performs soft combining after detection. On the reverse link, the multiple APs must detect and combine the terminal's transmission, and care must be taken by the AP schedulers not to place two terminals in the same resource block. This requires tight coordination between the APs.

The introduction of MIMO to next-generation systems for the most part does not influence the support of handoff, but it can allow for more flexibility. An example of this is the support of MIMO on the forward link where the multiple transmit antennas are on different APs. From the terminal perspective, the signal is processed in the same way as if the two (or more) transmit antennas were on the same AP, but there is an extra degree of flexibility on the network side. Of course, this requires the tight coordination of transmissions from multiple APs.

11.3 Handoff Management in Next-Generation Broadband Wireless Networks

11.3.1 Handoff Management

To realize ubiquitous wireless connectivity in future broadband wireless networks, which support mobility as fast as 250 km/h, effective mobility management is critical. Mobility management consists of location management and handoff management. As location may change while mobiles are idle, location management enables the wireless networks to locate a roaming Mobile Station (MS) and set up a connection. Handoff management is the process that changes a mobile's point of attachment to the network. A typical handoff process in a wireless network will involve the following entities: MS, serving access point (SAP), target access point (TAP), serving gateway (SGW), target gateway (TGW), and home agent (HA). The AP is the air link attachment point for the MS. It terminates the air link and link layer MAC protocols. The gateway (GW) is defined as the node that is the IP layer attachment point for the MS and serves as the first point for IP forwarding. In this chapter, we will investigate different architectures, for example, flat and hierarchical, and their impacts on the performance of handoff. The physical locations of the entities involved in the handoff could be different in different architectures. A handoff can be either a layer-2 or a layer-3 handoff. These are defined as follows:

■ Layer-2 handoff (L2HO) is often referred to as "micromobility" and only changes the air interface attachment point but keeps the IP attachment point unchanged. The process typically involves detection of changes in signal strength, releasing the connection to the SAP, and establishing a connection to the TAP. L2HO is transparent to the upper layer protocols, so it usually has small handoff latency and low packet loss.

■ Layer-3 handoff is often referred to as "macromobility" and changes the IP attachment point of a mobile user. During a layer-3 handoff, the MIP4/MIP6 protocol is used to update the HA with the care-of address (CoA) of the MS. During a layer-3 handoff, the MS must be registered and authenticated with the HA every time it moves from SGW to TGW, which introduces extra latency to the communication as usually HA is located far away from SGW.

11.3.2 WiMAX/802.20 Handoff Schemes

L2HO has a number of common attributes in all modern cellular data communications systems. In 802.20, WiMAX, 3GPP, and 3GPP2 systems, there are similar mechanisms used to accomplish L2HO, that range from a complete hard handoff (HHO) with no context sharing between BSs or APs, to complete SHO with full context sharing. In this section we look more closely at the common features of L2HO in these evolving systems. In general, there are five types of L2HO supported by different systems. These are (along with the systems that support them):

- HHO (WiMAX and 802.20)
- Optimized hard handoff (OHHO) (WiMAX)
- SHO (WiMAX)
- Fast Base Station switching (FBSS) based on downlink channel condition (WiMAX, 802.20)
- FBSS based on uplink channel condition (802.20)

In HHO, the air interface link is broken at all layers before being established again at the TAP. No context information is shared between serving and target BSs. For HHO, the handoff latency is typically on the order of 100 msec or more, due to the robust way in which messages are exchanged and communication re-established at the new AP. In exchange for a higher latency, this is typically the simplest and most robust type of handoff, and is often used for handoff between radio carriers as well as between systems operated by different service providers.

In OHHO, the physical radio link is broken before it is re-established at the TAP, as with HHO. In OHHO, however, part or all of the L2 and L3 context information is exchanged between the SAP and the TAP, to make the latency of the communication re-establishment shorter. This can include different levels of optimization (as in WiMAX), where the latency of the HO is decreased as more context information is exchanged. With enough information exchange, the latency of an OHHO can be less than 100 msec.

FBSS is yet a further level of optimization to decrease handoff latency. In the case of FBSS based on the downlink channel condition, there is still a break of the PHY layer channel before the link is established at the TAP, as in HHO and OHHO. However, PHY layer signaling is used for fast TAP selection indication. This results in handshake latency on the order of tens of milliseconds. FBSS requires the concept of an active set, which is a set of candidate APs to which the terminal is likely to handoff in the near future, based on a periodic measurement of channel quality. In FBSS, most layer-2 and layer-3 context information is retained among the candidate APs in the active set, to minimize data interruption. Lower MAC context information, such as Hybrid Automatic ReQuest (HARQ) state information, may not be retained or transferred when switching between APs of noncollocated cell sites. FBSS based on the downlink channel condition is supported by WiMAX, 802.20, and 1xEV-DO.

The case of FBSS based on uplink channel condition is similar to FBSS based on the downlink channel, except that it is the uplink channel measurement that is used as the basis for the handoff trigger. As with downlink-based FBSS, this mechanism requires an active set, and uses PHY layer signaling for target selection indication to reduce the handoff latency to a few tens of milliseconds. This scheme coexists with the downlink-based FBSS in 802.20, where the best AP for the downlink and uplink may be different.

The last general HO mechanism is SHO, in which a physical link to the TAP is established before the link to the SAP is broken. SHO requires an active set just as FBSS does, and requires the network to retain all layer-2 and layer-3 context information during the handoff. The resulting handoff is essentially seamless, and appears to be an uninterrupted data stream to the terminal. SHO can be used either on the downlink (supported by 802.16), or on the uplink (supported by both 802.16 and 1xEV-DO).

There are two main areas in which these handoff mechanisms differ. The first is in the level of context information that needs to be retained or transferred between APs. As the latency and data disruption caused by the handoff decreases, the level of context sharing increases, and this impacts the network architecture and backhaul engineering. In general, the more context information that is retained or transferred, the less overall handoff latency and data flow interruption at the expense of more costly backbone network. The second area of difference between the mechanisms is the over-the-air call flow that supports the handoff technique. In general, PHY layer signaling or lower MAC signaling allows for a faster handoff. There is also a tradeoff between reliability and latency. More reliable signaling (e.g., ARQ-protected layer-3 signaling) requires more processing and incurs more delay, while unprotected or lightly protected PHY layer signaling (e.g., PHY layer handoff switch indication) is very fast but also carries the risk of errors and false alarms.

The layer-3 handoff process of WiMAX and 802.20 can follow a typical mobile IPv6 handoff flow [6]. During the layer-3 handoff an MS changes its IP attachment point and needs layer-3 reanchoring. MIPv6 binding update and binding acknowledgment messages are required to update the HA the current location of MS. A suitable TAP is usually selected and a L2HO is completed before a layer-3 handover is initiated. The MS establishes a new layer-2 connection to that AP and acquires a valid CoA in the new subnetwork. Finally, it can initiate the layer-3 handover by sending a binding update message with the new CoA to its HA and possibly the correspondent node as well. In the next section, we will address how different network architectures impact the handoff processes described before.

11.4 Handoff Management with Different Network Architectures

In the past two decades, wireless networks went through a number of technology transformations. The first- and second-generation cellular networks were designed for circuited switched voice services. As packet data services were introduced in the third-generation networks, a number of packet data-specific nodes were added to the network to provide the connectivity to the IP world. The networks are highly hierarchical, which means that radio access nodes are connected to the IP network through one or more levels of management and controlling nodes, which increases the network latency. In recent years, Wi-Fi radio APs have gained popularity. Wi-Fi employs a flat network architecture, with no intermediate network nodes above the APs to connect to the IP network. However, high-speed mobility is not supported in Wi-Fi network. An interesting challenge is to find architectural options that improve system performance beyond 3G networks but at a reduced cost closer to Wi-Fi for mobility environments [7]. Depending on the deployment scenarios, the answer may not be unique.

11.4.1 Handoff in Traditional 3G Hierarchical Architecture

Traditional 3G network architectures have been highly hierarchical, designed to accommodate mobility users in outdoor macrocellular environments. This hierarchical architecture also fits well into typical deployment scenarios for 3G operators, where the large capacity carrier-grade network control nodes and GWs are located in a central office, interconnected to a large number of remote radio cell-sites through point-to-point backhaul T1/E1 links. In this star-like topology, there is no direct connection between radio cell-sites. All signaling and traffic have to go through the network nodes in the central office. The 3G Universal Mobile Telecommunication System (UMTS) network reference model [8] is a typical example of hierarchical network architecture. As shown in Figure 11.1, the access network consists of many radio cell-sites called node B, controlled by a radio network controller (RNC). Serving GPRS support node (SGSN), and gateway GPRS support node (GGSN) form a GPRS core network, providing connectivity to the IP network.

Hierarchical network architectures served well in traditional macrocellular mobility networks. It is highly scalable in geographical coverage area and system capacity. Complex signaling and protocol processing are concentrated and confined to a few nodes. As many multiple nodes are involved in processing a call, call setup time is generally longer in a hierarchical network, ranging from several seconds to tens of seconds. However, this is not a major issue for voice service as the call setup process is only invoked at the beginning of a call, unlike a packet data session, which sets up and tears down the connection frequently to preserve radio resources.

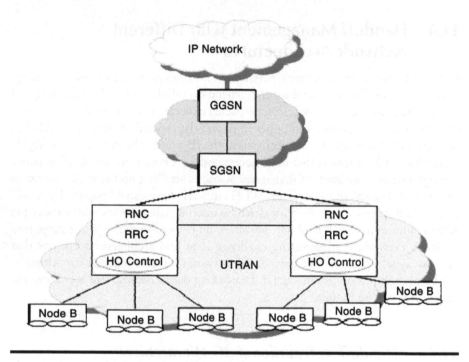

Figure 11.1   UMTS network reference model.

Hierarchical architecture is well suited for high-speed mobility networks. For a MS traveling at a high speed in a macrocellular environment, its radio link has to be frequently handed over from one radio site to another. A hierarchical handoff design confines the impact of frequent handoffs to a minimum number of nodes.

Figure 11.2 shows a high level handoff procedure in a hierarchical architecture network. The radio link is anchored at the GW, which is the RNC in a UMTS network. The MS is initially connected to an SAP, or a node B in UMTS. If the SAP has multiple radio sectors, MS moving among the sectors of the same SAP will be handled locally with no impact on the rest of the network. When the MS moves from its SAP to a TAP, the MS sends a HO_REQ (handoff request) message to the GW, indicating the need for an intra-GW handoff. The GW checks the resource availability with the TAP through RES_REQ (resource request) and RES_RSP (resource response) messages. Once resources are confirmed, the GW initiates a SESS_XFER (session transfer) process. As the radio session is anchored at the GW, most of the session information is stored at the GW without the need for being transferred. The GW then instructs the MS to execute the handoff through the HO_RSP (handoff response) message. The MS sends a HO_CMP (handoff complete) message to GW once the handoff is complete. As a radio session is anchored at the GW and a GW usually covers a large geographical area (e.g., a city), impacts of handoffs within such an area are confined only to layer-2. In case the MS moves

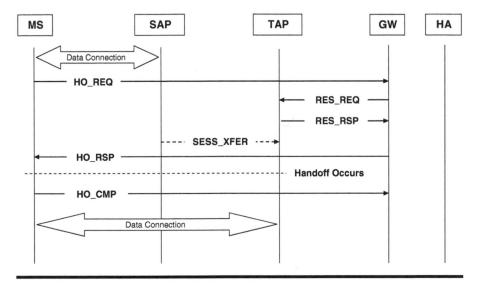

Figure 11.2   Handoff flow for a hierarchical architecture network.

into another GW, an inter-GW handoff is required. This can be accomplished by a layer-3 handoff with the assistance of network nodes above the GW, including mobile IP handoff with the help of a HA. It can be observed that in a hierarchical network, with a radio session anchored at the GW, mobility support is confined to within the RAN. HO_REQs are escalated to upper layer network nodes only when the MS travels beyond the RAN coverage area.

11.4.2   Handoff in Flat Network Architecture

Flat network usually means a network in which all stations can reach each other without going through another node. In broadband wireless design, flat network often refers to a network consisting of only APs. There are in fact two types of flat networks. One can be considered "true flat," where each AP is an independent IP node, containing no functionality to serve as a controller to other APs. Commercial 802.11 AP is an implementation of such true flat network architecture, as shown in Figure 11.3a. There is no coordination at the link layer between APs. As each hand-off is a layer-3 handoff, service could be disrupted by the slow mobile IP process. With these limitations, true flat network is only used in fixed hot spots with no real mobility requirements.

Another type of flat network is called "semi-flat," which is shown in Figure 11.3b. The AP in this case has the handoff control and the radio resource control (RRC) functionality built-in. From an architectural viewpoint, each AP can be considered as an integrated implementation of a complete 3G radio access network, with node B, RNC, and IP GW collapsed into one box. Each AP has direct access

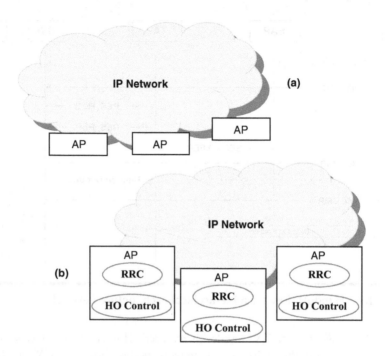

Figure 11.3   Flat network architectures.

to the IP network, and may serve as an AP, an RNC, or both to a particular MS. In a sense, the RNC and the IP GW functions are all consolidated into each AP, therefore a consolidated implementation of a hierarchical network. In the semi-flat architecture, the mobile can be anchored at a certain AP at the IP layer. Therefore, the handoff between APs does not have to change the IP anchor point and it becomes a L2HO.

There are two kinds of handoff procedures over a flat network. The true flat architecture shown in Figure 11.3a supports handoff without session anchoring at the AP and the semi-flat architecture shown in Figure 11.3b supports handoff with session anchoring at the AP. In both cases, the AP and the GW, which contains the RRC and the handoff control functions, are integrated into one node.

In the case of handoff without session anchoring (Fig. 11.4), the SAP and SGW are always colocated on the same PHY node. This link layer handoff flow is the same as in Figure 11.2. When the MS detects the need for handoff, it issues a HO_ REQ to the GW (in this case the SAP/GW). The SAP/GW verifies the resource with TAP, orders MS to perform handoff. Once the link layer handoff is completed, the MS initiates a mobile IP handoff to move the session anchoring point from the SAP/GW to the TAP/GW. The data traffic will go through the new AP and new GW (TAP/GW) once both layer-2 and layer-3 handoffs are completed. With this

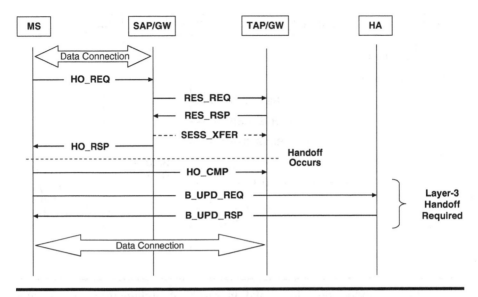

Figure 11.4 Handoff flow for a flat architecture without session anchoring.

handoff process, the serving GW location follows that of the MS. The obvious benefit is that the IP routing is always optimized. However, layer-3 handoff may disrupt services, and depending on how fast the MS moves, there may be too many layer-3 handoffs for the network to handle.

In the case of handoff with session anchoring (Fig. 11.5), the SAP and SGW can be located differently. The old GW serves as the anchoring point for an MS as long as its data session is active. There is no need for a layer-3 handoff after layer-2 handoff is complete. The session anchoring point remains at the SAP/GW, although the radio connection has been handed over from SAP to TAP. Thus, the link performance is improved without the disruption from layer-3 handoff. Such systems may support high-speed mobility as all handoffs are fast layer-2 handoffs. However, there are also drawbacks with this design. As the GW is anchored, user data and signaling traffic from TAP have to go through the anchoring GW before reaching the IP network. This significantly increases backhaul traffic. The cost may be prohibitive for access network consisting of APs which only have point-to-point links to a central office through T1/E1. This design may be less of an issue if there are direct low-cost broadband connections among APs. It should be noted that layer-3 handoff and route optimization can be applied after the data session transitions into idle mode.

Flat network architectures are considered attractive for a number of reasons. First of all, the network latency may be minimized as the number of nodes between the MS and the IP network is reduced to only one. With the functionality of multi-nodes integrated into one box, flat network avoids complex and costly internode interface and communications. Eliminating higher level nodes also avoids a single

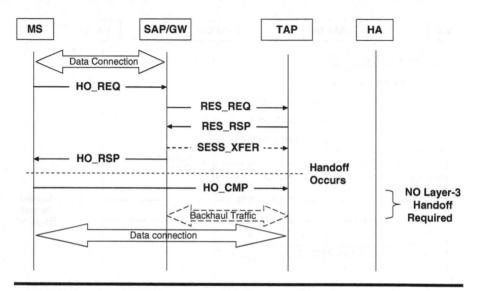

Figure 11.5 Handoff flow for a flat hierarchical architecture with session anchoring.

point of failure that may affect a large number of radio nodes. With IP intelligence pushed to the edge, the network is easier to scale up and down in size, by simply adding or removing the number of APs attached to the IP network. Finally, with the significant reduction of the number of nodes in the access network, there are less network components to manage.

However, some of the cost savings may not be applicable to all deployment environments. Wi-Fi APs, for example, have limited transmit power, which translates into low-cost radios, limited coverage, and reduced reliability requirements. All these contribute to the low cost of the Wi-Fi AP. These cost savings are not applicable to a carrier grade macrocellular network. The most significant challenge for the flat network architectures are the handoff support for high-mobility terminals, and system level radio resource management. As AP communicates with others at the IP layer in a flat network, it requires extra efforts to support link layer handoff and to share radio resource information among a group of APs.

11.4.3 Handoff with the Evolved Network Architectures

As we have discussed in the previous sections, selecting appropriate network architecture is critical for the handoff support in the next-generation broadband wireless mobility networks. There are many technical and economic tradeoffs and considerations, including performance, CAPital Expenditure (CAPEX) and OPerational

Expenditure (OPEX), and deployment environment such as the topology of the available backhaul infrastructure. Both hierarchical and flat network architectures have strong technical and economical merits, depending on the exact application environment.

There is a general trend in the industry to simplify the network to improve performance and to cut cost. There may not be one single answer that can meet all possible requirements. For IEEE 802.16e systems, WiMAX NWG (Network Working Group) looked into multiple architectural options (or profiles) to address different deployment scenarios [9]. As shown in Figure 11.6, Profile A can be considered a simplified hierarchical architecture, with only two nodes in the access network. This architecture balances the need for efficient high-speed handoff support and low network latency. Profile B can be implemented as a flat network node, capable of supporting handoff with or without session anchoring as we discussed in the previous section. The benefit and the cost of this architecture depend on the deployment environment and the available backhaul infrastructure. Profile C is a hybrid design of flat and hierarchical architectures, with handoff control and radio resource control distributed to the BSs or APs. However, there is a separate GW node in the network to perform nonmobility-related tasks. There are continuing debates in WiMAX NWG on the technical and business merits of each network profile. Such debates are essential for the industry to develop the best technology solution.

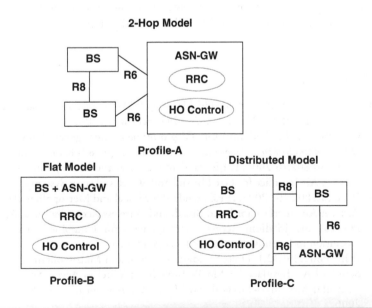

Figure 11.6 WiMAX Network Working Group access network reference models.

The important issue of network architecture evolution is also explored in other major international standards development forums. 3GPP and 3GPP2 are both looking at the two-hop hierarchical network architecture, while leaving the flexibility to go flat. With the availability of wireless broadband service to mobility users, one can expect that many real-time applications will become possible over the next-generation networks. High-performance handoff is essential to deliver the required quality of service (QoS) for real-time services. As network architecture has great impact on handoff solution and cost of network ownership, it is necessary to design the network that can address both handoff performance and network architecture concerns.

11.5   Conclusions

Handoff is a critical technology component for next-generation wireless broadband mobility networks. It is becoming even more important as the network increasingly supports real-time applications and services. Handoff complexity and performance are significantly affected by the underlying wireless access network architecture. In addition, network architecture selection affects the overall end-to-end performance of a mobility system. It also has a major impact on CAPEX and OPEX, depending on the existing deployment and the available backhaul infrastructure. This chapter addresses the key handoff issues, challenges, and architectural options that we face in designing the next generation of wireless broadband mobility systems. It contributes to the global efforts on the architecture design and mobility management by offering some technical focuses for further investigation.

References

1. G. L. Stuber et al., "Broadband MIMO-OFDM wireless communications," *IEEE Proceedings*, Vol. 92, No. 2, pp. 271–294, February 2004.
2. C. Dubuc et al., "A MIMO-OFDM prototype for next-generation wireless WANs," *IEEE Communications Magazine*, Vol. 42, No.12, pp. 82–87, December 2004.
3. IEEE Std 802.16-2004, IEEE Standard for Local and Metropolitan Area Networks, Part 16: Air Interface for Fixed Broadband Wireless Access Systems.
4. IEEE Std 802.16e-2005, IEEE Standard for Local and Metropolitan Area Networks, Part 16: Air Interface for Fixed Broadband Wireless Access Systems, Amendment 2: Physical and Medium Access Control Layers for Combined Fixed and Mobile Operation in Licensed Bands, and Corrigendum 1.
5. IEEE P802.20/D2.1, Draft Standard for Local and Metropolitan Area Networks—Standard Air Interface for Mobile Broadband Wireless Access Systems Supporting Vehicular Mobility—Physical and Media Access Control Layer Specification, May 2006.
6. IETF RFC 3775 "Mobility Support in IPv6."

7. J. De Vriendt, P. Laine, C. Lerouge, and Xiaofeng Xu, "Mobile network evolution: a revolution on the move," *IEEE Communications Magazine*, Vol. 40, No. 4, pp. 104–111, April 2002.
8. 3GPP TR 25.913 Technical Specification Group Radio Access Network; Requirements for Evolved UTRA (E-UTRA) and Evolved UTRAN (E-UTRAN) (Release 7), V7.3.0, March 2006.
9. WiMAX Forum Network Working Group, "WiMAX End-to-End Network Systems Architecture (Stage 3: Detailed Protocols and Procedures)," April 10, 2006.

Chapter 12

Distributed Architecture for Future WiMAX-Like Wireless IP Networks

Suman Das, Thierry E. Klein, Harish Viswanathan,
Kin K. Leung, Sayandev Mukherjee,
George E. Rittenhouse, Louis G. Samuel,
and Haitao Zheng

The migration of the radio access network to an all-Internet Protocol (IP) network, as proposed in the IEEE 802.16 (WiMAX) standard, provides significant advantages in terms of cost savings, scalability, and simplified network management for providing a variety of IP-based services in wireless networks. This paper proposes a novel, distributed radio network architecture to support all-IP services using orthogonal frequency division multiple access (OFDMA)-shared access. Specifically, the architecture distributes the so-called fast cell-site selection (FCSS) functionality among base-station routers (BSRs) so that data transmission for a call can be switched quickly from one BS with a weak radio link to another with a better link, as a means to enhance system capacity. A key objective of the new architecture with distributed radio resource management functions is to provide a unified approach to supporting multimedia IP applications. We present a distributed architecture for radio network control that specifically supports signaling message exchanges among network elements for FCSS in the presence of terminal mobility. We study the design and quality-of-service (QoS) requirements of the backhaul network for supporting FCSS.

12.1   Introduction

The Internet has evolved with the Internet Protocol (IP) as the single major transport mechanism for all data services. Mobile cellular networks on the other hand use specific protocols and network architectures that are primarily designed for efficiently carrying voice traffic. With voice penetration reaching saturation levels in many markets, for future growth service providers are now turning to a variety of data services that are currently supported in IP-based wireline networks. It is thus natural that mobile cellular network architectures and protocols also evolve to carry both the traditional voice and the new data services. An IP-based radio access network (RAN) akin to the Internet is likely to be the most favorable approach for future RANs. The advantages of an all-IP RAN are by now well understood and documented in the literature [1–3]. These include cost efficiency from economies of scale (because the radio network has the same network elements as the Internet, allowing for use of off-the-shelf components), improved reliability, separation of the control and transport planes, convenient implementation of new services, independent scaling of control and transport, and straightforward integration of multiple networks. A typical all-IP network architecture is illustrated in Figure 12.1. The BSs, which are now routers, are connected to the service provider's Intranet (also referred to as the backhaul network). In this chapter, we interchangeably use the terms BS, Base Station router (BSR), and cell site. A gateway connects the Intranet to the Internet. IP is the network protocol transporting user and control information within the Intranet. The control server provides the necessary call service control. An important characteristic of the network in the figure is that several radio network functionalities are integrated with the BS functionalities and are thus distributed across the network.

We propose the use of fast cell-site selection (FCSS) as a means to guarantee delay constraints for users particularly near the edges of cells, thereby enhancing Voice over IP (VoIP) capacity in the absence of soft handoff on the shared channels in a system. FCSS refers to the procedure that allows rapid switching of the transmission for a call from one BS with a weak radio link to another with a better link, thereby harnessing the advantage of the time-varying channel quality of the wireless link. Thus, FCSS is also referred to simply as cell switching here. A previous study on soft handoff suggests that, on the forward link of a system, soft handoff

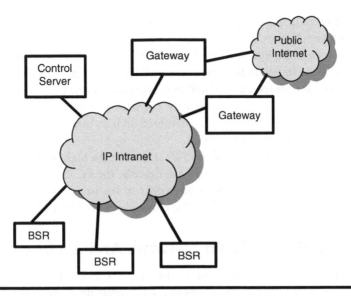

Figure 12.1   Generic All-IP network architecture with Base Station routers (BSRs).

can be avoided without significant loss in capacity, provided that FCSS can be used especially when multiple terminal antennas are used to provide link diversity. We exploit this fact and propose an architecture that allows fast switching between cells rather than soft handoff. This eliminates the need for tight synchronization between BSs and framing protocols between nodes to guarantee strict latency requirements (even when the application does not require it), as would otherwise be required for soft handoff. We propose here to use fast cell switching on the reverse link instead of frame selection or soft handoff to simplify the architecture. Receiving diversity at the BS mitigates the effect of not having soft handoff on the reverse link.

Several proposals for IP RAN [4,5] and for micro-mobility management have been published in the literature, including Hierarchical Mobile IP (HMIP) [6], HAWAII [7], Cellular IP [8], and Brain Candidate Mobility Protocol (BCMP) [9]. However, these proposals do not address the issue of ping-pong of a mobile between BSs in the case of rapid signal fluctuations from the different BSs. Seamless transfer of radio link protocol information was also not considered. In an orthogonal frequency division multiple access (OFDMA) system, universal frequency reuse is employed to maximize capacity and thus interference can come from an immediate neighboring BS. Further, the radio link from a BS that can best serve a given mobile, as reflected by its best Signal-to-Interference-plus-Noise Ratio (SINR) among adjacent BSs, could be rapidly changing even when the mobile is stationary because of time-varying channel fading. Thus handoffs are not only driven by user mobility but by signal propagation characteristics as well. Any mobility management architecture must be designed for a significant number of handoffs between a small set of BSs.

In a scenario with frequent handoffs, it becomes necessary to ensure that the hand-off process is fast, lossless, and efficient.

The main contributions of this chapter are: (i) a distributed architecture that supports FCSS, seamless radio link control (RLC) transfers over the radio links and the backhaul networks, and distributed registration and paging; (ii) quantifying the quality-of-service (QoS) design for the backhaul network to enable FCSS and establishing the feasibility of the presented architecture; and (iii) quantifying the gain in VoIP capacity from FCSS.

This chapter is organized as follows. We start with the high-level network architecture in Section 12.2. In Section 12.3 we describe the FCSS scheme in detail. In Section 12.4 we address the QoS requirements in the backhaul network to carry the control signals required for FCSS. We conclude in Section 12.5.

12.2   Distributed Radio Network Architecture

The current generation of cellular networks supports mobility management with associated requirements of low packet loss and low handoff delay through a centralized architecture. While it is possible to consider IP transport within the centralized architecture, a distributed architecture with decentralized control is preferable from the perspective of scalability and robustness. In the centralized architecture, the radio network controller (which typically serves a large number of BSs) performs the following functions: load and congestion control of individual cells, admission control in simplex and soft handoff modes, management and configuration of individual cells, mapping of traffic to appropriate physical channels, macro-diversity combining and distribution for soft handoff, outer loop power control for soft handoff on the reverse link, paging coordination and mobility management. Among these different functionalities, the ones specific to individual cells such as code allocation management, congestion control, and admission control in simplex mode can be straightfor-wardly distributed to the BSs as these functions do not require interactions among BSs. However, functionalities related to soft handoff, paging and mobility management require signaling between BSs when distributed and thus careful design of the architecture to facilitate these functions is required. In what follows, we describe how the proposed distributed architecture addresses these functionalities. Besides being distributed, the proposed architecture is all-IP in the sense that IP-based protocols are used for transport of data and signaling within the RAN.

Mobile IP [10] has been standardized for macro-mobility management in IP networks. Several extensions to mobile IP are being considered to support micro-mobility and low latency, low-loss handoffs in wireless networks. Examples include the aforementioned HAWAII, Cellular IP, and HMIP [6]. We use the framework of HMIP with route optimization in our mobility management proposal. While we have chosen HMIP as our framework, our enhancements to the mobility management can also be applied to other micro-mobility management protocols. HMIP is enhanced to support FCSS with seamless RLC transfers and header compression.

The proposed architecture is illustrated in Figure 12.2. A gateway foreign agent (GFA) is located at the boundary of the RAN and the Internet. Packets are hierarchically tunneled from a correspondent host (CH) to the BSs through the GFA. We define a network active set (NAS) consisting of the set of BSs between which the mobile can switch on a fast time scale. BSs are added or deleted from the NAS on a slow time scale, based on certain criteria for the link quality between the mobile and the added or deleted BS. Within the NAS, one of the BSs is called the primary foreign agent (PFA) while the other BSs are called the secondary foreign agents (SFA). Finally a database collecting all the user location information is connected or colocated with the GFA.

It is worth noting that PFA and SFA are logical entities for performing various network control functions and physically, they all are BSs. In the proposed architecture, one and only one BS in the active set is serving a mobile at a time and that is referred as the serving BS. As a result, it is possible that PFA, SFA, and the serving BS correspond to the same BS for a given mobile. Further, we note that different BSs can serve as the PFA for different mobiles, unlike in the traditional architecture where a single RNC performs the resource management for all mobiles. In addition, due to user mobility, different base stations may serve as the PFA for that particular user at different times during the connection. Thus the resource management function is distributed across the network. The NAS is similar to the active set in a code division multiple access (CDMA) system defined for soft handoff. However, in FCSS, only one BS within the active set transmits at any

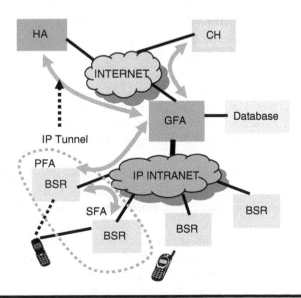

Figure 12.2 Mobility management structure. PFA, primary foreign agent; SFA, secondary foreign agent; GFA, gateway foreign agent.

given time, unlike in soft handoff for which several BSs simultaneously transmit to and receive from the same user. Nevertheless, all BSs in the NAS are assigned air-interface resources and maintain some RLC state information for the mobile so that they can immediately transmit if they become the serving BS.

The main features of the proposed architecture include:

1. The radio network control functionalities such as call admission control, and paging control, are distributed to the different BSRs in the network.
2. IP is used as the transport protocol to carry all data and signaling traffic between the different nodes.
3. Maintaining a GFA as in HMIP at the root of the domain which serves one or more than one location or paging area.
4. Maintaining a NAS of BSs for enabling FCSS for each mobile terminal.
5. A PFA serves as the mobility anchor and the PPP/PDCP initiation/termination point. Header compression, if enabled, is implemented at the PFA.
6. PFA multicasts forward-link user data to all the SFAs or selectively a subset of SFAs in the NAS. The subset can be chosen dynamically and intelligently by the PFA based on system loading, channel characteristics, and terminal mobility pattern.
7. Separate transmitting and receiving BSs for any given mobile: in general the cell switching can be independent on the forward and reverse links as the BS to which the mobile has the best channel quality on the forward and reverse links need not be the same on both links.
8. Packet forwarding mechanisms as in MIP route optimization [10] to ensure smooth relocation of the PFA as the mobile moves through the network. Additionally, whenever feasible, the PFA relocation is implemented in an opportunistic way when the mobile is in the dormant state and thus packets are not buffered in the network.
9. Maintaining Radio Resource Control (RRC) and call-processing signaling between the mobile and the network to enable mobility management without any layer-3 messages. This requires introduction of proxy registration messages into MIP so that the BSR can register with the home agent on behalf of the mobile node.
10. QoS for control signals and relevant data transfers on the backhaul network to enable FCSS using quasi-static multi-protocol label switching (MPLS) paths between BSs as described in more detail in Section 12.4.

We present a brief description of how a call is initiated and how it proceeds in the proposed architecture. Consider a mobile that powers up in the vicinity of a set of BSs. The mobile acquires the pilot signals from the BSs and uses the access channel to communicate with the BS from which it received the strongest signal to initiate a session. The BSR that receives the mobile's signal then performs admission control, and, if it admits the user, establishes resources for the mobile.

This BSR is designated to be the PFA for this mobile. Assume for the purpose of illustration that the mobile already has an IP address assigned to it that is topologically valid in the current network where it powered up. (If not, the mobile can then obtain a topologically valid IP address through a local DHCP and this MIP signaling will not be required.) The mobile then registers with this IP address to the BSR that in turn sends hierarchical MIP proxy registration messages to the GFA and the home-agent (HA) of the mobile. When the mobile receives a neighboring BSR pilot with signal strength above a certain threshold, it sends an RRC signal to request the addition of this BS to the NAS. This RRC signal is processed by the PFA; the PFA then adds the indicated BSR to the NAS and configures it as an SFA by HMIP proxy registration and response messages (with the mobile as the source and PFA as the next-level foreign agent). A correspondent host (CH) that sends a packet addressed to the mobile is first routed to the home network and intercepted by the HA. The HA then tunnels the packet to the GFA which in turn tunnels the packet to the PFA. The PFA performs header compression (if enabled) and forwards the packet to the serving BS for transmission over the air. As the mobile moves or as the signal strength changes, the serving BS can change rapidly according to the FCSS protocol described in Section 12.3. Periodically as the mobile moves over a larger distance, the PFA is relocated using context transfer protocols.

12.3 FCSS Protocol

The main objective of FCSS is to track channel fading of multiple cell sites and select the cell with the best channel quality to serve a terminal, thereby achieving cell site diversity and higher link throughput. FCSS is thus similar to hard handoff between BSs in that, at any given time, there is exactly one serving BS that supports a given terminal. However, the principal difference between FCSS and hard handoff is that for each terminal in the FCSS scheme, we propose to maintain an active set of BSRs that include the BSR with the best pilot SINR (as received at the terminal) and all other BSRs with link loss to the terminal within some threshold of that for the BSR with the best SINR. Because there is no active set for hard handoff, it takes much longer to switch from one BS to another. For the forward link, FCSS theoretically could lead to a substantial improvement in capacity compared with conventional hard handoff.

12.3.1 Control and Trigger of FCSS

We propose that each terminal is responsible for initiating cell switching for both forward and reverse links. Specifically, each terminal monitors channel quality for all links from the BSs in the active set, selects the best one according to pre-specified criteria, and broadcasts the selected cell identity via reverse link signaling. To reduce unnecessary cell switching (i.e., ping-pong effects), it is desirable to use a form of time-averaged channel quality to select the best BS.

Note that this approach has its shortcoming; namely, the switching decision by a terminal is based on channel quality without considering any other factors such as traffic loading and resource consumption. Utilizing a central controller (e.g., PFA) for decision-making would allow dynamic load balancing and cell coordination. However, it suffers from high delay in switching between cell sites due to excessive signaling. Hence, in addition to initiating cell switching by terminals, we also propose that, for example in case of traffic overload at the new cell site selected by a terminal, the PFA, upon receiving such a signaling message, can signal the terminal to cancel the chosen cell switch. In fact, this cancellation step is similar to that in the current systems. If the initial selection is rejected, it is envisioned that the terminal would select the next best cell site according to the specified criterion and inform the PFA of its selection. Alternatively, rather than potentially going through this iterative procedure for selecting the cell site, the terminal could provide an ordered list of cell sites to the PFA and let the PFA make the selection based on network-wide criteria, such as traffic loading.

In general, two different BSRs can be selected as the serving BSs for the forward and reverse links for a given terminal. As the channel qualities of the forward and reverse links typically fluctuate independently of each other in time, such a selection allows us to realize the FCSS gain to its fullest extent. Nevertheless, a single active set can be maintained for both forward and reverse links for operation simplicity. Because the terminal can monitor the channel quality for all forward links from the BSs, it is natural for the terminal to determine the best serving BS for its forward link. Then the selection is forwarded to the PFA to finalize the decision. As for the reverse links from the terminal, involved BSs have to assess the channel quality and decide among themselves (via a central decision-making entity such as the PFA) upon the best cell serving the reverse link for the terminal. On the other hand, to simplify the design and to reduce associated signaling overhead, a network designer may prefer to use the same BS to serve both links for a terminal. In this case, each terminal has one single active set of BSs to serve both its forward and reverse links. The best combined channel quality for both forward and reverse links is used as a criterion in determining the serving BS for a given terminal. Examples of such combined quality include: (i) a weighted sum of time-averaged pilot SINR for forward and reverse links and (ii) the minimum time-averaged pilot SINR for forward and reverse links between the terminal and a BS in the active set. As a terminal can readily assess the quality of all its forward links from various BSs, with information about reverse-link quality received from the BSs, a terminal can then select its best cell based on the specified criterion.

12.3.2 Operation Timeline of FCSS

We now describe the FCSS operation timeline, shown in Figure 12.3 and which is applicable to both forward and reverse links. Each mobile alternates between two operating states: active and suspending state. When a mobile is in the active state (represented by a white rectangle), it can decide, according to prespecified criteria such as signal strength, upon the best cell site (i.e., BSR) from which to receive data (respectively the best cell site to transmit data to). If the best cell site is different from the current serving cell, the decision is made to switch to a new cell, referred to as FCSS decision. The mobile leaves the active state and enters the suspending state (represented by a gray rectangle). During the suspending state, the mobile constantly monitors the channel quality to each BSR in the network active set and collects channel statistics. The duration of this state, referred to as suspension time, is a tunable parameter but should be long enough to provide adequate averaging of channel quality to avoid excessive and inaccurate cell switching. In a way, the tunable duration controls the speed and aggressiveness of the FCSS procedure. That is, a long suspension time provides stable estimation of the averaged channel quality but becomes less capable of tracking fast channel variations and utilizing cell-site diversity. A short suspension time, on the other hand, could lead to spurious cell switches. The choice of the appropriate suspension time depends on hardware implementations, mobile speed, quality of channel estimation as well as delay characteristics of the backhaul network. For low mobility and instantaneous channel estimation and prediction, it is feasible to track channel variations and switch between cells for capacity gain without overburdening mobile hardware. As the mobile's speed increases, the gain of FCSS diminishes as FCSS can no longer track fast channel variations for diversity gain.

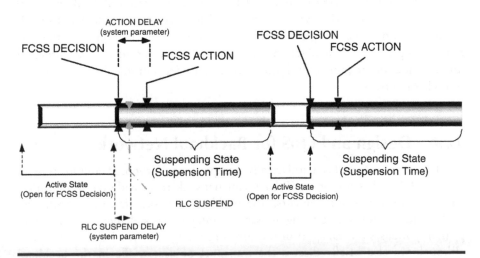

Figure 12.3 Fast cell-site selection (FCSS) operation timeline.

The operation of switching the mobile from one cell site to another is referred to as FCSS action. Ideally, each FCSS decision is immediately followed by the subsequent FCSS action. However, this is impossible in practical systems due to non-negligible signaling delays and hardware limitations. The time gap between an FCSS decision and its associated FCSS action is referred to as the action delay. A short action delay is beneficial to realize the gain of FCSS and to preserve the accuracy of the FCSS decision at the time of FCSS action.

After an FCSS decision, the present serving BS and the next serving BS should prepare for the FCSS action. Aside from being ready for transmission, one important factor is data and state synchronization at the Medium Access Control (MAC) level between the two BSRs, so as to avoid unnecessary packet losses or duplicate packet transmissions. One approach to achieve that, assuming the action delay is relatively small compared to the suspension time, is to suspend data transmission at both the MAC and RLC layers shortly after each FCSS decision and resume data transmission at the time of FCSS action. Negative impacts due to such suspension are expected to be small for cells serving multiple users as BSs could utilize this time period to serve other users. This is especially true for systems employing channel-aware scheduling algorithms. Furthermore, the system should be designed such that the time duration of transmission suspension due to FCSS is not much longer than action delay, which is the minimum amount of time to complete the FCSS operations. This is particularly so for scenarios having the same BS to serve forward and reverse links for a given terminal where both transmissions on both links are halted by the suspension.

As part of the FCSS action, the MAC protocols at both the serving BS and terminal are reset and a status report for each RLC entity associated with the terminal is generated and forwarded to the next serving BS to prepare for FCSS action. As the new serving BS does not have any information of MAC protocol state at the current serving BS, resetting MAC protocol state would avoid signaling overhead of transferring the entire MAC protocol information from the current serving BS to the new serving BS. A reset of the terminal MAC protocol however does not require to flush the reordering buffers but to deliver the content to higher layers.

12.4   Design and QoS for Backhaul Network

As IP networks have been universally accepted, the proposed architecture in Figure 12.2 makes use of an IP backhaul network to provide communication among various network elements. The backhaul network transports both user traffic, and control and signaling messages among network elements. This section primarily focuses on transport of the control messages for the FCSS operations, although we also briefly discuss the transport of user traffic in the backhaul network.

Despite abundant availability of IP network equipments, the proposed network architecture imposes a key challenge for the underlying IP network. On one hand, as the proposed architecture does not require its BSRs to be synchronized, the backhaul network can have relaxed QoS requirements for data transport in the proposed network. On the other hand, however, to support FCSS for improved network performance, associated signaling and control messages have very stringent requirements for packet delay, throughput, and packet-loss probability. Thus, the IP backhaul network has to provide adequate QoS to support such control message exchanges among various network elements. In this section, we examine several possible approaches and discuss their feasibility, and associated tradeoffs for application in the considered context. We also show that control traffic associated with FCSS is "smooth," thus verifying the feasibility of the proposed architecture without an excessive amount of bandwidth, as would be needed for very bursty control traffic in the backhaul network.

In general, there are two approaches for QoS in IP networks. The first approach is at the IP layer, namely, by use of Integrated Services (IntServ) and Differentiated Services (DiffServ) [11], while the second one is at layer-2, by use of Multi-Protocol Label Switching (MPLS) [12]. At a high level, IntServ uses Resource Reservation Protocol (RSVP) for explicit signaling and dynamic allocation of resources along the communication path of a given connection, as a means to guarantee end-to-end performance. Evidently, if there are frequent changes for a connection (e.g., due to handoff), the signaling overhead can be so significant that IntServ becomes unattractive in practice. On the other hand, DiffServ provides a number of service classes such as premium, assured, and best-effort classes. For DiffServ, packets are marked and classified at the edge router. Typically, QoS is provided by the router's scheduling mechanism on a hop-by-hop basis. Consequently, DiffServ does not guarantee absolute delay or throughput performance, but rather provides relative performance differentiation among various service classes. On the other hand, MPLS requires setting up a label-switch path (LSP) between each pair of network elements and the allocated bandwidth can be guaranteed along the entire path. Thus, desirable QoS for a prespecified offered traffic load can be achieved by proper bandwidth dimensioning of LSPs.

In terms of capability, IntServ could be appropriate for achieving the stringent QoS requirements for the FCSS operations. It is particularly so because the topology of the RAN does not change often. As far as control message exchanges among various network elements are concerned, the associated connections remain relatively static in time and the overhead associated with IntServ for connection changes can thus be avoided. Unfortunately, to our knowledge, IntServ has not been widely implemented to this date. Combining this with the fact that DiffServ cannot guarantee delay or throughput performance—a key requirement for the FCSS, we propose to use the MPLS approach for transport of control and signaling messages with appropriate network dimensioning to achieve the QoS requirements imposed by the FCSS operations.

Basically, one LSP is set up for each pair of BSRs and between a BSR and another network element such as a GFA. Control traffic associated with all calls between a given pair of network elements is multiplexed onto the single LSP between the corresponding network elements. Identity information for each call is included in the higher protocol layers and resolved at the receiving end. Signaling traffic load between each pair of network elements is estimated based on the expected size of control messages and frequency of such messages. In turn, the messaging frequency depends on call distribution, mobility characteristics, and radio conditions in neighboring cells in the actual deployed network. The signaling traffic load can be specified in terms of average, peak, or equivalent bandwidth [13] requirement for each pair of network elements. Based on the traffic load estimates, we propose applying existing tools [14] to obtain the set of required LSPs such that their allocated bandwidths are guaranteed. The actual setup of LSPs in real networks is achieved by the RSVP-TE protocol [12]. Existing MPLS dimensioning tools do not use target end-to-end delay as performance requirement for generating LSPs with guaranteed bandwidth allocation along the communication path. Thus, strictly speaking, end-to-end delay performance is not guaranteed in the MPLS network. However, with adequate bandwidth reserved for each LSP, QoS for control traffic can be satisfied with a high degree of confidence. Relatively speaking, it is more difficult for the DiffServ approach to provide such required QoS by prioritized packet scheduling on a hop-by-hop basis. As pointed out earlier, this approach is feasible particularly because the network topology does not change often. In fact, for best network performance, the MPLS network may be redimensioned occasionally or periodically based on actual traffic measurements, to adapt to slow changes or the periodic nature of traffic demands.

Based on simulation results, we show in Section 12.6 that control traffic associated with the FCSS operations is fairly smooth and does not exhibit a lot of burstiness. Thus, the MPLS approach with proper backhaul network dimensioning can meet the stringent delay requirement for FCSS. We have therefore demonstrated the feasibility of the proposed architecture to transport FCSS control traffic without the use of an excessive margin of backhaul bandwidth to handle traffic burstiness.

Now, let us comment on the backhaul network for the transport of user traffic. First, even if IntServ were commonly available in IP networks, we do not view it as an appropriate candidate for guaranteeing QoS for user traffic. This is so because IntServ incurs high signaling overhead and delay for connection setup and modification that cannot meet the needs of fast and frequent handoff and FCSS in the wireless networks. Second, although we have proposed to use the MPLS approach to meet the stringent QoS for control traffic, it is not immediately clear to us whether MPLS also represents the most suitable way to achieve QoS for user traffic. Indeed calls (especially those located at cell boundaries) under FCSS may possibly be bounced back and forth among a few BSs according to the fluctuation of link quality. In such cases, as a call is switched from one BS to the next, the RSVP-TE protocol has to be utilized to modify the associated LSPs. Rapid changes of BS under FCSS may incur unacceptable protocol overhead and delay, as LSPs have to be set and reset whenever the mobile terminal switched between BSRs.

On the other hand, we consider the widely implemented DiffServ service as a candidate for transport of user traffic in the backhaul network. Of course, the DiffServ approach requires a centralized bandwidth broker to serve as call admission control for ensuring QoS, which introduces additional complexity. Such a centralized broker may potentially be placed at the GFA. The tradeoff between the MPLS and DiffServ approaches for transport of user traffic in the context of FCSS is a subject for future investigation.

12.5 Conclusions

We have proposed a distributed architecture for all-IP wireless networks such as WiMAX. Functionalities associated with the FCSS are distributed among BSRs for improved network performance. In essence, the proposed architecture together with FCSS provides a unified air-interface and network architecture for supporting multimedia IP applications. We have also proposed a set of protocols and use of MPLS in the backhaul network to support FCSS operations. The architecture has the advantages of improved scalability, reliability, and reduced backhaul latencies and also provides cost savings because of the all-IP unified structure. In our internal simulation studies, we have verified the feasibility and the merits of the proposed architecture for the typical network condition.

Acknowledgments

The authors express their sincere thanks to Ajay Rajkumar, Dimitri Stiliadis, and T. V. Lakshman for their helpful discussion.

References

1. D. Wisely, P. Eardley, and L. Burness, *IP for 3G,* John Wiley & Sons Ltd., 2002.
2. S. Uskela, "Key concepts for evolution toward beyond 3G networks," *IEEE Wireless Communications Magazine,* vol. 10, no. 1, pp. 43–48, February 2003.
3. M. S. Corson, R. Laroia, A. O'Neill, V. Park, and G. Tsirtsis, "A new paradigm for IP-based cellular networks," *IT Pro,* pp. 20–29, November–December 2001.
4. J. Kempf and P. Yegani, "OpenRAN: a new architecture for mobile wireless internet radio access networks," *IEEE Communications Magazine,* May 2002, pp. 118–123.
5. Y. Cheng and W. Zhuang, "DiffServ Resource Allocation for Fast Handoff in Wireless Mobile Internet," *IEEE Communications Magazine,* May 2002, pp. 130–136.

6. E. Gustafsson, A. Jonsson, and C. E. Perkins, "Mobile IPv4 regional registration," Internet-draft, draft-ietf-mobileip-reg-tunnel-07.txt (work in progress) October 2002.

7. R. Ramjee, K. Varadhan, L. Salgarelli, S. Thuel, S. Wang, and T. LaPorta, "HAWAII: a domain-based approach for supporting mobility in wide-area wireless networks," in *IEEE/ACM Transactions on Networking*, vol. 10, no. 3, pp. 396–410, 2002.

8. A. Campbell, J. Gomez, S. Kim, A. Valko, C. Wan, and Z. Turanyi, "Design, implementation, and evaluation of Cellular IP," *IEEE Personal Communications Magazine*, August 2002, pp. 42–49.

9. IST EU Project BRAIN (Broadband Radio Access over IP) Deliverable 2.2, "BRAIN architecture specifications and models, BRAIN functionality and protocol specification," available at www.ist-brain.org.

10. C. Perkins, ed., "IP mobility support for IPV4, revised," IETF RFC 2002, September 2001.

11. J. F. Kurose and K. W. Ross, *Computer Networking: A Top-Down Approach Featuring the Internet*, Addison Wesley, New York , 2001.

12. S. Harnedy, *The MPLS Primer: An Introduction to Multiprotocol Label Switching*, Prentice Hall, NJ, 2002.

13. R. Guerin, H. Ahmadi, and M. Naghshineh, "Equivalent capacity and its application to bandwidth allocation in high-speed networks, *IEEE JSAC*, vol. 9, pp. 968–981, September 1991.

14. M. Aukia, P. V. Kodialam, T. V. Koppol, H. Lakshman, B. Sarin, and B. Suter, "RATES: a server for MPLS traffic engineering," *IEEE Network Magazine*, 2000.

Chapter 13

Centralized Scheduling Algorithms for 802.16 Mesh Networks

Petar Djukic and Shahrokh Valaee

13.1 Introduction

IEEE 802.16 protocol [1,2] specifies two different modes of operation. The first mode of operation is the point-to-multipoint (PMP) mode. In the PMP mode, each 802.16 access point has a dedicated broadband connection to the Internet. Wireless terminals connect to the access points on their first hop and their traffic goes to the Internet through the access point's broadband connection. The second mode of operation is the mesh mode, where access points are interconnected with wireless links. In the mesh mode, wireless terminals connect to the access points on their first hop. Then, the wireless mesh carries their traffic to the point-of-presence (POP) where it goes to the Internet (Fig. 13.1). Mesh networks decrease the cost of running the access points, as all access points in the same area share a single broadband connection.

Current mesh networks use 802.11a technology to interconnect the mesh backbone [3,4]. However, 802.11a technology is a decade old and was not designed for mesh networks. In particular, 802.11 Medium Access Control (MAC) lacks the

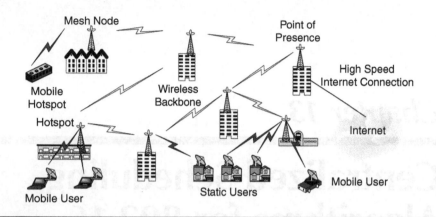

Figure 13.1 Mesh networks.

extensions to provide quality of service (QoS) in multihop wireless environments [5]. IEEE 802.16 introduces QoS in the mesh with time division multiple access (TDMA) MAC technology [1]. In 802.16, access points negotiate their end-to-end bandwidth with the POP. IEEE 802.16 MAC enforces end-to-end QoS by scheduling wireless transmissions on the links connecting the access point to the POP.

IEEE 802.16 mesh protocol specifies two TDMA scheduling protocols: centralized and decentralized scheduling protocols. The Base Station (POP) uses the centralized scheduling protocol to establish networkwide schedules. In the centralized scheduling protocol, mesh nodes request bandwidth from the Base Station (BS). The BS uses the end-to-end bandwidth requests to assign bandwidth to every link in the network, so that the links form a routing tree terminating at the BS. The BS then multicasts the assignment to all mesh nodes. Mesh nodes use the link bandwidth assignment and a common scheduling algorithm to determine a global transmission schedule. In the decentralized scheduling protocol, mesh nodes negotiate bandwidth assignments with their neighbors without centralized coordination.

This chapter reviews the 802.16 centralized scheduling protocol and research on scheduling algorithms for this protocol. Section 13.2 reviews the 802.16 mesh protocol, including the centralized and decentralized scheduling protocols. Section 13.3 gives a detailed description of the 802.16 centralized scheduling protocol. Section 13.4 reviews the current research in centralized scheduling algorithms and compares several published algorithms with each other.

13.2 802.16 Medium Access Control

In this section, we first describe the physical (PHY) layer used in 802.16 and compare it with the PHY layer used in 802.11a. Both PHY layers use orthogonal frequency division multiplexing (OFDM) in the 5-GHz frequency band. We then describe the 802.16 MAC and its centralized and decentralized scheduling protocols.

13.2.1 802.16 PHY Layer

IEEE 802.16 is a TDMA-based MAC protocol [1], built on an OFDM PHY layer. OFDM transforms blocks of bits into constant duration symbols carried on a set of frequency orthogonal pilot carriers. In 802.16, OFDM symbols are grouped into frames of equal length and frames repeat over time (Fig. 13.2). The duration of OFDM symbols, T_s, depends on the total bandwidth occupied by the carriers: at 20-MHz bandwidth $T_s = 12.5$ μs and at 10-MHz bandwidth $T_s = 25$ μs. As OFDM symbol duration for 10-MHz bandwidth is twice as long as the OFDM symbol duration for 20-MHz bandwidth, the raw bit-rate at 10-MHz bandwidth is half of the raw bit-rate at 20-MHz bandwidth (Table 13.1).

IEEE 802.16 uses OFDM in the 5-GHz frequency band. 802.16 can use hardware with 10-MHz total bandwidth in the licensed frequency bands and hardware with 20-MHz bandwidth in the license-exempt frequency band. IEEE 802.11a PHY layer uses OFDM in the 5-GHz license-exempt frequency band with hardware operating at 20-MHz bandwidth. Table 13.1 compares 802.16 raw bit-rate at 10- and 20-MHz bandwidths to 802.11a raw bit-rate. The two PHY layers have comparable raw bit-rates, however 802.16 provides a TDMA MAC layer, which guarantees QoS [6,7].

13.2.2 802.16 Framing

IEEE 802.16 frames OFDM symbols for two reasons. First, the frame boundaries are used to synchronize the mesh. Second, the frame structure allows the division of control and data traffic into subframes (Fig. 13.2). The control subframe is divided into transmission opportunities, which are seven OFDM symbols long.

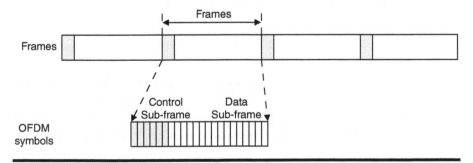

Figure 13.2 **802.16 time division multiple acess (TDMA). OFDM, orthogonal frequency division multiplexing.**

Table 13.1 Comparison of 802.11a and 802.16 Bit-Rates

Modulation	Data Bits/Symbol		Bit-Rate Mbits/Second		
	802.11	*802.16*	*802.11*	*802.16 10 MHz*	*802.16 20 MHz*
BPSK-1/2	24	96	6.0	3.84	7.68
BPSK-3/4	36	X	9.0	X	X
QPSK-1/2	48	192	12.0	7.68	15.36
QPSK-3/4	72	288	18.0	11.52	23.04
16QAM-1/2	96	384	24.0	15.36	30.72
16QAM-3/4	144	576	36.0	23.04	46.08
64QAM-2/3	192	768	48.0	30.72	61.44
64QAM-3/4	216	864	54.0	34.56	69.12

In each control subframe transmission, three OFDM symbols are guard symbols and four OFDM symbols are used for data transmissions at the lowest rate of BPSK-1/2. The size of the 802.16 control subframe is 7×MSH–CTRL–LEN OFDM symbols, where MSH-CTRL-LEN is a parameter specified by the network operator. Details of the coordination mechanism in the control subframe are available in the 802.16 standard [1], as well as in our summary of the standard [2].

Data transmissions have timing similar to the timing of control packets; each data transmission has two or three OFDM guard symbols and a variable number of data-carrying OFDM symbols. However, 802.16 mandates that the total number of OFDM symbols in each data subframe transmission be a multiple of the number of slots in one data subframe transmission opportunity. The number of OFDM symbols in each data subframe transmission opportunity is determined by dividing the number of OFDM symbols in the data subframe with 256:

$$\texttt{DataTxOppSize} = \left\lceil \frac{T_f/T_s - 7 \times \texttt{MSH-CTRL-LEN}}{256} \right\rceil, \qquad (13.1)$$

where $\lceil \cdot \rceil$ is the ceiling function, T_f is the duration of the 802.16 frame, and T_s is the OFDM symbol duration. However, there may be fewer than 256 transmission opportunities in the data subframe, depending on the frame size and the size of the control subframe. The actual number of transmission opportunities in the data subframe is:

$$\texttt{DataTxOppNum} = \left\lfloor \frac{T_f/T_s - 7 \times \texttt{MSH-CTRL-LEN}}{\texttt{DataTxOppSize}} \right\rfloor. \qquad (13.2)$$

For example, if the frame duration is $T_f = 10$ ms, the bandwidth of the OFDM is 20 MHz ($T_s = 12.5$ μs) and MSH–CTRL–LEN 10, the size of a data transmission opportunity is DataTxOppSize = 3 and there are DataTxOppNum = 243 transmission opportunities in the data subframe. The standard restricts the number of transmission opportunities in the data sub-frame to at most 256 because the duration fields in the scheduling control packets are eight bits long.

The guard symbols and the transmission opportunity size affect the granularity of bandwidth assignments. For example, if transmission opportunities are three OFDM symbols long and each transmission uses three guard symbols, the smallest data packet size at BPSK-1/2 modulation is 36 bytes in two transmission opportunities (one transmission opportunity is the overhead of three OFDM symbols), the next size for a data packet is 72 bytes in three transmission opportunities and then 108 bytes in four transmission opportunities. If the frame size is $T_f = 10$ ms, this granularity implies that bandwidth should be assigned in the increments of 28.8 kbps (36 bits = 288 bits).

13.2.3 Assignment of Transmission Opportunities

IEEE 802.16 specifies two different protocols used to negotiate assignment of transmission opportunities in the data subframe: centralized and decentralized scheduling protocols. The centralized scheduling protocol assigns transmission opportunities in the first MSH–CSCH–DATA–FRACTION × DataTxOppNum transmission opportunities of the data subframe, where MSH–CSCH–DATA–FRACTION is a percentage of data subframe transmission reserved for centralized scheduling, specified by the network operator. The decentralized scheduling protocol assigns transmission opportunities in the last (1–MSH–CSCH–DATA–FRACTION) × DataTxOppNum transmission opportunities of the data subframe.

In the centralized scheduling protocol, mesh nodes monitor the traffic from their wireless terminals and use this information to request end-to-end bandwidth from the BS. The BS uses the end-to-end bandwidth requests from the mesh nodes and assigns link bandwidths to all links in the network. The assignment of link bandwidths creates a routing tree in the mesh. After assigning link bandwidths, the BS multicasts the assignment through the mesh. The mesh nodes use the link bandwidth assignment to determine all starting times and link durations in the frame (a common, global, schedule). We describe the centralized scheduling protocol in detail in the next section.

In the decentralized scheduling protocol, neighboring mesh nodes negotiate local schedules. First, a node wishing to change its allocation of transmission opportunities sends a request for transmission opportunities to its neighbors. One or more of the neighbors correspond with a range of available transmission opportunities. The node chooses a subrange of these transmission opportunities and confirms that it will use them with a third message. The 802.16 standard does not specify any algorithm for the distributed scheduling protocol. We have proposed a distributed scheduling algorithm [8] that can be adapted for this purpose. Our algorithm uses a distributed Bellman-Ford algorithm to iteratively find the TDMA schedule from link bandwidths. The algorithm determines TDMA schedules with a partial knowledge of the network topology, available from 802.16 neighbor tables.

The centralized and distributed scheduling give rise to two different QoS levels in the mesh network. Links scheduled with the centralized scheduling protocol have guaranteed bandwidth, granted by the BS and known throughout the network. The hop-by-hop bandwidth guarantee in the centralized scheduling routing tree allows end-to-end QoS guarantees for the traffic flows traversing the tree. On the other hand, links scheduled with the decentralized scheduling protocol have transient behavior and their bandwidth depends on the grants from the node's neighbors. The uncertainty in link bandwidth translates to the best-effort QoS for end-to-end connections using the links scheduled with the decentralized scheduling protocol. We have shown how to take advantage of the QoS provided by 802.16 with network layer solutions in [2]. In that work, we propose a convergence sublayer, connecting the network layer, and the 802.16 MAC layer, that allows the Internet Protocol (IP) DiffServ architecture [9] to take advantage of the two 802.16 QoS service levels.

Finding the minimum number of centralized scheduling transmission opportunities in the data subframe is an important provisioning question for 802.16 mesh networks. MSH-CSCH-DATA-FRACTION should be minimized so that as much bandwidth as possible be available for best-effort traffic and enough bandwidth can be allocated for the services requiring guaranteed bandwidth. We have proposed an algorithm that minimizes the number of slots needed to schedule the traffic in the centralized scheduling part of the data frame [10]. The algorithm finds the smallest value of MSH-CSCH-DATA-FRACTION that supports the link bandwidths, subject to the limit on the maximum TDMA delay in the network. TDMA delay occurs when an outgoing link on a mesh node is scheduled to transmit before an incoming link in the path of a packet [11].

13.3   802.16 Centralized Scheduling

In this section, we discuss the 802.16 scheduling protocol in detail. First, we describe the 802.16 mesh control packets of the BS and the mesh nodes use to negotiate end-to-end bandwidths. Then, we propose a framework for bandwidth assignment. We compare several scheduling algorithms with each other in the next section.

13.3.1   802.16 Centralized Scheduling Protocol

The centralized scheduling protocol uses MSH-CSCH and MSH-CSCF mesh control packets. The BS uses MSH-CSCF packets to distribute the network topology and MSH-CSCH packets to assign end-to-end bandwidth. Mesh nodes use MSH-CSCH packets to request end-to-end bandwidth.

Centralized scheduling MSH-CSCH and MSH-CSCF packets are transmitted in the control subframe [1,2]. Scheduling of MSH-CSCH and MSH-CSCF control packets is performed by following a breadth-first traversal of a globally known tree topology. The BS starts by broadcasting the global tree topology with MSH-CSCF packets. When a node receives a MSH-CSCF message, it learns the multicast routing tree, as well as which node is broadcasting the MSH-CSCF packet. Given this information, the node calculates when it should transmit its MSH-CSCF packet. After the last node in the tree receives the MSH-CSCF packet from its parent, all nodes know the network topology, as well as the transmission schedule for MSH-CSCH and MSH-CSCF packets. If the topology changes, subsequent MSH-CSCF messages notify the mesh nodes of the changes.

The nodes request bandwidth from the BS by sending MSH-CSCH messages to their parents in the tree. After all of the requests reach the BS, the BS uses them to calculate the bandwidth for each link in the network and multicasts granted end-to-end bandwidths to mesh nodes with new MSH-CSCH messages. If by changing the link bandwidths, the BS also changes the routing tree for the network, it multicasts routing changes with MSH-CSCF messages before it multicasts MSH-CSCH messages. When a mesh node receives a MSH-CSCH message, it uses the routing tree, known from prior MSH-CSCF messages, and the assignment of end-to-end bandwidths to find the assignment of link bandwidths for the entire network. Each mesh node uses the link bandwidth assignment to find a global transmission schedule. The new schedule takes place in the first frame after the last node in the tree receives the MSH-CSCH message from its parent.

The 802.16 centralized scheduling protocol forces three requirements on centralized scheduling algorithms. First, the assignment of link bandwidths should result in a tree whose root node is the BS. Second, link bandwidths should be assigned so that end-to-end bandwidth requests are satisfied. If the end-to-end bandwidth requests are not fully satisfied, the assignment of end-to-end bandwidths should be fair. Third, the number of transmissions for each link should be limited to one per frame as the overhead of each transmission is large.

The breakdown of algorithm requirements gives a convenient way to split the scheduling problem. First, the BS decides on a tree of links that should carry the traffic. Then, the BS assigns link bandwidths using the routing tree and the end-to-end bandwidth requests. The BS multicasts the routing tree with MSH-CSCF messages and end-to-end bandwidth grants with MSH-CSCH messages. After receiving the latest tree topology and end-to-end bandwidth grants, mesh nodes find link bandwidth assignments for the entire network. The BS and the mesh nodes use the same algorithm to assign link bandwidths from end-to-end bandwidths. Finally, each mesh node uses a scheduling algorithm to find a global transmission schedule from the link bandwidths and the network topology (the routing tree). The scheduling algorithm decreases link bandwidths, if necessary, to obtain a valid schedule.

Next, we give a simple method for the centralized assignment of link bandwidths. The BS uses this method to find link bandwidths from a routing tree and a set of mesh node bandwidth requests. The mesh nodes use this method to find link bandwidths from the routing tree and the set of mesh node bandwidth grants received in the MSH-CSCH messages. We review several scheduling algorithms in the next section.

13.3.2 Centralized Bandwidth Assignment

IEEE 802.16 standard uses an example to describe how to find link bandwidths from end-to-end bandwidths. In this section, we propose a formal method to find link bandwidth assignments, after the BS receives end-to-end bandwidth requests. We also provide a reverse of the method to find end-to-end bandwidths given a 802.16 TDMA schedule. We use the reverse method in the next section to compare the performance of 802.16 scheduling algorithms.

We model the mesh network with a topology graph connecting the nodes in the wireless range of each other. We assume that if the two nodes are in the range of each other, they establish mutual links in the MAC layer, so the mesh can be represented with a connectivity graph $G(V, E, f_t)$, where $V = \{v_1, ..., v_n\}$ is the set of nodes, $E = \{e_1, ..., e_m\}$ are directional links between neighboring nodes, and $f_t: E \rightarrow V \times V$ assigns links to pairs of nodes. The connectivity map f_t enforces the fact that all links are directional, so for a link $e_k \in E, f_t(e_k) = (v_i, v_j)$ means that node v_i uses link e_k to transmit data to node v_j.

We model channel quality with link bit-rates. Bit-rates depend on the modulation, which depends on signal-to-noise ratio (SNR) for the link. The SNR is divided into several discrete levels and each is associated with its maximum bit-rate. We define link bit-rate as the number of bits transmitted in an OFDM symbol, which is represented with $b: E \rightarrow \{96, 192, ..., 864\}$, where 96 is the number of bits carried in a symbol with the BPSK-1/2 modulation and 864 is the number of bits carried in a symbol with 64 QAM-3/4 (Table 13.1).

The routing algorithm chooses a subset of links, forming two spanning trees in the network. The first tree is associated with the uplink traffic and it consists of the links carrying the uplink traffic (to the BS). The second tree is associated with the downlink traffic and contains links carrying downlink traffic (to the mesh nodes). Each mesh node is connected to the BS with two, acyclic, directed paths. The first path is the uplink path directed from the mesh node to the base station and the second path is the downlink path directed from the BS to the mesh node. The uplink and the downlink paths are represented with separate sets: $\mathcal{P}^{up} = \{P_2^{up}, ..., P_n^{up}\}$ and $\mathcal{P}^{down} = \{P_2^{down}, ..., P_n^{up}\}$ where there are $(n-1)$ paths on the uplink, one for each node, and $(n-1)$ paths on the downlink, also one for each node. We use the convention that v_1 is the BS. There are no paths to the BS as the traffic from its wireless terminals does not traverse the wireless mesh. Each path is a set of links that the packets use to traverse the mesh, so $\forall P_i^{up} \in \mathcal{P}^{up}, P_i^{up} \subseteq E$ and $\forall P_i^{down} \in \mathcal{P}^{down}, P_i^{down} \subseteq E$.

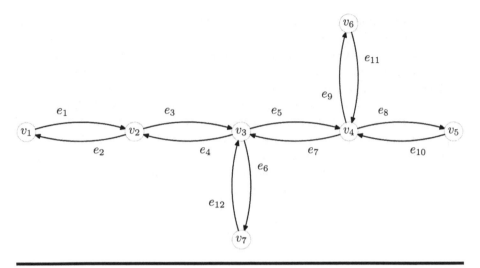

Figure 13.3   Simple example of a mesh network.

We represent the node uplink and downlink demands, in bits per second, with $g^{up}: V \to \mathbb{R}$ and $g^{down}: V \to \mathbb{R}$, respectively.* Because each node corresponds to an uplink and a downlink path, the uplink and downlink requests of each node are associated with the uplink and downlink paths for the node. The required bandwidth on each link, $r: E \to \mathbb{R}$, can be found by adding up the traffic on each path traversing the link:

$$r_j = \sum_{i=2}^{n} g_i^{up} I(e_j \in P_i^{up}) + \sum_{i=2}^{n} g_i^{down} I(e_j \in P_i^{down}), \qquad (13.3)$$

where $I(\cdot)$ is the indicator function, which is 1 when its argument is true and 0 when its argument is false.

Because links are unidirectional, uplink and downlink trees are disjoint, hence links never carry both the uplink and the downlink traffic. This means that one of the sums in Equation 13.3 is always zero. For example, for the topology in Figure 13.3, $r_4 = g_3^{up} + g_4^{up} + g_5^{up} + g_6^{up} + g_7^{up}$ and $r_3 = g_3^{down} + g_4^{down} + g_5^{down} + g_6^{down} + g_7^{down}$.

Given the required rates on each link, the number of OFDM symbols the link uses in the frame is:

$$d_j = \left\lceil \frac{r_j T_f / b_j + h}{\texttt{DataTxOppSize}} \right\rceil \times \texttt{DataTxOppSize}, \qquad (13.4)$$

---

* Although we are using the notation that bandwidth demands real numbers, in reality the bandwidths are discrete as they must be rounded to fit into 802.16 MSH–CSCF packets.

where h is the number of overhead slots ($h = 2$ or $h = 3$), b_j is the number of bits in each OFDM symbol, T_f is the frame duration and the formula makes sure the number of slots is divisible by `DataTxOppSize`. The number of transmission opportunities required by the link in each frame is $d_j/$`DataTxOppsize`.

Ideally, a centralized scheduling algorithm assigns the number of OFDM symbols to each link corresponding to Equation 13.4. However, the scheduling algorithms we review in the following section do not always assign the exact number of required slots to the links. In such cases, it is useful to know the actual end-to-end bandwidth assigned to each mesh node given a specific schedule. We find the end-to-end rates in two steps. First, we find the actual rate assigned to link e_j, given the number of slots assigned to the link, d_j:

$$\hat{r}_j = (d_j - h)\frac{b_j}{T_f} \tag{13.5}$$

Second, we use the actual link bandwidth, to find the achievable end-to-end rates. We assume that the nodes use a mechanism such as weighted fair queueing (WFQ) [12]. With WFQ, all end-to-end connections can get a proportional share of bandwidth on the links they traverse. For example, if the end-to-end flows are weighted according to their requested bandwidth, the share of uplink bandwidth that node v_i gets on link e_j is $g_i^{\text{up}} \hat{r}_j/r_j$ as r_j is the total bandwidth of all connections traversing the link, Equation 13.3. Similarly, the share of downlink bandwidth that node v_i gets on link e_j is $g_i^{\text{down}} \hat{r}_j/r_j$. The achievable end-to-end bandwidth is found by considering the minimum bandwidth the end-to-end connection gets on all the links on the path:

$$\hat{g}_i^{\text{up}} = \min_{e_j \in P_i^{\text{up}}} \frac{\hat{r}_j}{r_j} g_i^{\text{up}} \tag{13.6}$$

and

$$\hat{g}_i^{\text{down}} = \min_{e_j \in P_i^{\text{down}}} \frac{\hat{r}_j}{r_j} g_i^{\text{down}} \tag{13.7}$$

The reason for taking the minimum over all links in the path is to ensure that all queue lengths on the path are bounded.

13.4 Centralized Scheduling Algorithms

In this section, we compare three different 802.16 centralized scheduling algorithms with each other. The first algorithm is suggested in the 802.16 standard [1] and does not use spatial reuse to increase the capacity in the network. We derive the second scheduling algorithm from a load-balancing node scheduling algorithm proposed for 802.16 [13]. The new algorithm is equivalent to the original algorithm [13], when the uplink and downlink traffic are scheduled separately. However, our adaptation allows scheduling of uplink and downlink traffic at the same time, thus increasing spatial reuse. We propose the third algorithm based on the Bellman-Ford TDMA scheduling algorithm [11].

Each algorithm takes the number of OFDM slots each link should transmit in the frame (link durations) as the input and produces a transmission schedule. Link durations are readily available from the routing tree and end-to-end node bandwidths (Equations 13.3 and 13.4). The algorithms follow a common procedure to find a transmission schedule. First, each algorithm finds a link ranking, $R: E \rightarrow \mathbb{Z}$. The ranking determines the order of transmissions in the frame; links with a lower rank transmit before the links with a higher rank. For example, we obtained the schedule in Figure 13.4 with the 802.16 ranking mechanism, which ranks links according to a breadth-first traversal of the tree. The links in the figure have the following ranking: $R_1 = R_2 = 0$, $R_3 = R_4 = 1$, $R_5 = R_6 = R_{12} = R_7 = 2$, and $R_8 = R_9 = R_{10} = R_{11} = 3$. We elaborate on this ranking mechanism later in the section.

After an algorithm assigns link rankings, it proceeds to assign transmission opportunities to links. In the case of the 802.16 algorithm, the starting transmission opportunity of each link is the first unused transmission opportunity. In the case of the load-balancing algorithm, the starting transmission opportunity is the first transmission opportunity that allows the link to transmit without overlapping with any of its conflicting links. In the case of the Bellman-Ford algorithm, the starting transmission opportunity is based on the link's minimum distance in the conflict graph. The distance in the conflict graph is based on the ranking and the link duration, so links with a lower ranking transmit before the links with a higher ranking. The 802.16 algorithm and the Bellman-Ford algorithm also include a final step in which the link transmission times are scaled down to fit in the centralized scheduling part of the data subframe.

In the rest of the section, we use the network topology from Figure 13.3. The arrows in the figure correspond to links connecting nodes in the range of each other. The Modulation on all links in the network is BPSK-1/2. We set the frame duration to ten milliseconds, giving a total of 800 OFDM symbols in each frame and MSG–CTRL–LEN=10, making the number of OFDM symbols in the data sub-frame 730 (there are $7 \times$ MSH–CTRL–LEN $= 70$ OFDM symbols in the control subframe). With these network parameters, the number of OFDM symbols in each transmission opportunity is DataTxOppSize $= 3$ and there are a total of DataTxOppNum $= 243$ transmission opportunities in the data subframe. We set MSH–CSCH–DATA–FRACTION $= 80$ percent, leaving 194 centralized scheduling transmission opportunities in every frame.

We assign link bandwidths so that uplink and downlink requests of each mesh node are 245 kbps. To satisfy the end-to-end connection bandwidths, link rates should be $r_1 = r_2 = 1470$ kbps, $r_3 = r_4 = 1225$ kbps, $r_5 = r_7 = 490$ kbps, and

$r_6 = r_8 = r_9 = r_{10} = r_{11} = r_{12} = 245$ kbps. Using the 802.16 network parameters, we find the total duration required by each link with Equation 13.4 as $d_1 = d_2 = 156$, $d_3 = d_4 = 129$, $d_5 = d_7 = 54$, and $d_6 = d_8 = d_9 = d_{10} = d_{11} = d_{12} = 27$ OFDM symbols.

Next we give the details of how each scheduling algorithm works and we also make observations about algorithm performance. The numerical results from our simulations are summarized in the tables that follow.

13.4.1 802.16 Algorithm

The 802.16 scheduling algorithm finds a link ranking during a breadth-first traversal of the routing tree. The first visited link, in the traversal of the tree, is assigned the lowest rank. The link traversed next is assigned a higher rank and so on, until all links are assigned a ranking. After ranking the links, the algorithm assigns transmission opportunities. The link with the lowest rank is assigned transmission opportunities at the beginning of the data subframe. The link with the next highest rank is assigned the subsequent transmission opportunities and so on, until all links are scheduled. In case that two links have the same rank, the link with the smaller identifier transmits first.

If the total number of assigned transmission opportunities is larger than the number of transmission opportunities reserved for centralized scheduling, the algorithm scales down link durations. After the scaling, the number of transmission opportunities assigned to link e_j is:

$$\hat{D}_j = \frac{d_j}{\texttt{DataTxOppSize}} \frac{\texttt{DataTxOppNum} \times \texttt{MSH} - \texttt{CSCH} - \texttt{DATA} - \texttt{FRACTION}}{N_g} \quad (13.8)$$

where $d_j/\texttt{DataTxOppSize}$ is the original allocation of transmission opportunities for the link, $\texttt{MSH-CSCH-DATA-FRACTION} \times \texttt{DataTxOppNum}$ is the number of centralized scheduling transmission opportunities and N_g is the total number of transmission opportunities needed by the schedule. Because the resulting number of transmission opportunities is rational, the 802.16 standard specifies how to round the duration times. First, the links are ordered by the increasing size of the rational part of their scaled duration. Second, for pairs of links, one at the beginning of the ordered list and the other at the end of the ordered list, the algorithm rounds down the link at the beginning of the list and rounds up the link at the end of the list. The algorithm proceeds inward, towards the middle of the list, until all durations are integers.

We show the schedule obtained with the 802.16 breadth-first algorithm in Figure 13.4. The schedule corresponds to the network scenario in Figure 13.3. The 802.16 algorithm schedules the links one after another according to their ranking. The algorithm does not use spatial reuse so the total length of the resulting schedule exceeds the number of slots available for centralized scheduling. The links were scaled by about 35 percent so that the schedule can fit in the frame. Tables 13.2

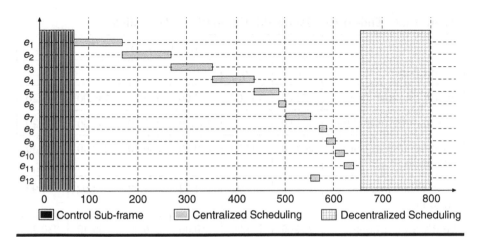

Figure 13.4 Schedule obtained with the 802.16 scheduling algorithm and breadth-first traversal of the routing tree. (From IEEE standard for local and metropolitan area networks. Part 16; air interface for fixed broadband wireless access systems, 2004.)

and Table 13.3 show the end-to-end bandwidths resulting from this schedule. We observe that the scaling algorithm preserves proportional fairness between end-to-end flows. This is a consequence of the linear relationship between link bandwidths and end-to-end bandwidths (Equation 13.3).

Table 13.2   End-to-End Bandwidth Comparison (Uplink)

Algorithm	Bandwidth (kbps)					
	v_2	v_3	v_4	v_5	v_6	v_7
802.16 (breadth first)	153.60	153.60	153.60	144.00	144.00	144.00
802.16 (minimum delay)	153.60	153.60	153.60	144.00	144.00	144.00
Load balancing, one iteration	244.80	244.80	240.00	230.40	230.40	28.80
Load balancing, two iterations	244.80	155.52	155.52	155.52	155.52	155.52
Bandwidth optimal	244.80	244.80	240.00	230.40	230.40	230.40
Bandwidth optimal, $D_{max} = 10$ ms	163.20	161.28	161.28	144.00	144.00	144.00
Bellman-Ford (minimum delay)	163.20	161.28	161.28	144.00	144.00	144.00

Table 13.3 End-to-End Bandwidth Comparison (Downlink)

Algorithm	Bandwidth (kbps)					
	v_2	v_3	v_4	v_5	v_6	v_7
802.16 (breadth first)	153.60	153.60	153.60	115.20	144.00	115.20
802.16 (minimum delay)	153.60	153.60	153.60	115.20	144.00	115.20
Load balancing, one iteration	244.80	244.80	240.00	230.40	230.40	230.40
Load balancing, two iterations	244.80	1.92	1.92	1.92	1.92	1.92
Bandwidth optimal	244.80	244.80	240.00	230.40	230.40	230.40
Bandwidth optimal, $D_{max} = 10$ ms	163.20	161.28	161.28	144.00	144.00	144.00
Bellman-Ford (minimum delay)	163.20	161.28	161.28	144.00	144.00	144.00

The breadth-first ranking does not result in the best transmission order. For example, the return path TDMA delay between v_1 and v_4 for the schedule in Figure 13.4 is approximately 30 ms (Table 13.4). The TDMA delay occurs when an outgoing link on a mesh node is scheduled to transmit before an incoming link in the path of a packet [11]. For the schedule in Figure 13.4, the delay of approximately 10 ms occurs at nodes v_2, v_3, and v_4.

Table 13.4 Return Trip Propagation Delay

Algorithm	Minimum Delay (msec)	Maximum Delay (msec)
802.16 (breadth first)	8.76	28.76
802.16 (minimum delay)	8.76	8.76
Load balancing, one iteration	8.05	28.05
Load balancing, two iterations	8.05	28.05
Bandwidth optimal	8.05	18.05
Bandwidth optimal, $D_{max} = 10$ ms	8.69	8.69
Bellman-Ford (minimum delay)	8.69	8.69

We decrease the TDMA delay resulting from the 802.16 ranking procedure with the minimum TDMA delay ranking algorithm [11]. The minimum delay ranking algorithm ranks the links in a way that guarantees that the total return path TDMA delay is one frame. Initially, the algorithm sets the rank of all the links to zero. The algorithm then examines each return path to the BS, link-by-link, and assigns a rank to each link as a function of the distance from the root of the routing tree. For links in a return path $\mathcal{P} = \{e_i,..., e_k, e_l...e_j\}$, where e_i is the starting link at the BS and e_j is the terminating link at the BS, the rank is assigned as follows:

$$R_k = \max \{R_k, R_l + 1\}, \forall e_k, e_l \in \mathcal{P} : e_l \rightsquigarrow e_k, \qquad (13.9)$$

where $e_l \rightsquigarrow e_k$ means that e_k follows e_l in the path. For the example in Figure 13.3, we have $R_1 = 0$, $R_3 = 1$, $R_5 = R_6 = 2$, $R_8 = R_9 = R_{12} = 3$, $R_{10} = R_{11} = 4$, $R_7 = 5$, $R_4 = 6$, and $R_2 = 7$.

We show the schedule obtained with the 802.16 scheduling algorithm and the minimum delay ranking in Figure 13.5. The total length of the schedule obtained in this way is the same as the total length of the schedule obtained with the 802.16 algorithm, so end-to-end bandwidths resulting from either algorithm are the same (Tables 13.2 and 13.3). The advantage of using the new ordering is that the maximum return path delay is decreased (Table 13.4). The ranking forces the links to transmit in the order that results in TDMA delay on any path of less than ten milliseconds. If we follow links in the topology graph (Fig. 13.3) in the order of transmissions in Figure 13.5, we see that every node in the tree is visited twice before the end of the frame. This is not the case with the schedule in Figure 13.4, where visiting all the nodes twice requires three frames in some cases.

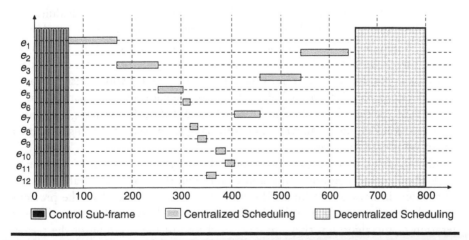

Figure 13.5 Schedule obtained with the 802.16 scheduling algorithm and minimum delay ranking.

13.4.2 Load-Balancing Algorithm

We use the 802.16 node load-balancing algorithm [13] for link scheduling. The link scheduling version of the algorithm uses the same procedure as the node scheduling algorithm, but it schedules links instead of nodes. When the uplink and downlink traffic are scheduled on their own, nonoverlaping, parts of the data subframe, node scheduling is equivalent to link scheduling. For the example in Figure 13.3, the downlink schedule for nodes v_2, v_3, v_4, v_5, v_6, v_7 corresponds to the schedule for links e_1, e_3, e_5, e_6, e_8, e_9. For the uplink, node schedule for v_2, v_3, v_4, v_5, v_6, v_7 corresponds to the link schedule for e_2, e_4, e_7, e_{10}, e_{11}, e_{12}. However, the link scheduling algorithm increases spatial reuse in the network, as it allows the uplink and downlink to be scheduled simultaneously.

The load-balancing algorithm works in iterations. At the begining of each iteration, the algorithm calculates the link ranks according to the link satisfaction with the schedule from the previous iteration. The satisfaction is directly proportional to the ratio of the link bandwidth achieved with the schedule and the bandwidth assigned to the link by the BS. For link e_j, the satisfaction is:

$$s_j = \frac{\hat{r}_j}{r_j},$$ (13.10)

where $\hat{r}_j$ is obtained from the previous schedule with Equation 13.5 and r_j is the required bandwidth on the link. The satisfaction index determines the ranking of the links; links with the lowest satisfaction index have the highest rank and the links with the highest satisfaction index have the lowest rank. If two links have the same satisfaction, the link with the higher link identifier is assigned the higher rank.

After establishing the ranking, the load-balancing algorithm schedules the links. First, the algorithm schedules the link with the smallest rank at the beginning of the data subframe. Then, the algorithm schedules the rest of the links in the order of their rank. Each link is scheduled as close to the beginning of the data subframe as possible, so that its transmission does not overlap with transmissions of its conflicting links. Scheduling links in this way takes advantage of spatial reuse in the network, however links scheduled toward the end of the frame may get a smaller number of OFDM slots than what they were assigned by the BS. In this case, their transmissions are truncated.

The links that were truncated get more OFDM symbols in the next iteration of the load-balancing algorithm. In the next iteration, the algorithm recalculates the satisfaction indexes of the links. The links that were truncated in the previous schedule (previous iteration of the algorithm) have the lowest index, so in the next iteration of the algorithm they are scheduled toward the begining of the frame. Scheduling the links sooner in the frame ensures that their transmissions are longer, making their bandwidths higher than in the previous frame. On the other

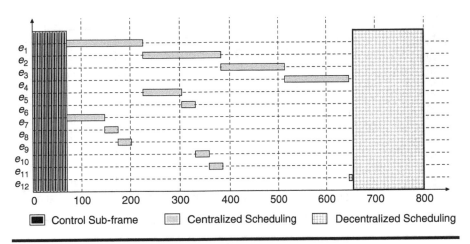

Figure 13.6 Schedule obtained with the load-balancing algorithm, one iteration.

hand, links that were not truncated in the previous iteration have the highest satisfaction index, so they are scheduled toward the end of the frame where they may be
truncated.

We show a schedule obtained in the first iteration of the load-balancing algorithm in Figure 13.6. In the first iteration, the satisfaction of all links is the same as
they start with the $\hat{r}_j = 0$, $\forall e_j \in E$. In this case, the link rank corresponds to link
index, so link e_{12} is scheduled last and its transmission is truncated. Tables 13.2 and
13.3 show the end-to-end bandwidths resulting from this schedule. We see that the
uplink bandwidth of node v_7 is lower than its assigned bandwidth as e_{12} is its bottleneck link.

The load-balancing algorithm increases the bandwidth of link e_{12} in the second
iteration. At the beginning of the second iteration, link e_{12} has the lowest satisfaction,
so it is scheduled first in the frame (with no loss of bandwidth). Link e_3 has the highest
satisfaction index, so it is scheduled last and is truncated. We show the schedule after
the second iteration in Figure 13.7. Tables 13.2 and 13.3 show the end-to-end bandwidths resulting from this schedule. We see that the downlink bandwidths of nodes
v_3, v_4, v_5, v_6, and v_7 are significantly decreased as e_3 is their bottleneck link. In the next
iteration, the algorithm should increase the bandwidth on e_3, so this is only a temporary loss of bandwidth for the end-to-end connections traversing e_3.

The load balancing does not take into account delay. The maximum return path
delay is approximately 30 ms after both iterations (Table 13.4).

13.4.3 Bellman-Ford Algorithm

The Bellman-Ford scheduling algorithm is based on our previous work [11]. The
algorithm has two steps. First, the algorithm finds a ranking with good TDMA

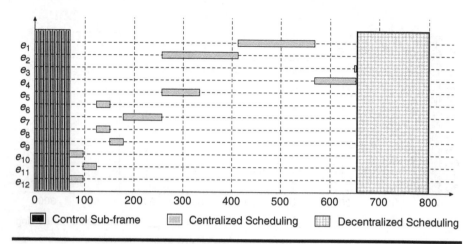

Figure 13.7   Schedule obtained with the load-balancing algorithm, two iterations.

delay properties. Finding a ranking with good TDMA delay properties is a computationally hard problem [11]. Nevertheless, we propose a branch-and-bound search method for bandwidth optimal transmission orders, as well as heuristics for good rankings [11]. A ranking is bandwidth optimal if link bandwidths resulting from its schedule cannot be increased with a schedule from another ranking. Second, the algorithm finds a transmission schedule with the Bellman-Ford algorithm. The link transmission times are calculated from the minimum distances in the conflict graph for the network. The conflict graph has links as vertexes and conflicts between the links as arcs and links conflict if they cannot transmit at the same time. As the schedule is based on the distances in the conflict graph, the conflicting link transmissions do not overlap [11].

Figure 13.8 shows a schedule resulting from a bandwidth optimal ranking with no limit on the return path TDMA delay. We use the branch-and-bound method to find the bandwidth optimal ranking [11]. We show the end-to-end bandwidths resulting from the schedule in Tables 13.2 and 13.3. We observe that the end-to-end bandwidths resulting from this schedule closely match the required end-to-end bandwidths. The final end-to-end bandwidths are different from the required bandwidths as links are allocated OFDM symbols in multiples of `DataTxOppSize`, which dictates the minimum granulation of bandwidth assignments. As there is no limit on the return path TDMA delay, this schedule causes a maximum return path TDMA delay of approximately 20 ms (Table 13.4).

Figure 13.9 shows a schedule resulting from a bandwidth optimal ranking, when the maximum return path TDMA delay is limited to ten milliseconds. We use the branch-and-bound method to find this ranking as well. This schedule is different from a schedule that can be obtained by fixing the ranking to the minimum delay ranking with Equation 13.9; it is bandwidth optimal. Tables 13.2 and 13.3 show end-to-end bandwidths resulting from the schedule. We note that end-to-end

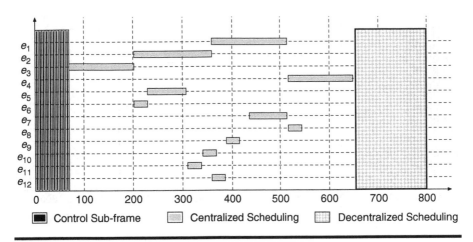

Figure 13.8   Bandwidth optimal schedule.

bandwidths are significantly lower than the end-to-end bandwidth resulting from the bandwidth optimal schedule with no delay restrictions (30 percent lower), however the bandwidths are also higher than the end-to-end bandwidths resulting from the 802.16 schedule (8 percent higher). The advantage of the schedule in Figure 13.9 is that it limits the return path delay to ten milliseconds, which is half of the maximum delay resulting from the bandwidth optimal schedule with no delay restrictions.

Finally, Figure 13.10 shows the schedule resulting from the minimum delay ranking obtained with Equation 13.9. The end-to-end bandwidths resulting from this schedule are the same as the end-to-end bandwidths obtained with the bandwidth optimal schedule with the delay limit (Tables 13.2 and 13.3), showing that the simple heuristic is effective.

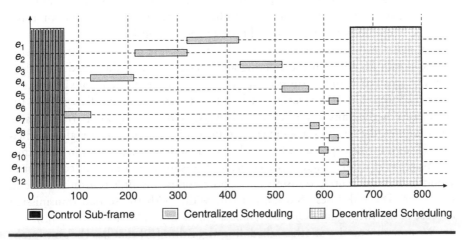

Figure 13.9   Bandwidth optimal schedule, delay limited to $D_{max} = 10.0$ msec.

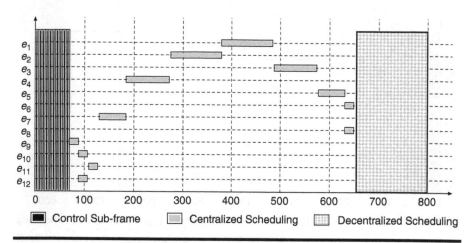

Figure 13.10   Bellman-Ford algorithm, fixed minimum delay ranking.

13.4.4   Other Approaches

TDMA scheduling has been a research topic for some time, however most of that research is not appropriate for the 802.16 mesh protocol. We identify two main reasons for the disconnect between the previous research in TDMA scheduling protocols and 802.16. First, many of the previous approaches do not model all conflicts that exist in TDMA wireless networks [14–16]. In these works, the authors assume that only the links sharing a node in common conflict (primary conflicts). For example, in Figure 13.3 this conflict model ignores conflicts such as the one between the links e_1 and e_5. It is possible to resolve secondary conflicts by assigning orthogonal frequencies to links in secondary conflicts. However, finding this frequency assignment is hard [17,18]. Also, the 802.16 standard does not completely resolve the use of multiple frequencies in 802.16, meaning that any solution using multiple frequencies requires new additions to the standard.

Second, many of the algorithms in current literature do not take transmission overhead into account. In the coloring approaches [14–16], the scheduling algorithm assigns colors to transmission opportunities in the frame before it begins to color the links. After the coloring, each link has a set of colors different from the colors used by the links it conflicts with. Because colors correspond to transmission opportunities, the resulting assignment of transmission opportunities is also conflict-free. However, this approach allows links to transmit multiple times in the frame. For example, if a link assigned two colors for transmissions and these colors correspond to nonconsecutive transmission opportunities, the link transmits twice in the frame. The link loses 28.8 kbps of bandwidth with the second transmission because of the three guard OFDM symbols (36 bytes = 288 bits every 10 ms).

We now summarize another scheduling algorithm [19] that has the same problem as the coloring algorithms. The algorithm assigns transmission opportunities in rounds. In each round, the algorithm assigns one transmission opportunity to links that need it to satisfy their bandwidth requests. Links may not receive any transmission opportunities in the round if the assignment introduces conflicts with other links in the round. The reason why we could not implement this scheduling algorithm without significant changes is that we cannot limit the number of times each link transmits in the frame. This can have a very negative impact on the resulting end-to-end bandwidths. For example, the algorithm may schedule a link to transmit all of its data in nonconsecutive data transmissions, thus allocating no bandwidth to the link, if `DataTxOppSize` ≤ 3.

13.5 Conclusion

We have given a summary of the 802.16 mesh protocol, with focus on centralized scheduling algorithms. We have compared three centralized link scheduling algorithms with each other. The first algorithm is proposed in the 802.16 standard [1]. We have obtained the second algorithm from a load-balancing node scheduling algorithm [13]. The third algorithm is based on the Bellman-Ford algorithm and was proposed for TDMA scheduling [11].

The Bellman-Ford algorithm has the best properties. Unlike the 802.16 algorithm, it takes advantage of spatial reuse and unlike the load-balancing algorithm it always allocates bandwidth proportional to assigned link bandwidths. The Bellman-Ford algorithm also takes TDMA delay into account, so it can find link schedules that result in small end-to-end TDMA delay.

References

1. IEEE standard for local and metropolitan area networks. Part 16: air interface for fixed broadband wireless access systems (2004).
2. P. Djukic and S. Valaee. 802.16 mesh networking. In eds. S. Ahson and M. Ilysa, *Handbook of Wimax,* Chapter 8, pp. 149–177, CRC Press, Boca Raton (2007).
3. J. Camp, J. Robinson, C. Steger, and E. Knightly. Measurement driven deployment of a two-tier urban mesh access network. Technical Report TREE0505, Rice University (December 2005).
4. Nortel Network. Wireless mesh network—extending the reach of wireless LAN, securely and cost-effectively. http://www.nortelnetworks.com/solutions/wlan/ (November 2003).
5. S. Xu and T. Saadawi. Does the IEEE 802.11 MAC protocol work well in multihop wireless ad hoc network, *IEEE Communication Magazine,* 39(6), 130–137 (June 2001).
6. P. Djukic and S. Valaee. 802.16 MCF for 802.11a based mesh networks: a case for standards re-use. In *23rd Queen's Biennial Symposium on Communications* (2006).
7. P. Djukic and S. Valaee. Towards guaranteed QoS in mesh networks: Emulating WiMAX mesh over WiFi hardware. In *The Fourth Workshop on Wireless Ad hoc and Sensor Networks WWASN* (2007).

8. P. Djukic and S. Valaee. Distributed link scheduling for TDMA mesh networks. In *Proceedings of ICC* (2007).

9. Y. Bernet, P. Ford, R. Yavatkar, F. Baker, L. Zhang, M. Speer, R. Braden, B. Davie, J. Wroclawski, and E. Felstaine. A framework for integrated services operation over Diffserv networks. RFC 2998 (Informational) (November 2000). URL http://www.ietf.org/rfc/rfc2998.txt.

10. P. Djukic and S. Valaee. Quality-of-service provisioning in multi-service TDMA mesh networks. In *Proceedings of 20th International Teletraffic Congress* (2007).

11. P. Djukic and S. Valaee. Link scheduling for minimum delay in spatial re-use TDMA. In *INFOCOM* (2007).

12. S. Keshav. *An Engineering Approach to Computer Networking* (Probability, Random Variables and Stochastic Processes, Addison-Wesley, Boston, MA, 1997).

13. D. Kim and A. Ganz. Fair and efficient multihop scheduling algorithm of IEEE 802.16 BWA systems. In *Broadnets*, pp. 895–901 (2005).

14. B. Hajek and G. Sasaki. Link scheduling in polynomial time. *IEEE Transactions on Information Theory*, 34(5), 910–917 (September 1988).

15. T. Salonidis and L. Tassiulas. Distributed dynamic scheduling for end-to-end rate guarantees in wireless ad hoc networks. In *MobiHoc*, pp. 145–156 (2005).

16. M. Kodialam and T. Nandagopal. Characterizing achievable rates in multi-hop wireless mesh networks with orthogonal channels. *IEEE/ACM Transactions on Networking*, 13(4), 868–880 (2005).

17. W. K. Hale, Frequency assignment: theory and applications. *Proceedings of the IEEE*, 68(12), 1497–1514 (December 1980).

18. M. Alicherry, R. Ghatia, and L. Li. Joint channel assignment and routing for throughput optimization in multi-radio wireless mesh networks. In *Mobicom* (2006).

19. H.-Y. Wei, S. Ganguly, R. Izamailov, and Z. Haas. Interference-aware IEEE 802.16 WiMax mesh networks. In *VTC Spring'05* (2005).

Chapter 14

Load-Balancing Approach to Radio Resource Management in Mobile WiMAX Network

---

Stanislav Filin, Sergey Moiseev, Mikhail Kondakov, and Alexandre Garmonov

This chapter presents a comprehensive set of radio resource management algorithms for the Mobile WiMAX network, including call admission control, adaptive transmission, horizontal handover, and vertical handover algorithms. These algorithms are based on our system load model of the Mobile WiMAX network. Each algorithm individually and all algorithms as a whole satisfy Quality-of-Service (QoS) requirements and maximize the network capacity. The sections of this chapter are divided into two groups, that is, the basic ones and the extensions. In the basic sections, we give a brief overview of the Mobile WiMAX system, describe our system load model of the Mobile WiMAX network, present call admission control, dynamic bandwidth allocation, adaptive transmission, and horizontal handover algorithms, and show how all radio resource management algorithms operate together. The first extension is devoted to the QoS-guaranteed load-balancing vertical handover algorithm in the 4G heterogeneous wireless network. In the second extension, which is the last section, we describe our vision of the future research on improvement of the Mobile WiMAX network.

14.1   Introduction

IEEE standards 802.16-2004 [1] and 802.16e-2005 [2] specify the requirements for the Medium Access Control (MAC) and physical (PHY) layers of the WiMAX and Mobile WiMAX systems. These systems are attracting huge interest as a promising solution for delivering fixed and mobile broadband wireless access services. The standards have incorporated key technologies, such as Quality-of-Service (QoS) mechanisms, Adaptive Modulation and Coding (AMC), power control, selective and hybrid Automatic Repeat Request (ARQ), Orthogonal Frequency Division Multiplexing (OFDM) and Orthogonal Frequency Division Multiple Access (OFDMA), support of Adaptive Antenna Systems (AAS), and Multiple-Input Multiple-Output (MIMO) transmission. This provides great potential for satisfying users and operators needs.

From a user's perspective, a wireless system should deliver his or her data with the required QoS level, while an operator aims at maximizing network capacity and revenue. These two goals are achieved by Radio Resource Management (RRM) algorithms. These algorithms play a key role in the Mobile WiMAX system. However, they are quite complex, which is largely due to a lot of degrees of freedom and optimization opportunities.

RRM in the OFDMA-based Mobile WiMAX network includes Call Admission Control (CAC), adaptive transmission, and handover algorithms. The CAC algorithm handles system overloading and satisfies users' QoS by limiting the number of users in the system. Adaptive transmission algorithms enable QoS-guaranteed opportunistic data transmission over a wireless medium. They include scheduling, AMC, power control, and time-frequency resource allocation. A seamless horizontal handover guarantees continuous service by assigning new serving Base Stations (BSs) to a user during his or her mobility and system load variations. In the 4G heterogeneous wireless network, including the Mobile WiMAX network, a seamless vertical handover algorithm is added to the set of RRM algorithms.

The Mobile WiMAX network has a number of distinct features that complicate the use of conventional RRM algorithms. Users may have multiple service flows with different traffic and QoS requirements, be in different receiving conditions, and use different coding and modulation schemes and transmission power values. Consequently, it is difficult to determine the time-frequency and power resources required for the given number of users. This is the main challenge of the CAC algorithm. In adaptive transmission algorithms, the key problems are computational intensity due to a large number of degrees of freedom, complicated frame structure, and complex MAC and PHY layers processing procedures. Traditional horizontal handover algorithms are based on the received signal level or Signal-to-Interference-plus-Noise Ratio (SINR). However, such handover algorithms neither guarantee QoS requirements nor maximize the capacity in the Mobile WiMAX OFDMA network. Moreover, the SINR-based approach is inconvenient for vertical handover algorithms because different networks use different transmission technologies and operate in different SINR ranges. Therefore, RRM algorithms taking into account the distinct features of the Mobile WiMAX OFDMA network should be provided.

We present a comprehensive set of RRM algorithms for the Mobile WiMAX OFDMA network. These algorithms satisfy users' QoS requirements and maximize the network capacity. We define the network capacity as the maximum achievable network throughput when QoS requirements are satisfied for all the users served. Our algorithms are based on our system load model of the Mobile WiMAX OFDMA network. This model takes into account different traffic, QoS requirements, and receiving conditions of the users and efficiently combines time-frequency and power resources, as well as downlink (DL) and uplink (UL) resources in the expression for the system load. We use a load-balancing approach in all RRM algorithms, including CAC, adaptive transmission, horizontal handover, and vertical handover algorithms. We demonstrate the advantages of our algorithms over the conventional ones by means of system level simulation. Each algorithm individually and all algorithms as a whole satisfy QoS requirements and maximize the network capacity.

The sections of this chapter may be divided into two groups, that is, the basic ones and the extensions.

The next six sections form the basic group. First, we give a brief overview of the Mobile WiMAX system. Our main focus is on the features of the Mobile WiMAX network that are essential for the RRM algorithms. Secondly, we describe our system load model of the Mobile WiMAX network. We show an approach for optimal estimation of the system load and present fast practical estimation algorithms. Then, we proceed to the CAC algorithm and the dynamic bandwidth allocation algorithm. Next, we describe our QoS-guaranteed load-balancing adaptive transmission algorithms. We give a general formulation of the adaptive transmission conditional optimization problem in the Mobile WiMAX system. Our load-balancing adaptive transmission algorithm solves this problem with high spectral efficiency and low computational complexity. Then, the horizontal handover algorithm in the Mobile WiMAX network is discussed. Our load-balancing handover algorithm guarantees QoS and has a considerable network capacity gain over the SINR-based handover algorithm. Finally, we summarize our RRM algorithms and describe how they all operate together.

The last two sections are the extensions. The first extension is devoted to the QoS-guaranteed load-balancing vertical handover algorithm in the 4G heterogeneous wireless network. Our system load model of the Mobile WiMAX network may be extended to the system load model of any wireless network. This enables applying our horizontal handover algorithm to the 4G heterogeneous network. In the last section we describe our vision of the future research on improvement of the Mobile WiMAX network. We present several research directions and give key ideas for them.

14.2 Mobile WiMAX System Description

The Mobile WiMAX system is based on the OFDMA PHY layer of the IEEE standards 802.16-2004 [1] and 802.16e-2005 [2]. The Mobile WiMAX system profile is developed by the WiMAX Forum and is described in [3,4].

The Mobile WiMAX network is supposed to have a cellular topology. Each cell will have three sectors. Possible frequency reuse factor values are one and three. Consequently, the Mobile WiMAX network comprises some sectors and some users. The sectors transmit data to the users in the DL and the users transmit data to the sectors in the UL. The sectors in the DL and the users in the UL have the maximum transmission power constraints. Each user may have several DL service flows and several UL service flows (see Fig. 14.1). A service flow is a flow of data packets from a service.

Different service flows may have different traffic arrival rates. Each service flow has a set of the QoS requirements. The set of the QoS requirements includes three key parameters, that is, the minimum data rate, the maximum data block delay, and the maximum data block reception error probability.

Each sector performs transmission in the DL and reception in the UL using frames. The logical structure of the frame in the Mobile WiMAX system is shown in Figure 14.2. Here we consider the Time Division Duplex (TDD) frame structure as a more general case. The frame has a fixed duration and comprises the DL and the UL subframes. The position of the frame boundary between the DL and UL

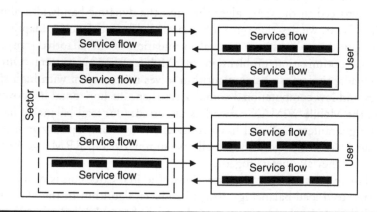

Figure 14.1   Downlink and uplink service flows.

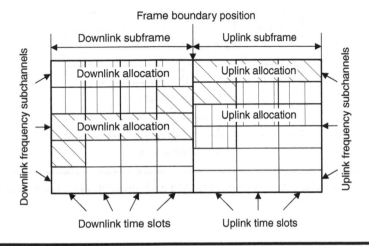

Figure 14.2   Logical frame structure.

subframes may be adaptively changed on the frame-by-frame basis thus taking into account the DL and UL load variations.

In the time domain the DL and UL subframes are divided into the time slots, each formed by one or several OFDM symbols. In the frequency domain the subcarriers of the DL and UL subframes are partitioned into the frequency subchannels, each containing a group of subcarriers. Time slots and frequency subchannels form the time-frequency resource of the frame.

A part of the time-frequency resource of the DL or UL subframe is allocated to each DL or UL service flow for transmission. We assume that the service flows are placed into the free part of a subframe from left to right. This allows allocating more transmission power per subcarrier, when a subframe is not fully loaded.

The transmit and receive processing in the Mobile WiMAX system is as follows. Data blocks of a service flow arrive from the upper layers at the MAC layer, where they are converted into data packets using fragmentation and packing operations (see Fig. 14.3). Also, cyclic redundancy check and ARQ mechanisms can be used. The set of data packets of the service flow arrives at the PHY layer, where it is converted into coding blocks. Each coding block is coded and decoded independently. After decoding, coding blocks are converted back into data packets. Data packets are converted back into data blocks. We call all these procedures MAC and PHY layers processing.

At the PHY layer, coding blocks are modulated using, for example, quadrature amplitude modulation. Modulation symbols are mapped onto subcarriers. In the Mobile WiMAX network subcarriers are subject to frequency-selective fading and interference. Two options exist in the Mobile WiMAX system to cope with these negative factors, that is, frequency diversity and multi-user diversity.

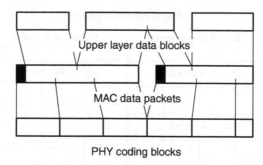

Figure 14.3   Medium Access Control (MAC) and physical (PHY) layers processing.

When the frequency diversity is used, the subcarriers of the given user are interleaved in the frequency domain within the entire signal bandwidth. In this case, even only a part of the subcarriers is allocated to the user, his or her receiving conditions are characterized by the receiving conditions, averaged over the OFDM symbol. In the Mobile WiMAX system, the frequency diversity is implemented by converting the logical frame structure into the physical one by means of subcarrier permutation. Subcarriers of each frequency subchannel are mapped onto the physical subcarriers that are interleaved within the entire signal bandwidth. This enables fading and interference averaging.

When the multi-user diversity is used, the subcarriers of the given user are not interleaved in the frequency domain. In other words, subcarriers of a logical frequency subchannel remain adjacent in the physical frame structure. In this case, the receiving conditions for the given user vary within the frequency subchannels. For the given frequency subchannel the receiving conditions are different for different users. Multi-user diversity gain may be obtained by allocating frequency subchannels to the users with the best receiving conditions.

RRM algorithms require knowing the coding block error rate as a function of the receiving conditions. The set of the receiving conditions commonly includes only one element, that is, average SINR. We proposed to estimate coding block error rate as a function of two variables, that is, the average SINR and the normalized SINR root-mean-square value in [5–7]. Our approach guarantees satisfying QoS requirements and provides considerable throughput gain compared with the traditional and the effective SNR approaches.

The following degrees of freedom are available for the RRM algorithms in the Mobile WiMAX network. Frame boundary position may be adaptively changed on the frame-by-frame basis. AMC and power control may be performed individually for each service flow. In the case of multi-user diversity, frequency subchannels may be adaptively allocated to the service flows. Users may be adaptively distributed among sectors.

The Mobile WiMAX system has two additional features that may improve its performance. The first one is the hybrid ARQ mechanism that has greater performance than the traditional ARQ mechanism. The second one is the MIMO technology. The Mobile WiMAX system supports several MIMO schemes, including smart antenna, space-time coding, and spatial multiplexing.

14.3 System Load Model

A system load model is traditionally used in CAC algorithms. System load characterizes the degree of system resources consumption. When all system resources have been consumed, new users are not admitted to the system.

Several system load models currently exist. In [8,9], system load is equal to the number of users per sector. In the Mobile WiMAX network the users may have different sets of traffic with different QoS requirements, be in different receiving conditions, and use different coding and modulation schemes. Consequently, they may consume quite different amounts of the system resources. Hence, the system load in the Mobile WiMAX network is not equal to the number of users.

The authors of Ref. [10,11] define the system load as the number of channels used in a sector. In the Mobile WiMAX OFDMA network the channels correspond to the time-frequency resource of the frame, where the size of the time resource is equal to the number of time slots and the size of the frequency resource is defined by the number of frequency subchannels. However, power resource should also be considered. We can refer to a case when a part of the time-frequency resource is not consumed although all the transmission power is used. Then the system is fully loaded but unused channels are available. On the other hand, all time-frequency resources could be consumed although some power resource is available. In this case, although all channels are consumed, we can select the coding and modulation schemes with higher transmission rates using the available power resource to provide the time-frequency resource for new users. As transmission rates in the Mobile WiMAX network are 4.5 times different, we can obtain about 80 percent of the time-frequency resource. Summing up, the number of channels used does not characterize the system load in the Mobile WiMAX network.

Throughput is used as a system load metric in [12,13]. This metric is noninformative. It does not characterize the consumed resource numerically, especially when the system is overloaded.

In the Code Division Multiple Access (CDMA) networks the system load is usually connected with interference [14,15]. Although this is a good idea for the CDMA networks, time-frequency and power resources should be considered in the Mobile WiMAX OFDMA network.

We proposed a system load model for the Mobile WiMAX OFDMA network in [16]. We formulate the requirements for the system load, which serve as the basis for our system load model development.

We formulate three requirements to the system load model:

■ System load model shall explicitly include the system resources. Only shared system resources shall be included.
■ System load shall be normalized to all available system resources to compare the system loads of different networks.
■ System load shall be equal to the minimum required amount of the system resources. It shall be calculated under the condition that the QoS requirements are satisfied for all users.

Based on the requirements, we define the system load as follows:

■ System load is the minimum amount of the normalized shared system resources required for the users.
■ The minimum amount of the required system resources is normalized to the total amount of the system resources.
■ The amount of the required system resources is calculated under the condition that the QoS requirements are satisfied for all users.

We use this definition to develop the system load model for the Mobile WiMAX network. Our system load model includes the following system loads:

■ UL load
■ DL load
■ Sector load
■ Network load

To calculate each of these system loads, we use the following approach. First, we write the expression for the amount of the normalized shared system resources, consumed by all users, as a function of the adaptation parameters. Then, we minimize this expression over the adaptation parameters under the constraint on the individual system resources, while satisfying the QoS requirements for all users.

14.3.1 UL Load

Let us consider the UL. In the UL, the shared system resource is the time-frequency resource of the UL subframe. Transmission power of each user is an individual system resource. Adaptation parameters are the set of the assigned uplink frequency subchannels, coding and modulation schemes, and transmission power values.

The normalized consumed UL resource s^{UL} can be expressed as

$$s^{UL} = \frac{S^{UL}}{S^{UL}_{max}}, \tag{14.1}$$

where S^{UL} is the UL time-frequency resource, consumed by all users, and S^{UL}_{max} is the total time-frequency resource of the UL subframe.

To obtain the UL load u^{UL}, we need to solve the following conditional optimization problem

$$u^{UL} = \min_{q^{UL}, p^{UL}, f^{UL}} (s^{UL}),$$ (14.2)

subject to

$$P_i^{UL} \leq P_{max,i}^{UL},$$ (14.3)

$$p_{i,j,\ell}^{UL} \geq p_{i,j,\ell}^{QoS},$$ (14.4)

where q^{UL} is the set of the coding and modulation schemes assigned to the users in the UL subframe, p^{UL} is the set of the transmission power values assigned to the users in the UL subframe, and f^{UL} is the set of the frequency subchannels assigned to the users in the UL subframe. P_i^{UL} and $P_{max,i}^{UL}$ are the transmission power and the maximum transmission power values of the user i, $p_{i,j,\ell}^{UL}$ is the transmission power value selected for the UL service flow j of the user i on the UL subchannel ℓ, $p_{i,j,\ell}^{QoS}$ is the minimum transmission power value that shall be assigned to the UL service flow j of the user i on the UL subchannel ℓ to satisfy QoS requirements for this service flow.

14.3.2 DL Load

Let us now consider the DL. In the DL, the shared system resources are the time-frequency resource of the DL subframe and the DL transmission power. Adaptation parameters are the set of the assigned DL frequency subchannels, coding and modulation schemes, and transmission power values.

The normalized consumed DL time-frequency resource s^{DL} can be written as

$$s^{DL} = \frac{S^{DL}}{S^{DL}_{max}},$$ (14.5)

where S^{DL} is the DL time-frequency resource consumed by all users and S^{DL}_{max} is the total time-frequency resource of the DL subframe.

The normalized consumed DL power ρ^{DL} can be given by

$$\rho^{DL} = \frac{P^{DL}}{P^{DL}_{max}},$$ (14.6)

where P^{DL} is the DL transmission power consumed by all users and P^{DL}_{max} is the maximum DL transmission power constraint.

To obtain the DL load u^{DL} we need to combine the shared DL system resources, that is, the time-frequency resource s^{DL} and the power ρ^{DL}. In our opinion, the DL load shall have the following properties:

- DL load u^{DL} (s^{DL}, ρ^{DL}) shall be the increasing function of its arguments s^{DL} and ρ^{DL}.
- u^{DL} (s^{DL}, ρ^{DL}) = 0 if and only if s^{DL} = 0 and ρ^{DL} = 0.
- If s^{DL} < 1 and ρ^{DL} < 1, then u^{DL} (s^{DL}, ρ^{DL}) < 1.
- If s^{DL} = 1 and ρ^{DL} < 1, then u^{DL} (s^{DL}, ρ^{DL}) = 1, that is, when the DL time-frequency resource is fully used, then the DL is fully loaded although the DL power resource is not fully used.
- If s^{DL} < 1 and ρ^{DL} = 1, then u^{DL} (s^{DL}, ρ^{DL}) = 1, that is, when the DL power resource is fully used, then the DL is fully loaded although the DL time-frequency resource is not fully used.
- If s^{DL} = ρ^{DL}, then u^{DL} (s^{DL}, ρ^{DL}) = s^{DL} = ρ^{DL}, that is, the DL time-frequency resource and the DL power resource are equivalent.
- Let us have m identical DL service flows with identical receiving conditions.

  – DL load value u^{DL} = α, where 0 < α < 1, means that if we take m/α service flows, identical to the service flows considered, then the DL will be fully loaded, that is, u^{DL} = 1.
  – DL load value u^{DL} = β, β > 1, means, that if we take β DL subframes, identical to the considered DL subframe, then the DL will not be overloaded.
  – If we iteratively add these service flows to the DL subframe until we get the fully loaded DL, then we will get the maximum number of the served service flows. In other words, calculating the DL load shall lead to the maximum system throughput for the given service flows.

A linear function cannot satisfy all these properties. There is a nonlinear function, that is, max(s^{DL}, ρ^{DL}) which satisfies all these requirements. Consequently, we define the DL resource consumption function U^{DL} as

$$U^{DL} = \max(s^{DL}, \rho^{DL}). \tag{14.7}$$

To obtain the DL load u^{DL}, we need to solve the following conditional optimization problem

$$u^{DL} = \min_{q^{DL}, p^{DL}, f^{DL}} (U^{DL}), \tag{14.8}$$

subject to

$$p_{i,j,\ell}^{DL} \geq p_{i,j,\ell}^{QoS}, \tag{14.9}$$

where $\mathbf{q}^{DL}$ is the set of the coding and modulation schemes assigned to the users in the DL subframe, $\mathbf{p}^{DL}$ is the set of the transmission power values assigned to the users in the DL subframe, and $\mathbf{f}^{DL}$ is the set of the frequency subchannels assigned to the users in the DL subframe; $p_{i,j,\ell}^{DL}$ is the transmission power value selected for the DL service flow j of the user i on the DL subchannel ℓ, $p_{i,j,\ell}^{QoS}$ is the minimum transmission power value that shall be assigned to the DL service flow j of the user i on the DL subchannel ℓ to satisfy QoS requirements for this service flow.

14.3.3 Sector Load and Network Load

Let us consider a sector. The shared system resources are the UL and the DL resources. A new adaptation parameter, that is, frame boundary position t between the DL and UL subframes is added to the UL and DL adaptation parameters. Hence, we define the load of the sector k as

$$u_k = \min_t [\max(u_k^{UL}, u_k^{DL})]. \tag{14.10}$$

Finally, let us consider the entire OFDMA network comprising K sectors. Network load $\mathbf{u}$ is characterized by the set of its sector loads

$$\mathbf{u} = \{u_1, u_2, ..., u_K\}. \tag{14.11}$$

14.3.4 Fast System Load Estimation

The optimal solution to optimization problems (Equations 14.2 through 14.4) for the UL, (Equations 14.8 and 14.9) for the DL, and (Equation 14.10) for the sector is exhaustive search. Because it is inapplicable in practice, several suboptimal algorithms can be used to estimate system load. Different algorithms will have different accuracy and computational complexity. Here, in a top–down manner we describe our approach to fast system load estimation.

Sector load (Equation 14.10) can be found in two steps. During the first step, the position of the frame boundary is selected such that the time-frequency resource of the frame is divided between the DL and UL subframes proportionally to the DL and UL traffic loads. This frame boundary position does not take into account the receiving conditions of the users. For this frame boundary position we estimate the DL and UL loads to take the receiving conditions into account.

During the second step, we calculate the position of the frame boundary such that the time-frequency resource of the frame is divided between the DL and UL subframes proportionally to the DL and UL loads, found during the first step. For the new frame boundary position, we estimate the DL and UL loads as well as the sector load.

The sector load value, found during the second step, may be used as the final sector load estimate. Alternatively, additional greedy search for the optimal frame boundary position may be performed in the vicinity of the position, found during the second step.

Approaches to estimate the DL and UL loads are different for the frequency and multi-user diversity.

We use the following approach to estimate the DL load in case of frequency diversity. In the DL, the shared system resources are the time-frequency resource of the DL subframe and the DL transmission power. Adaptation parameters are the set of the assigned coding and modulation schemes and transmission power values. Consequently, the optimization problem (Equations 14.8 and 14.9) can be written as

$$u^{\mathrm{DL}} = \min_{\mathbf{q}^{\mathrm{DL}}, \mathbf{p}^{\mathrm{DL}}} (U^{\mathrm{DL}}), \tag{14.12}$$

subject to

$$p_{i,j}^{\mathrm{DL}} \geq p_{i,j}^{\mathrm{QoS}}. \tag{14.13}$$

We solve the optimization problem (Equations 14.12 and 14.3) iteratively. The number of iterations is equal to the number of all the DL service flows of all the users.

During the first iteration, we select one user and one service flow of this user and solve the optimization problem

$$q_{1,1}^{\mathrm{DL,opt}} = \arg \min_{q_{1,1}^{\mathrm{DL}}} [\max(s_{1,1}^{\mathrm{DL}}, \rho_{1,1}^{\mathrm{DL}})] \tag{14.14}$$

using the full enumeration and taking $p_{1,1}^{\mathrm{DL,opt}} = p_{1,1}^{\mathrm{DL}}(q_{1,1}^{\mathrm{DL}})$.

In Equation 14.14, values of $s_{1,1}^{\mathrm{DL}}$ and $\rho_{1,1}^{\mathrm{DL}}$ are calculated given that only service flow (1,1) is placed into the DL subframe. Examples of finding transmission power value $p_{1,1}^{\mathrm{DL}}$ as a function of coding and modulation scheme number $q_{1,1}^{\mathrm{DL}}$, QoS requirements, and receiving conditions for the frequency diversity are given in [17–19].

After the optimization problem (Equation 14.14) is solved, we fix coding and modulation scheme $q_{1,1}^{\mathrm{DL,opt}}$ and transmission power value $p_{1,1}^{\mathrm{DL,opt}}$ for service flow (1,1).

During the second iteration, we again select one user and one service flow of this user and solve the optimization problem

$$q_{2,2}^{DL,opt} = \arg\min_{q_{2,2}^{DL}}[\max(s_{1,1}^{DL,opt} + s_{2,2}^{DL}, \rho_{1,1}^{DL,opt} + \rho_{2,2}^{DL})] \qquad (14.15)$$

using the full enumeration and taking $p_{2,2}^{DL,opt} = p_{2,2}^{DL}(q_{2,2}^{DL})$.

In Equation 14.15, values of $s_{1,1}^{DL,opt}$ and $\rho_{1,1}^{DL,opt}$ are calculated during the first iteration, values of $s_{2,2}^{DL}$ and $\rho_{2,2}^{DL}$ are calculated given that service flow (1,1) is already placed into the DL subframe with coding and modulation scheme $q_{1,1}^{DL,opt}$ and transmission power value $p_{1,1}^{DL,opt}$, and service flow (2,2) is additionally placed into the DL subframe.

We continue the iterations until all the DL service flows of all the users are considered. Recall that during each iteration we solve the optimization problem only for one service flow of one user.

When a DL subframe is overloaded, that is, we are unable to place all the DL service flows scheduled for transmission into it, we use the following procedure. We place as many service flows as possible into the DL subframe subject to $u^{DL} \leq 1$. Then, we virtually add the second DL subframe, the same as the first DL subframe, below the first DL subframe. We place as many remaining service flows as possible into the second DL subframe subject to $u^{DL} \leq 1$. This condition is checked only for the second DL subframe. We repeat this procedure until all DL service flows have been placed. Then, we calculate the DL load over all DL subframes used. Here, we add all time-frequency resources used of all DL subframes used to obtain s^{DL}. Moreover, we add all transmission power used in all DL subframes used to obtain ρ^{DL} (therefore we add each new DL subframe below the DL subframes already used).

Different criteria can be used to select one user and one service flow during each iteration. We use the following criterion for the frequency diversity. We select users starting from the one with the worst receiving conditions. For each selected user we sequentially consider all its service flows. We start from the service flow with the maximum traffic load.

We use the following approach to estimate the DL load in case of multi-user diversity. Again, the shared system resources are the time-frequency resource of the DL subframe and the DL transmission power. Adaptation parameters are the set of the assigned DL subchannels, coding and modulation schemes, and transmission power values. Consequently, we need to solve the optimization problem (Equations 14.8 and 14.9). Again, we solve this problem iteratively.

During the first iteration, we select one DL service flow and one DL subchannel and solve the optimization problem

$$q_{1,1,1}^{DL,opt} = \arg\min_{q_{1,1,1}^{DL}}[\max(s_{1,1,1}^{DL}, \rho_{1,1,1}^{DL})] \qquad (14.16)$$

using the full enumeration and taking $p_{1,1,1}^{DL,opt} = p_{1,1,1}^{DL} (q_{1,1,1}^{DL})$.

In Equation 14.16, values of $s_{1,1,1}^{DL}$ and $\rho_{1,1,1}^{DL}$ are calculated given that only service flow (1,1) is placed into subchannel (1) of the DL subframe. Examples of finding transmission power value $p_{1,1,1}^{DL}$ as a function of coding and modulation scheme number $q_{1,1,1}^{DL}$, QoS requirements, and receiving conditions for the multi-user diversity are given in [20].

After optimization problem (Equation 14.16) is solved, we fix subchannel (1), coding and modulation scheme $q_{1,1,1}^{DL,opt}$, and transmission power value $p_{1,1,1}^{DL,opt}$ for service flow (1,1).

During the second iteration, we again select one DL service flow and one DL subchannel and solve the optimization problem

$$q_{2,2,2}^{DL,opt} = \arg \min_{q_{2,2,2}^{DL}}[\max(s_{1,1,1}^{DL,opt} + s_{2,2,2}^{DL}, \rho_{1,1,1}^{DL,opt} + \rho_{2,2,2}^{DL})] \qquad (14.17)$$

using the full enumeration and taking $p_{2,2,2}^{DL,opt} = p_{2,2,2}^{DL} (q_{2,2,2}^{DL})$.

In Equation 14.17, values of $s_{1,1,1}^{DL,opt}$ and $\rho_{1,1,1}^{DL,opt}$ are calculated during the first iteration, values of $s_{2,2,2}^{DL}$ and $\rho_{2,2,2}^{DL}$ are calculated given that service flow (1,1) is already placed into subchannel (1) of the DL subframe with coding and modulation scheme $q_{1,1,1}^{DL,opt}$ and transmission power value $p_{1,1,1}^{DL,opt}$, and service flow (2,2) is additionally placed into subchannel (2) of the DL subframe.

We continue the iterations until all the DL service flows of all the users are considered. Recall that during each iteration we solve the optimization problem only for one service flow of one user in one subchannel. In case of overloading, we use the same approach as we used for the frequency diversity.

Different criteria can be used to select one subchannel and one service flow during each iteration. We use the following criterion for the multi-user diversity. We select the subchannel and service flow with the best receiving conditions.

We use the following approach to estimate the UL load in case of frequency diversity. In the UL, the shared system resource is the time-frequency resource of the UL subframe. Adaptation parameters are the set of the assigned coding and modulation schemes and transmission power values.

In the UL, the maximum transmission power constraints (Equation 14.3) are of a distributed nature. Consequently, in case of frequency diversity the optimization problem (Equations 14.2 through 14.4) can be divided into several optimization subproblems

$$u_i^{UL} = \min_{q_{i,j}^{UL}, p_{i,j}^{UL}} (s_i^{UL}), \qquad (14.18)$$

subject to

$$P_i^{UL} \le P_{max,i}^{UL}, \qquad (14.19)$$

$$p_{i,j}^{\text{UL}} \geq p_{i,j}^{\text{QoS}}, \tag{14.20}$$

where

$$u^{\text{UL}} = \sum_{i} u_i^{\text{UL}}. \tag{14.21}$$

We use the following iterative algorithm to solve the optimization subproblem (Equations 14.18 through 14.20).

First, we assign the coding and modulation schemes with the maximum transmission rates and transmission power values $p_{i,j}^{\text{UL}} = p_{i,j}^{\text{UL}}(q_{i,j}^{\text{UL}})$ (as described in Refs. [17–19]) to all the service flows of a user.

Then we iteratively decrease transmission rates of these service flows until the constraint (Equation 14.19) is satisfied. Let us note that when the transmission rate is decreased, the transmission power value $p_{i,j}^{\text{UL}}$ required to satisfy the QoS requirements is also decreased.

When the optimization subproblems (Equations 14.18 through 14.20) are solved for all the users, we find the UL load using Equation 14.21. As optimization subproblems (Equations 14.18 through 14.20) are solved independently for each user in case of frequency diversity, the order does not matter.

We use the following approach to estimate the UL load in case of the multi-user diversity. Again, the shared system resource is the time-frequency resource of the UL subframe. Adaptation parameters are the set of the assigned UL subchannels, coding and modulation schemes, and transmission power values.

In case of multi-user diversity, the optimization problem (Equations 14.2 through 14.4) cannot be divided into optimization subproblems. However, the maximum transmission power constraints (Equation 14.3) are of a distributed nature. Consequently, we solve the optimization problem (Equations 14.2 through 14.4) user-by-user. We start from the user with the best receiving conditions.

For a given user, we first assign the coding and modulation scheme with the maximum transmission rate to all its service flows. Then, we place these service flows into the free part of the UL subframe starting from the subchannels with the best receiving conditions. For each service flow and for each frequency subchannel we select the transmission power value $p_{i,j,\ell}^{\text{UL,opt}} = p_{i,j,\ell}^{\text{UL}}(q_{i,j,\ell}^{\text{UL}})$ as described in [20].

For the multi-user diversity receiving conditions are different on different subchannels. Consequently, we sequentially place users in the same UL subframe while solving the optimization problem (Equations 14.2 through 14.4) for them.

When an UL subframe is overloaded, we use the following procedure. We place as many service flows as possible into the first UL subframe. Then, we virtually add the second UL subframe, the same as the first UL subframe, to the right of the first UL subframe (to keep the maximum UL transmission power constraints adequate). We place as many service flows as possible into the second UL subframe. We repeat this procedure until all UL service flows have been placed. Then, we calculate the UL load over all UL subframes used. Here, we add all time-frequency resources used of all UL subframes used to obtain s^{UL}.

14.3.5 Summary

The described system load model of the Mobile WiMAX network plays a key role in the RRM algorithms. It takes into account all distinct features of the Mobile WiMAX network, including the time-frequency and power resources, QoS requirements, AMC and power control, TDD and adaptation of the frame boundary between the DL and UL subframes, different users with different traffic and receiving conditions.

Our system load model includes UL load, DL load, sector load, and network load. To calculate each of these system loads, we use the following approach. First, we write the expression for the amount of the normalized shared system resources, consumed by all users, as a function of the adaptation parameters. Then, we minimize this expression over the adaptation parameters under the constraint on the individual system resources, while satisfying the QoS requirements for all users.

We formulated the conditional optimization problems to perform the optimal estimation of the system load. Also, we described the fast practical algorithms of the system load estimation, both for the frequency and multi-user diversity.

14.4   Call Admission Control

Any wireless network is resource-constrained and can serve a limited number of users with a given QoS level. Hence, a CAC algorithm that decides whether new users should be admitted to the network is required. The goals of the CAC are satisfying QoS requirements for the admitted users, maximizing operator's revenue by maximizing the network capacity, and supporting fairness and priorities among the users [21,22].

A CAC algorithm admits a new user to the network based on a given criterion. The admission criteria may be different. The known CAC schemes are based on SINR, interference, bandwidth, load, or system capacity [21,22]. Several publications are available that consider the CAC in the OFDM network [23,24]. The algorithms of these publications admit new users based on the analysis of the current status of the queues of the active users.

In the Mobile WiMAX network the most suitable scheme is the one maximizing network capacity while satisfying QoS requirements for all the admitted users. Such scheme maximizes operator's revenue and guarantees user's satisfaction. We propose such CAC algorithm based on our system load model.

We admit a new user to the sector, when the following conditions are satisfied

$$u_{\text{new}} \leq 1, \qquad (14.22)$$

$$p_{\text{new},j}^{\text{UL}} \geq p_{\text{new},j}^{\text{QoS}}. \qquad (14.23)$$

Condition 14.22 checks that the new user does not overload the sector. Condition 14.23 checks that the new user is within the coverage area of the sector, that is, the QoS requirements are satisfied for all his or her service flows.

In our model system load is equal to the minimum required system resources. Consequently, our CAC algorithm maximizes the sector capacity in the Mobile WiMAX network.

14.4.1 Dynamic Bandwidth Allocation

The CAC algorithm does not admit new users to a sector when this leads to the sector overloading. Let us consider two neighboring sectors. The first sector is almost fully loaded, while the second sector has available resources. If the new users attempt to access the first sector, they will not be admitted. In this case it is advantageous to transfer some available resources from the second sector to the first sector. The idea of adaptive sharing of the system resources by the sectors is called dynamic bandwidth allocation.

The Mobile WiMAX network has a convenient possibility for the dynamic bandwidth allocation. Adjacent sectors may use different groups of subcarriers within a common signal bandwidth. For example, one-third of all the subcarriers may be used in each of the three sectors. When the adjacent sectors have a common RRM, the subcarriers can be adaptively distributed among these sectors. Thus, we can take into account different traffic loads in different sectors.

We propose using our system load model for the optimum dynamic bandwidth allocation in the Mobile WiMAX network. When the subcarriers are allocated to the adjacent sectors in such a way that their sector loads are equal, the total capacity of these sectors is maximized.

14.5 Adaptive Transmission

While optimizing the performance of a wireless communication system, it is advantageous to use adaptive transmission approach [25–27] and cross-layer optimization approach [28–30]. The adaptive transmission approach enables adaptation of the PHY layer transmission parameters to the changing receiving conditions. Examples of the adaptive transmission are power control [31–34] and adaptive modulation [35,36]. The cross-layer optimization approach enables opportunistic data transmission over a wireless medium by joint adaptation of transmission parameters of two or more layers of the wireless system.

In OFDMA, adaptive transmission algorithms include scheduling, AMC, power control, and adaptive resource allocation. A scheduler guarantees QoS and fairness and can also handle user priorities by making a decision on how much data and of which service flows will be transmitted in the current frame. To satisfy the required QoS level, different pairs of coding and modulation scheme number and transmission power value can be used. The selection of a pair for data transmission is based on the target function. For example, when the total transmission power is minimized, a coding and modulation scheme with the minimum transmission rate and a corresponding transmission power are selected. However, when the total allocated time-frequency resource is minimized, a coding and modulation scheme with the maximum transmission rate and a corresponding transmission power are selected. In the OFDMA with the multi-user diversity, adaptive resource allocation algorithm plays an important role. Receiving conditions are different for different users on the same frequency subchannel. Moreover, they are different for the same user on different frequency subchannels. Users can be assigned to the subchannels with the best receiving conditions, thereby multi-user diversity gain is obtained. When the cross-layer optimization approach is used, scheduling, AMC, power control, and adaptive resource allocation algorithms should have the same target function and should be jointly optimized.

Optimization of the OFDMA systems is a subject of a considerable literature. Time-frequency resource minimization [37,38], transmission power minimization [39–41], throughput maximization [42–45], and utility function optimization [29,30,46–50] are traditionally performed. However, most of the known algorithms do not consider the distinct features of the Mobile WiMAX network.

First, MAC and PHY layers processing is not taken into account. Data blocks of a service flow arrive from the upper layers at the MAC layer, where they are converted into data packets using fragmentation and packing operations. Also, cyclic redundancy check and ARQ mechanisms can be used. The set of data packets of the service flow arrives at the PHY layer, where it is converted into coding blocks. Each coding block is coded and decoded independently. QoS requirements are specified for the data blocks, while the coding blocks are transmitted and received. Hence, MAC and PHY layers processing should be taken into account to enable QoS-guaranteed data transmission [17–20].

Moreover, most of the known algorithms do not perform joint DL and UL optimization. A common problem for all the adaptive transmission algorithms in the OFDMA networks is their computational complexity due to a large number of degrees of freedom.

In the following subsections we describe our load-balancing transmission algorithm for both the frequency and multi-user diversity. We show the efficiency of our algorithm using system level simulation. We compare our load-balancing algorithm with our other QoS-guaranteed algorithms of [17,18,20]. Algorithms of [17,18] minimize the total consumed time-frequency resource and the total transmission power, respectively. They are both for the frequency diversity. Algorithm of [20] minimizes time-frequency resource for the multi-user diversity. Algorithms of [17,18,20] also take into account the distinct features of the Mobile WiMAX network, but they are not load-balancing.

14.5.1 *Frequency Diversity*

In case of frequency diversity, the adaptation parameters are position of the frame boundary between the DL and UL subframes, coding and modulation schemes, and transmission power values of the DL and UL service flows. Positions of the service flows within the DL and UL subframes may be selected arbitrary.

In our previous QoS-guaranteed algorithms [17,18] and in our load-balancing algorithm we assume that the scheduler selects a set of the DL and UL service flows to be transmitted in the current frame. We perform optimization for these service flows on the frame-by-frame basis. Moreover, we optimize each sector of the Mobile WiMAX network independently. Under these assumptions, the general formulation of the adaptive transmission optimization problem in the Mobile WiMAX network with the frequency diversity is as follows.

Find:

■ Position of the frame boundary between the DL and UL subframes
■ Coding and modulation schemes
■ Transmission power values

that optimize the target function subject to

■ Maximum DL and UL transmission power constraints
■ Constraints on the time-frequency resource available in the DL and UL subframes
■ QoS requirements

In [17] our target function was minimization of the total used time-frequency resource, while in [18] we minimized the total power transmitted in the DL and UL subframes. Here our target is to maximize the sector capacity.

The formulated adaptive transmission problem is a nonlinear conditional optimization problem. Target function and conditions are nonlinear functions of the optimization parameters. The optimal solution to the problem is the exhaustive search, which is impractical.

Different nonoptimal algorithms, which search for one local extreme value of the target function, are possible to solve the formulated problem. However, they all have two key difficulties. First, all the mentioned target functions have a lot of extreme values; some of them are far from the optimum. Secondly, computational complexity of the algorithms is of great practical interest. The better an algorithm overcomes these difficulties, the more practical value it has.

The simplification of the optimization problem is that for the given position of the frame boundary, optimization in the DL and UL may be performed independently, both in the optimal and nonoptimal algorithms. Consequently, the optimization algorithm may be divided into two procedures, that is, frame boundary position optimization and the DL and UL optimization.

When the DL or UL is optimized, we need to select a coding and modulation scheme and a transmission power value for each DL or UL service flow. Let us consider one service flow. For an arbitrary selected coding and modulation scheme, transmission power value is determined by the QoS requirements, receiving conditions, and MAC and PHY layers processing of this service flow [17–20]. However, for any coding and modulation scheme available in the Mobile WiMAX system there exists a transmission power value such that the QoS requirements are satisfied. Consequently, for each service flow we have a set of pairs of coding and modulation scheme and transmission power value that satisfy the QoS requirements of this service flow.

The next step is to select a particular pair of coding and modulation scheme and transmission power value for each service flow such that the maximum DL and UL transmission power constraints and the constraints on the time-frequency resource available in the DL and UL subframes are satisfied. Here, we have a tradeoff between the time-frequency resource used and the transmission power used. If we increase the transmission rate of a service flow, we need less time-frequency resource and more transmission power. If we decrease the transmission rate of a service flow, the situation is the opposite.

In the DL, both time-frequency and power resources are shared by all the DL service flows of all the users. In the UL, the time-frequency resource is shared by all the UL service flows of all the users, although power resources are individual resources of the users. However, all the UL service flows of a given user share its power resource.

When we minimize the total used time-frequency resource in [17], we use the following procedure for the DL optimization. First, we select the coding and modulation scheme with the maximum transmission rate for all the DL service flows. In other words, we start from the unconditional minimum of the total DL time-frequency resource used. Then we iteratively decrease transmission rates of the service flows until the maximum DL transmission power constraint is satisfied. If during any iteration the constraint on the time-frequency resource available in the DL subframe is not satisfied, we discard one service flow and start the procedure from the very beginning. In the UL the optimization procedure is very similar.

When we minimize the total power transmitted in the DL and UL subframes in [18], we use the following procedure for the DL optimization. First, we select the coding and modulation scheme with the minimum transmission rate for all the DL service flows. In other words, we start from the unconditional minimum of the total DL transmitted power. Then we iteratively increase transmission rates of the service flows until the constraint on the time-frequency resource available in the DL subframe is satisfied. If during any iteration the maximum DL transmission power constraint is not satisfied, we discard one service flow and start the procedure from the very beginning. In the UL the optimization procedure is very similar.

In [17,18], the target function is a sum of the DL and UL components, which are found during the DL and UL optimization. Consequently, the frame boundary position between the DL and UL subframes may be found, for example, by greedy search. In this case, the initial position is of great importance. In [17,18] we start from the position in the middle of the frame. Another value of the initial position is when the time-frequency resources of the DL and UL subframes are proportional to the DL and UL traffic loads. To the best of our knowledge the latter value of the initial position is the best one when the load-balancing approach is not used.

Both algorithms of [17,18] have two key drawbacks. They start from the initial point of the DL and UL optimization, where a part of the constraints are not satisfied in the general case. Moreover, they use a complicated iterative procedure to satisfy all the constraints. The latter drawback has two consequences. First, the algorithms may converge to a point that is far from the optimum. Secondly, the iterative procedure may require a lot of iteration and, thus, it may have high computational complexity.

Applying our load-balancing approach for adaptive transmission leads to the fast and near-optimum solution of the formulated conditional optimization problem. The target function of our load-balancing QoS-guaranteed adaptive transmission algorithm is sector capacity maximization. We use our load-balancing approach for both the frame boundary position optimization and the DL and UL optimization.

The procedure of the frame boundary position optimization is the same as the one used for fast calculation of the system load. First, the position of the frame boundary is selected such that the time-frequency resource of the frame is divided between the DL and UL subframes proportionally to the DL and UL traffic loads. For this frame boundary position we estimate the DL and UL loads using our fast system load estimation algorithm.

Then, we select the position of the frame boundary such that the time-frequency resource of the frame is divided between the DL and UL subframes proportionally to the DL and UL loads, found for the previous position of the frame boundary. This frame boundary position may be used as the final one. Alternatively, additional greedy search may be performed.

In the DL, the shared system resources are the time-frequency resource of the DL subframe and the DL transmission power. We use a very short iterative procedure to optimize the DL. The number of iterations is less than or equal to the number of the DL service flows to be transmitted.

The DL optimization procedure is analogous to the fast procedure of the DL load estimation. During each iteration we select one DL service flow and select a pair of coding and modulation scheme and transmission power value for it such that the DL load is minimized, as described in Equations 14.14 and 14.15. We select the DL service flows starting from the one with the best receiving conditions. We stop the procedure when all the DL service flows have been considered or when all the DL shared system resources have been consumed.

In the UL, the only shared system resource is the time-frequency resource of the UL subframe. The UL optimization procedure is analogous to the fast procedure of the UL load estimation.

We sequentially consider users starting from the user with the best receiving conditions. We select the coding and modulation scheme with the maximum transmission rate for all the UL service flows of this user. Then, we decrease transmission rate of its service flows until its maximum transmission power constraint is satisfied. Finally, we distribute the remaining power of this user among its service flows that use the ARQ mechanism.

Our load-balancing QoS-guaranteed adaptive transmission algorithm is very fast and converges to the point where the sector throughput is near the maximum sector capacity.

The system level simulation results show the efficiency of the load-balancing adaptive transmission algorithm. In the simulation, we consider the Mobile WiMAX network comprising seven cells, where six cells surround the central cell. Each cell has three sectors and a frequency reuse factor of three. The cell radius is 1000 m and the sector bandwidth is 10 MHz.

Figure 14.4 shows the sector throughput as a function of the traffic load for the load-balancing algorithm, the total used time-frequency resource minimization algorithm [17], and the total power transmitted in the DL and UL subframes minimization algorithm [18]. The load-balancing algorithm has a 0.3 b/sec/Hz spectral efficiency gain when the sector is almost fully loaded.

Figure 14.5 shows the simulation time as a function of the traffic load for three algorithms considered. The load-balancing algorithm has a considerable computational efficiency gain.

14.5.2 Multi-User Diversity

In case of multi-user diversity, the adaptation parameters are position of the frame boundary between the DL and UL subframes, coding and modulation schemes, transmission power values, and positions (frequency subchannels) of the DL and UL service flows.

In our previous QoS-guaranteed algorithm [20] and in our load-balancing algorithm we again assume that the scheduler selects a set of the DL and UL service flows to be transmitted in the current frame and we optimize each sector of the

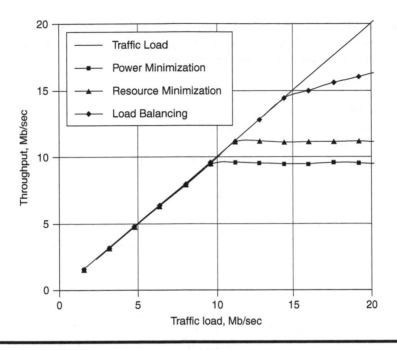

Figure 14.4   Sector throughput as a function of traffic load.

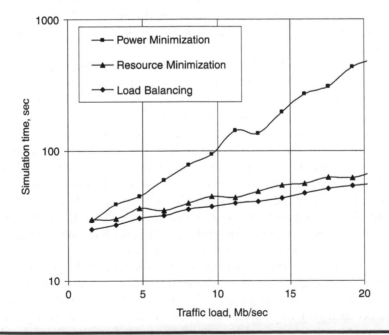

Figure 14.5   Simulation time as a function of traffic load.

Mobile WiMAX network independently. Under these assumptions, the general formulation of the adaptive transmission optimization problem in the Mobile WiMAX network with the multi-user diversity is as follows.

Find:

- Position of the frame boundary between the DL and UL subframes
- Coding and modulation schemes
- Transmission power values
- Positions of the DL and UL service flows within the frequency subchannels

that optimize the target function subject to

- Maximum DL and UL transmission power constraints
- Constraints on the time-frequency resource available in the DL and UL subframes
- QoS requirements

Compared with the frequency diversity case, an additional degree of freedom is added in the multi-user diversity. Receiving conditions are different for different users within the given frequency subchannel. Hence, adaptive frequency subchannel allocation is possible to enhance the system performance. OFDMA with the multi-user diversity maximizes the system capacity, when appropriate adaptive transmission algorithms are used [51]. However, this additional degree of freedom considerably increases computational complexity of the adaptive transmission algorithms.

As for the frequency diversity, the given position of the frame boundary optimization in the DL and UL may be performed independently. Consequently, the optimization algorithm may be divided into two procedures, that is, frame boundary position optimization and the DL and UL optimization.

When we minimize the total used time-frequency resource in [20], we use the following procedure for the DL optimization. First, we select the coding and modulation scheme with the maximum transmission rate for all the DL service flows. Then we iteratively decrease transmission rates of the service flows until the maximum DL transmission power constraint is satisfied. If during any iteration the constraint on the time-frequency resource available in the DL subframe is not satisfied, we discard one service flow and start the procedure from the very beginning. In the UL the optimization procedure is very similar.

Compared with the frequency diversity, we need to find the placement of the service flows within the frame, that is, select frequency subchannels for the service flows. During each iteration we first select one service flow and one frequency subchannel such that the transmission power value required to satisfy the QoS requirements is minimum for this service flow in this frequency subchannel. The minimum is found among all service flows and all frequency subchannels. Then we select the next service flow and the next frequency subchannel using the same approach and so on.

The procedure of the frame boundary position optimization is the same as for the frequency diversity.

If we add the placement procedure to the algorithm of minimization of the total power transmitted in the DL and UL subframes for the frequency diversity [18], then we will obtain the power minimization algorithm for the multi-user diversity.

Both algorithms for the multi-user diversity have two key drawbacks, the same as for the frequency diversity. They start from the initial point of the DL and UL optimization, where a part of the constraints are not satisfied in the general case. Moreover, they use a complicated iterative procedure to satisfy all the constraints. The latter drawback adds a lot of computational complexity, when the multi-user diversity is considered.

Applying our load-balancing approach for adaptive transmission leads to the fast and near-optimum solution of the formulated conditional optimization problem. The target function of our load-balancing QoS-guaranteed adaptive transmission algorithm is sector capacity maximization.

The procedure of the frame boundary position optimization is the same as the one used for the fast calculation of the system load and for the frequency diversity case.

In the DL, the shared system resources are the time-frequency resource of the DL subframe and the DL transmission power. We use a very short iterative procedure to optimize the DL.

The DL optimization procedure is analogous to the fast procedure of the DL load estimation. During each iteration we select one DL service flow and one DL frequency subchannel and we also select a pair of coding and modulation scheme and transmission power value for them such that the DL load is minimized, as described in Equations 14.16 and 14.17. We select the DL service flows and frequency subchannels starting from the service with the best receiving conditions in the selected frequency subchannels. We stop the procedure when all the DL service flows have been considered or when all the DL shared system resources have been consumed.

In the UL, the only shared system resource is the time-frequency resource of the UL subframe. The uplink optimization procedure is analogous to the fast procedure of the uplink load estimation.

We sequentially consider users starting from the user with the best receiving conditions. For a given user, we first assign the coding and modulation scheme with the maximum transmission rate to all its service flows. Then, we place these service flows into the free part of the UL subframe starting from the subchannels with the best receiving conditions.

Our load-balancing QoS-guaranteed adaptive transmission algorithm is very fast and converges to the point where the sector throughput is near the maximum sector throughput.

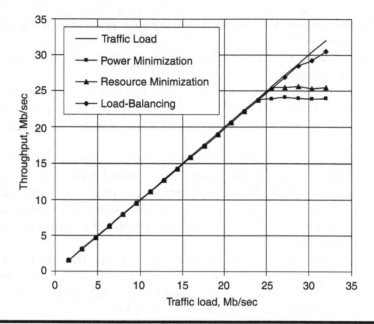

Figure 14.6 Sector throughput as a function of traffic load.

The system level simulation results show the efficiency of the load-balancing adaptive transmission algorithm. In the simulation, we consider the Mobile WiMAX network comprising seven cells, where six cells surround the central cell. Each cell has three sectors and a frequency reuse factor of three. The cell radius is 1000 m and the sector bandwidth is 10 MHz.

Figure 14.6 shows the sector throughput as a function of the traffic load for the load-balancing algorithm, the total used time-frequency resource minimization algorithm [20], and the algorithm minimizing the total power transmitted in the DL and UL subframes. The load-balancing algorithm has a 0.3 b/sec/Hz spectral efficiency gain when the sector is almost fully loaded.

Figure 14.7 shows the simulation time as a function of the traffic load for three algorithms considered. The load-balancing algorithm has a considerable computational efficiency gain of several orders.

14.5.3 Summary

We presented the general formulation of the adaptive transmission conditional optimization problem in the Mobile WiMAX system, both for the frequency and multi-user diversity. This problem formulation is different from the ones currently present in the literature. It is quite general. The only limitations are that

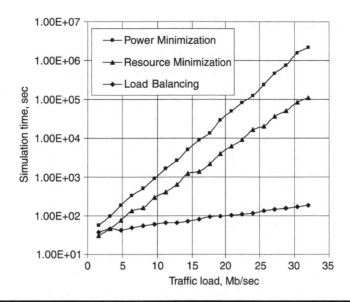

Figure 14.7 Simulation time as a function of traffic load.

optimization is performed frame-by-frame and independently for each sector. However, these limitations are driven by practical considerations.

An optimal solution of the formulated problem is the exhaustive search, which is inapplicable in practice. Different nonoptimal algorithms are possible to solve this conditional optimization problem. However, they all have two key difficulties. First, all the mentioned target functions have a lot of extreme values; some of them are far from the optimum. Secondly, computational complexity of the algorithms is of great practical interest. The better an algorithm overcomes these difficulties, the more practical value it has.

Although the system load model is not traditionally employed in the adaptive transmission algorithms, its usage results in a very efficient algorithm that provides a spectral efficiency gain and is considerably less computationally intensive.

Both for the frequency and multi-user diversity the spectral efficiency gain is about 0.3 b/sec/Hz. The computational efficiency gain is considerable. It is greater for the multi-user diversity and reaches a value of several orders.

When our target function is the sector capacity maximization, the load-balancing algorithm outperforms the algorithms of the total used time-frequency resource and the total transmitted power minimization, both for the frequency and multi-user diversity. We advise that this algorithm be used when the sector is almost fully loaded.

However, when the sector is not fully loaded, the total transmission power minimization algorithm may be useful. When the total transmission power is minimized, the interference to other sectors is decreased. This improves the receiving conditions in the neighboring sectors.

14.6   Horizontal Handover

Handover algorithms first appeared in cellular networks with mobile users. When moving, a user passes from the serving sector's coverage area to the coverage area of another sector. As the receiving conditions of this user in the serving sector degrade, we come to a point when the user can no longer maintain a connection in his or her serving sector. Therefore, it appears reasonable to hand over this user to the sector, to the coverage area of which the person currently belongs [52–55].

The receiving conditions are characterized by the received signal level or SINR. Consequently, traditional handover algorithms are based on the received signal level or SINR [56–58]. The user is handed over to the sector with the maximum signal level or SINR value. This scheme may be expanded by adding thresholds to decrease the number of ping-pong events and signaling load and to keep the call dropping probability low.

In the Mobile WiMAX network the horizontal handover should be seamless [59]. Seamless horizontal handover is a handover that continuously guarantees the required QoS for all users' service flows while the person is active in the network. To guarantee QoS requirements, DL and UL receiving conditions and the sector load of the serving sector should be taken into account. The horizontal handover algorithm guarantees QoS requirements by assigning a new serving sector to the user when the receiving conditions or the sector load change. Traditional horizontal handover algorithms do not take the load into account. Hence, they cannot guarantee QoS requirements in the Mobile WiMAX network.

We proposed the load-balancing QoS-guaranteed horizontal handover algorithm that maximizes the capacity of the Mobile WiMAX network in [60].

The primary goal of our handover algorithm is to distribute the load of the overloaded sectors among other sectors, that is, to eliminate overloading. In some sectors it may be impossible to eliminate overloading because of specific distribution of the users. For these sectors the primary goal is to decrease their loads as much as possible. The handover algorithm that achieves this goal reaches the network capacity under the saturation conditions.

In the real network, the handover algorithm requires some execution time. Thus, increasing the traffic load may result in the overloading of the sectors that are almost fully loaded. Hence, the secondary goal of the handover algorithm is to balance the load in the nonoverloaded sectors.

Consequently, we define the optimal handover algorithm as the algorithm that minimizes the maximum sector load in all subsets of the Mobile WiMAX network sectors set, starting from the subset with the maximum number of sectors and ending with the subsets with the minimum number of sectors.

We use the optimization procedure consisting of $K-1$ steps, where K is the number of sectors in the Mobile WiMAX network. During the first step we select the serving sectors for all network users to minimize the maximum sector load among all K sectors. Then, for the users of the sector with the maximum sector load this sector becomes a new serving sector. This sector and all its users are excluded from further optimization. During the second step the remaining users and $K-1$ sectors are optimized in the same way. During the last step we minimize the maximum sector load for the two remaining sectors. After this optimization procedure we initiate a horizontal handover procedure for the users, whose serving sector number has been changed.

The proposed horizontal handover algorithm maximizes the network capacity and guarantees QoS requirements when the network is not overloaded. We show the efficiency of our algorithm using the system level simulation. The simulated Mobile WiMAX network includes seven cells. Six cells surround the central cell. The frequency reuse factor is seven, the cell radius is 300 m, and the signal bandwidth is 10 MHz in each cell.

Figures 14.8 through 14.10 illustrate the advantages of our algorithm compared with the traditional SINR-based algorithm in the Mobile WiMAX network.

Figures 14.8 shows the maximum sector load among seven sectors as a function of frame number. It indicates that the traditional handover algorithm occasionally leads to overloading, that is, to the network condition when the QoS requirements are not satisfied for the users. Our load-balancing handover algorithm keeps the maximum sector load less than one, thus guarantees meeting the QoS requirements.

Figure 14.9 shows the load of the central sector as a function of the network traffic load. The SINR-based handover algorithm leads to overloading under the network traffic load value of about 22 Mb/sec, whereas our load-balancing algorithm results in the overloading condition under the network traffic load value of about 70 Mb/sec.

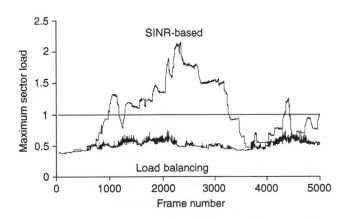

Figure 14.8 Maximum sector load as a function of frame number. SINR, Signal-to-Interference-plus-Noise-Ratio. (From Moiseev, S.N., Filin, S.A., Kondakov, M.S., Garmonov, A.V., Savinkov, A.Y., Park, Y.S., Yim, D.H., Lee, J.H., Cheon, S.H., and Han, K.T., *IEEE Global Communications Conference (GLOBECOM 2006)*. With permission.)

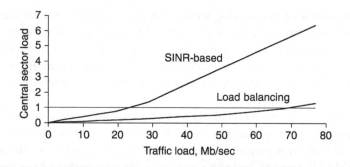

Figure 14.9 Load of central sector as a function of network traffic load. SINR, Signal-to-Interference-plus-Noise-Ratio. (From Moiseev, S.N., Filin, S.A., Kondakov, M.S., Garmonov, A.V., Savinkov, A.Y., Park, Y.S., Yim, D.H., Lee, J.H., Cheon, S.H., and Han, K.T., *IEEE Global Communications Conference (GLOBECOM 2006)*. With permission.)

Figure 14.10 shows the network throughput, that is, the throughput of seven sectors as a function of the network traffic load. The SINR-based handover algorithm reaches the maximum throughput value of about 22 Mb/sec, while our load-balancing algorithm gains the maximum throughput value of about 70 Mb/sec.

Therefore, the load-balancing approach enables development of the efficient horizontal handover algorithm in the Mobile WiMAX network. This algorithm guarantees QoS requirements and maximizes network capacity. Our load-balancing algorithm provides a considerable gain in the network capacity over the traditional SINR-based algorithm.

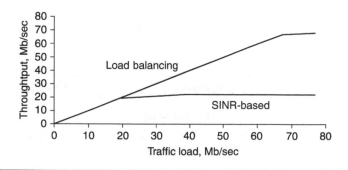

Figure 14.10 Network throughput as a function of network traffic load. SINR, Signal-to-Interference-plus-Noise-Ratio. (From Moiseev, S.N., Filin, S.A., Kondakov, M.S., Garmonov, A.V., Savinkov, A.Y., Park, Y.S., Yim, D.H., Lee, J.H., Cheon, S.H., and Han, K.T., *IEEE Global Communications Conference (GLOBECOM 2006)*. With permission.)

14.7   Summary of Mobile WiMAX RRM Algorithms

Let us describe how CAC, horizontal handover, dynamic bandwidth allocation, and adaptive transmission algorithms work together in the Mobile WiMAX network. The block diagram of these algorithms is shown in Figure 14.11.

When a new user arrives in the network, CAC algorithm is invoked. It makes decision whether to admit the user to the system. The algorithm takes into account the current network load and the new user characteristics, such as receiving conditions, traffic load, and QoS requirements of the user's service flows.

If the new user is admitted to the Mobile WiMAX network, horizontal handover and dynamic bandwidth allocation algorithms are invoked. Both of them distribute the load among the sectors of the network. The horizontal handover algorithm distributes the users among the sectors of the network in such a way that the maximum sector load in all subsets of the Mobile WiMAX network sectors set is minimized. Sometimes, the positions of the users within the sectors are such that it is more advantageous to distribute the system resources among these sectors. This is done by means of the dynamic bandwidth allocation algorithm. The horizontal handover and dynamic bandwidth allocation algorithms work in parallel with close coordination.

The adaptive transmission algorithm is invoked in each frame in each sector of the Mobile WiMAX network. It performs scheduling, AMC, power control, and adaptive allocation of frequency subchannels.

During transmission, some users may experience very bad receiving conditions for quite a while due to coverage holes. For these users, their QoS requirements are not satisfied, although they still consume system resources. This situation is indicated to the CAC algorithm, which removes these users from the network.

In our load-balancing RRM framework each algorithm individually and all algorithms as a whole satisfy QoS requirements and maximize the network capacity. They have the same target and operate in close cooperation.

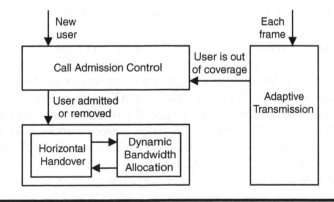

Figure 14.11   Block diagram of Mobile WiMAX radio resource management algorithms.

14.8   Vertical Handover

Multiple wireless networks based on different transmission technologies, belonging to different operators, having different coverage areas, and developed to provide different services are currently deployed. The 4G wireless network is a heterogeneous network, combining the existing wireless networks [61].

To increase the efficiency of the 4G heterogeneous network, it is beneficial to use a common RRM. It enables joint optimization of the entire 4G heterogeneous network. The key algorithm of the common RRM is a vertical handover algorithm. While the horizontal handover algorithm should guarantee QoS requirements, the vertical handover algorithm should be applied to a wider set of tasks [62]. The vertical handover algorithm should optimize the characteristics of the 4G heterogeneous wireless network, for example, maximize its capacity. The vertical handover algorithm solves these tasks by selecting a serving sector for a user among all sectors of all networks of the 4G heterogeneous wireless network. The vertical handover algorithm should take into account QoS, system load, monetary cost, security, user preferences, power consumption, and operator benefits [63,64].

The Mobile WiMAX network will be a part of the 4G heterogeneous wireless network. We propose the load-balancing QoS-guaranteed vertical handover algorithm for the 4G heterogeneous wireless network including the Mobile WiMAX network. Our algorithm maximizes the capacity of the 4G heterogeneous wireless network.

To develop the load-balancing vertical handover algorithm, we need the expressions for the sector load for all parts of the heterogeneous wireless network. Moreover, these sector loads should be comparable. Consequently, they should satisfy our requirements for the system load model of the Mobile WiMAX network:

- System load model should include shared system resources only.
- System load should be equal to the minimum amount of the system resource needed to satisfy QoS requirements for all the users served.
- System load should be normalized to the available system resource.

When the expressions for the sector load are available, the sectors of different networks may be considered as the sectors of the same heterogeneous wireless network. For this 4G heterogeneous wireless network we can use the proposed horizontal handover algorithm for the Mobile WiMAX network [60].

We illustrate the efficiency of the proposed vertical handover algorithm using system level simulation. The simulated 4G heterogeneous wireless network is shown in Figure 14.12. It includes seven Mobile WiMAX sectors, six IEEE 802.11b sectors, and three cdma2000 1xEV-DO sectors.

In the Mobile WiMAX network the sector radius is 300 m, the frequency reuse factor is seven, and the signal bandwidth is 10 MHz in each sector. In the IEEE 802.11b network the sector radius is 100 m, in the cdma2000 1xEV-DO network it equals 2000 m.

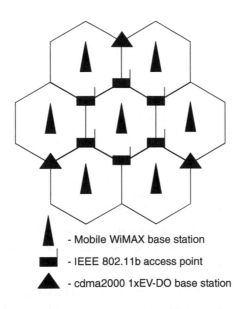

Figure 14.12   Simulated 4G heterogeneous wireless network.

For the IEEE 802.11b and cdma2000 1xEV-DO networks we use the system load models that satisfy our requirements. To calculate the system load in the IEEE 802.11b network, the ratio of the traffic load to the throughput should be minimized [65]. The minimization involves the maximization of the IEEE 802.11b saturation throughput, which can be performed as described in [66–68]. In the cdma2000 1xEV-DO network, the time and interference resources in the DL need to be combined and the interference resource in the UL should be used.

We compare our vertical handover algorithm with the SINR-based algorithm. In the SINR-based algorithm, we select the sector with the maximum SINR value given that this SINR value is higher than the threshold determined by the QoS requirements.

First, we show the maximum sector load among all sectors of the 4G heterogeneous network as a function of the system time in Figure 14.13. This figure indicates that the SINR-based vertical handover algorithm occasionally leads to overloading, that is, to the network condition when the QoS requirements are not satisfied for the users. Our load-balancing vertical handover algorithm keeps the maximum sector load less than one and, hence, guarantees meeting the QoS requirements.

Then, we show the maximum sector load among all sectors of the 4G heterogeneous network and the network throughput as the functions of the traffic load in Figures 14.14 and 14.15.

Figure 14.14 shows the maximum sector load as a function of the network traffic load. The SINR-based handover algorithm leads to overloading under the network

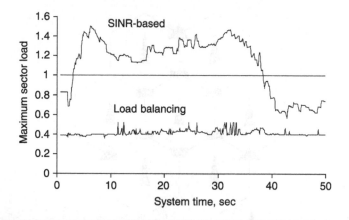

Figure 14.13 Maximum sector load as a function of system time. SINR, Signal-to-Interference-plus-Noise-Ratio.

traffic load value of about 25 Mb/sec, while our load-balancing algorithm results in the overloading condition under the network traffic load value of about 90 Mb/sec.

Figure 14.15 shows the network throughput as a function of the network traffic load. The SINR-based handover algorithm reaches the maximum throughput value of about 25 Mb/sec, while our load-balancing algorithm gains the maximum throughput value of about 90 Mb/sec.

Our vertical handover algorithm guarantees QoS requirements and maximizes the capacity of the 4G heterogeneous wireless network. It provides a considerable capacity gain over the SINR-based algorithm.

14.9 Future Research Issues

Some issues have been left out of our discussion. They have to do with future research and may further improve the performance of the Mobile WiMAX network. Let us briefly describe these issues and key ideas for each of them.

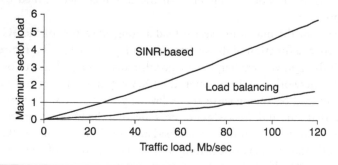

Figure 14.14 Maximum sector load as a function of network traffic load. SINR, Signal-to-Interference-plus-Noise-Ratio.

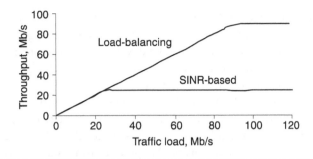

Figure 14.15 Network throughput as a function of network traffic load. SINR, Signal-to-Interference-plus-Noise-Ratio.

The first direction for the future research on RRM algorithms is the MIMO technology. Space-time coding and receive diversity techniques increase network coverage, whereas spatial multiplexing in the DL and collaborative spatial multiplexing in the UL improve spectral efficiency. Adaptive switching between these two MIMO techniques is a promising RRM algorithm. The idea is to use space-time coding for users with bad receiving conditions and to use spatial multiplexing for users with good receiving conditions.

Another MIMO technique, that is, smart antenna with interference cancellation improves spatial separation of the users. Also, adaptive transmission algorithms for the multi-user MIMO are of great interest. These algorithms are particularly important for the UL collaborative spatial multiplexing.

The second direction for the future research is a fractional reuse factor. Let us consider three adjacent sectors. We may use the same frequency for the users inside these sectors, and three different frequencies for the users around the edges of these sectors. In this case, the frequency reuse factor will be between one and three. It is expected that this scheme will have higher spectral efficiency than the fixed frequency reuse factor of one and three. The OFDMA technology and frame structure of the Mobile WiMAX network provide convenient capability for implementing the fractional reuse factor.

Another direction is time-frequency scheduling. In the neighboring sectors using the same frequency, different users cause different levels of interference to each other. Consequently, the mutual interference may be reduced by specific allocation of service flows within the time-frequency resource of the frame. Two users that cause the maximum interference to each other may be placed in the areas of nonoverlapping time-frequency resource. This approach will increase coverage and improve spectral efficiency of the Mobile WiMAX network.

So far the scheduling algorithm and other adaptive transmission algorithms have been considered independently. The scheduling algorithm selects the DL and UL service flows to be transmitted in the current frame. Then, AMC, power control, and adaptive frequency subchannel allocation algorithms are applied to the selected service flows. Consequently, the adaptive transmission algorithm including scheduling and all other algorithms may be developed. This algorithm will operate over a moving window of several frames. We believe that such algorithms can further improve the Mobile WiMAX system's performance.

An important issue in the Mobile WiMAX network is overhead. Some times it can take a half of the system capacity or even more. The key components of the overhead are the DL and UL Medium Access Protocol (MAP) messages, bandwidth requests, and acknowledgments. Adaptive transmission algorithms taking into account the overhead could greatly improve the Mobile WiMAX performance. Also, applying AMC and power control for the DL and UL MAP transmission is a promising approach.

One of the planned applications for the Mobile WiMAX network is Voice over IP (VoIP). However, the Mobile WiMAX system is not optimized for it. Transmission of many small packets leads to a lot of overhead. Consequently, development of the adaptive transmission algorithm specifically for the VoIP transmission may improve the VoIP capacity of the Mobile WiMAX network. One of the specific features of the VoIP traffic is that it is bi-directional. Hence, if the user cannot transmit a VoIP service flow in one direction, there is no need to waste the system resources for transmission of its VoIP service flow in the reverse direction.

Cross-layer optimization of the Mobile WiMAX network for supporting the end-to-end QoS is another interesting direction for research. Under this approach, the Mobile WiMAX network can be optimized for specific services or sets of services.

Finally, we believe that using our system load model of the Mobile WiMAX network will greatly improve the network planning tools. In the Mobile WiMAX network, the receiving conditions do not characterize the network coverage. Traffic load, QoS requirements, and available degrees of freedom should also be taken into account. Our system load model takes all these factors into account. Moreover, our general system load model, presented in the previous section, will be beneficial for planning the 4G heterogeneous wireless network.

References

1. *IEEE Standard for Local and Metropolitan Area Networks—Part 16: Air Interface for Fixed Broadband Wireless Access Systems*, IEEE Standard 802.16–2004, October 2004.
2. *Amendment to IEEE Standard for Local and Metropolitan Area Networks—Part 16: Air Interface for Fixed Broadband Wireless Access Systems—Physical and Medium Access Control Layers for Combined Fixed and Mobile Operation in Licensed Bands*, IEEE Standard 802.16e–2005, December 2004.
3. WiMAX Forum (June, 2006). Mobile WiMAX-Part I: A technical overview and performance evaluation. [Online]. Available: http://www.wimaxforum.org.

4. WiMAX Forum (May, 2006). Mobile WiMAX—Part II: A comparative analysis. [Online]. Available: http://www.wimaxforum.org.
5. S. N. Moiseev, S. A. Filin, M. S. Kondakov, A. V. Garmonov, A. Y. Savinkov, Y. S. Park, D. H. Yim, J. H. Lee, S. H. Cheon, and K. T. Han, "Practical propagation channel robust BLER estimation in the OFDM/TDMA and OFDMA broadband wireless access networks," *IEEE Vehicular Technology Conference (VTC 2006 Fall)*, September 2006.
6. S. N. Moiseev, S. A. Filin, M. S. Kondakov, A. V. Garmonov, D. H. Yim, J. Lee, S. Chang, and Y. S. Park, "Analysis of the statistical properties of the interference in the IEEE 802.16 OFDMA network," *IEEE Wireless Communications and Networking Conference (WCNC 2006)*, vol. 4, pp. 1830–1835, April 2006.
7. S. N. Moiseev, S. A. Filin, M. S. Kondakov, A. V. Garmonov, D. H. Yim, J. Lee, S. Chang, and Y. S. Park, "Analysis of the statistical properties of the SINR in the IEEE 802.16 OFDMA network," *IEEE International Conference on Communications (ICC 2006)*, June 2006.
8. N. Damji and T. Le-Ngoc, "Dynamic DL OFDM resource allocation with interference mitigation and macro diversity for multimedia services in wireless cellular systems," *IEEE Wireless Communication and Networking Conference (WCNC 2005)*, vol. 3, no. 1, pp. 1298–1304, March 2005.
9. E. Yanmaz and O. K. Tonguz, "Dynamic load balancing and sharing performance of integrated wireless networks," *IEEE Journal on Selected Areas in Communications*, vol. 22, no. 5, pp. 862–872, June 2004.
10. J. Karlsson and B. Eklundh, "A cellular mobile telephone system with load sharing— an enhancement of directed retry," *IEEE Transactions on Communications*, vol. 37, no. 5, pp. 530–535, May 1989.
11. S. Kim and P. K. Varshney, "Adaptive load balancing with preemption for multimedia cellular networks," *IEEE Wireless Communications and Networking Conference (WCNC 2003)*, vol. 4, no. 1, pp. 1680–1684, March 2003.
12. S. Das, H. Viswanathan, and G. Rittenhouse, "Dynamic load balancing through coordinated scheduling in packet data systems," *IEEE International Conference on Computer Communication (INFOCOM 2003)*, vol. 22, no. 1, pp. 786–796, March 2003.
13. H. Velayos, V. Aleo, and G. Karlsson, "Load balancing in overlapping wireless LAN cells," *IEEE International Conference on Communications (ICC 2004)*, vol. 27, no. 1, pp. 3833–3836, June 2004.
14. W.-Y. Yeo and D.-H. Cho, "An analytical model for reverse link rate control in cdma2000 1xEV-DO systems," *IEEE Communications Letters*, vol. 9, no. 3, pp. 270–272, March 2005.
15. J. Muckenheim and U. Bernhard, "A framework for load control in 3rd generation CDMA networks," *IEEE Global Telecommunications Conference (GLOBECOM 2001)*, vol. 6, no. 1, pp. 3738–3742, November 2001.
16. S. N. Moiseev, S. A. Filin, M. S. Kondakov, A. V. Garmonov, A. Y. Savinkov, Y. S. Park, D. H. Yim, J. H. Lee, S. H. Cheon, and K. T. Han, "System load model for the OFDMA network," *IEEE Communications Letters*, vol. 10, no. 8, pp. 620–622, August 2006.
17. S. A. Filin, S. N. Moiseev, M. S. Kondakov, A. V. Garmonov, D. H. Yim, J. Lee, S. Chang, and Y. S. Park, "QoS-guaranteed cross-layer adaptive transmission algorithms for the IEEE 802.16 OFDMA system," *IEEE Wireless Communications and Networking Conference (WCNC 2006)*, vol. 2, pp. 964–971, April 2006.

18. S. A. Filin, S. N. Moiseev, M. S. Kondakov, A. V. Garmonov, A. Y. Savinkov, Y. S. Park, D. H. Yim, J. H. Lee, S. H. Cheon, and K. T. Han, "QoS-guaranteed cross-layer adaptive transmission algorithms with selective ARQ for the IEEE 802.16 OFDMA system," *IEEE Vehicular Technology Conference (VTC 2006 Fall)*, September 2006.

19. S. N. Moiseev, S. A. Filin, M. S. Kondakov, A. V. Garmonov, A. Y. Savinkov, Y. S. Park, D. H. Yim, J. H. Lee, S. H. Cheon, and K. T. Han, "Optimal average number of data block transmissions for the ARQ mechanism in the IEEE 802.16 OFDMA system," *IEEE International Symposium on Personal, Indoor and Mobile Communications (PIMRC 2006), September 2006.*

20. S. A. Filin, S. N. Moiseev, M. S. Kondakov, A. V. Garmonov, D. H. Yim, J. Lee, S. Chang, and Y. S. Park, "QoS-guaranteed cross-*layer transmission algorithms* with adaptive frequency subchannels allocation in the IEEE 802.16 OFDMA system," *IEEE International Conference on Communications (ICC 2006)*, June 2006.

21. M. H. Ahmed, "Call admission control in wireless networks: a comprehensive survey," *IEEE Communications Surveys*, vol. 7, no. 1, pp. 50–69, first quarter 2005.

22. D. Niyato and E. Hossain, "Call admission control for QoS provisioning in 4G wireless networks: issues and approaches," *IEEE Network*, vol. 19, no. 5, pp. 5–11, September 2005.

23. D. Niyato and E. Hossain, "Connection admission control algorithms for OFDM wireless networks," *IEEE Global Telecommunications Conference (GLOBECOM 2005)*, vol. 24, no. 1, pp. 2453–2457, November 2005.

24. W. S. Jeon and D. G. Jeong, "Combined connection admission control and packet transmission scheduling for mobile Internet services," *IEEE Transactions on Vehicular Technology*, vol. 55, no.5, pp. 1582–1593, September 2006.

25. A. J. Goldsmith and S. G. Chua, "Variable-rate, variable-power MQAM for fading channels," *IEEE Transactions on Communications*, vol. 45, pp. 1218–1230, October 1997.

26. S. T. Chung and A. J. Goldsmith, "Degrees of freedom in adaptive modulation: a unified view," *IEEE Transactions on Communications*, vol. 49, pp. 1561–1571, September 2001.

27. A. J. Goldsmith and P. Varaiya, "Capacity of fading channels with channel side information," *IEEE Transactions on Information Theory*, vol. 43, pp. 1986–1992, November 1997.

28. V. Srivastava and M. Motani, "Cross-layer design: a survey and the road ahead," *IEEE Communications Magazine*, vol. 43, no. 12, pp. 112–119, December 2005.

29. G. Song and Y. (G.) Li, "Cross-layer optimization for OFDM wireless networks. Part I: Theoretical framework," *IEEE Transactions on Wireless Communications*, vol. 4, no. 2, pp. 614–624, March 2005.

30. G. Song and Y. (G.) Li, "Cross-layer optimization for OFDM wireless networks. Part II: Algorithm development," *IEEE Transactions on Wireless Communications*, vol. 4, no. 2, pp. 625–634, March 2005.

31. J. Zander, "Performance of optimum transmitter power control in cellular radio systems," *IEEE Transactions on Vehicular Technology*, vol. 41, no. 1, pp. 57–62, February 1992.

32. J. Zander and M. Frodigh, "Comment on Performance of optimum transmitter power control in cellular radio systems," *IEEE Transactions on Vehicular Technology*, vol. 43, no. 3, p. 636, August 1994.

33. J. Zander, "Distributed cochannel interference control in cellular radio systems," *IEEE Transactions on Vehicular Technology*, vol. 41, no. 3, pp. 1057–1071, August 1992.
34. S. A. Grandhi, R. Vijayan, D. J. Goodman, and J. Zander, "Centralized power control in cellular radio systems," *IEEE Transactions on Vehicular Technology*, vol. 42, no. 4, pp. 466–468, November 1993.
35. T. Keller and L. Hanzo, "Adaptive multicarrier modulation: A convenient framework for time frequency processing in wireless communications," *IEEE Proceedings of the IEEE*, vol. 88, no. 5, pp. 611–640, May 2000.
36. T. Keller and L. Hanzo, "Adaptive modulation techniques for duplex OFDM transmission," *IEEE Transactions on Vehicular Technology*, vol. 49, no. 5, pp. 1893–1906, September 2000.
37. I. Koutsopoulos and L. Tassiulas, "Carrier assignment algorithms in wireless broadband networks with channel adaptation," *IEEE International Conference on Communications (ICC 2001)*, no. 1, pp. 1401–1405, June 2001.
38. M. Acena and S. Pfletschinger, "A spectrally efficient method for subcarrier and bit allocation in OFDMA," *IEEE Vehicular Technology Conference (VTC 2005 Spring)*, vol. 3, pp. 1773–1777, May–June 2005.
39. C. Y. Wong, R. S. Cheng, K. B. Letaief, and R. D. Murch, "Multiuser OFDM with adaptive subcarrier, bit, and power allocation," *IEEE Journal on Selected Areas in Communications*, vol. 17, no. 10, pp. 1747–1758, October 1999.
40. D. Kivanc, G. Li, and H. Liu, "Computationally efficient bandwidth allocation and power control for OFDMA," *IEEE Transactions on Wireless Communications*, vol. 2, no. 6, pp. 1150–1158, November 2003.
41. Y. J. Zhang and K. B. Letaief, "Adaptive resource allocation and scheduling for multiuser packet-based OFDM networks," *IEEE International Conference on Communications (ICC 2004)*, vol. 27, no. 1, pp. 2949–2953, June 2004.
42. H. Yin and H. Liu, "An efficient multiuser loading algorithm for OFDM-based broadband wireless systems," *IEEE Global Communications Conference (GLOBECOM 2000)*, no. 1, pp. 103–107, November 2000.
43. G. Li and H. Liu, "Dynamic resource allocation with finite buffer constraint in broadband OFDMA networks," *IEEE Wireless Communications and Networking Conference (WCNC 2003)*, vol. 4, no. 1, pp. 1037–1042, March 2003.
44. Y. J. Zhang and K. B. Letaief, "Multiuser subcarrier and bit allocation along with adaptive cell selection for OFDM transmission," *IEEE International Conference on Communications (ICC 2002)*, vol. 25, no. 1, pp. 861–865, April 2002.
45. Y. J. Zhang and K. B. Letaief, "Multiuser adaptive subcarrier-and-bit allocation with adaptive cell selection for OFDM systems," *IEEE Transactions on Wireless Communications*, vol. 3, no. 5, pp. 456–468, September 2004.
46. G. Song and Y. (G.) Li, "Utility-based joint physical-MAC layer optimization in OFDM," *IEEE Global Communications Conference (GLOBECOM 2002)*, vol. 21, no. 1, pp. 680–684, November 2002.
47. G. Song, Y. (G.) Li, "Adaptive resource allocation based on utility optimization in OFDM," *IEEE Global Communications Conference (GLOBECOM 2003)*, vol. 22, no. 1, pp. 586–590, December 2003.
48. G. Song, Y. (G.) Li, L. J. Cimini Jr., and H. Zheng, "Joint channel-aware and queue-aware data scheduling in multiple shared wireless channels," *IEEE Wireless Communications and Networking Conference (WCNC 2004)*, vol. 5, no. 1, pp. 1922–1927, March 2004.

49. Z. Jiang, Y. Ge, and Y. (G.) Li, "Max-utility wireless resource management for best-effort traffic," *IEEE Transactions on Wireless Communications*, vol. 4, no. 1, pp. 100–111, January 2005.

50. L. T. H. Lee, C.-J. Chang, Y.-S. Chen, and S. Shen, "A utility-approached radio resource allocation algorithm for DL in OFDMA cellular systems," *IEEE Vehicular Technology Conference (VTC 2005 Spring)*, vol. 3, pp. 1798–1802, May–June 2005.

51. G. Li and H. Liu, "On the optimality of the OFDMA network," *IEEE Communications Letters*, vol. 9, no. 5, pp. 438–440, May 2005.

52. M. Gudmundson, "Analysis of handover algorithms," *IEEE Vehicular Technology Conference (IEEE VTC 1991 Spring)*, pp. 537–542, May 1991.

53. R. Vijayan and J. M. Holtzman, "A model for analyzing handoff algorithms," *IEEE Transactions on Vehicular Technology*, vol. 42, no. 3, pp. 351–356, August 1993.

54. N. Zhang and J. M. Holtzman, "Analysis of handoff algorithms using both absolute and relative measurements," *IEEE Transactions on Vehicular Technology*, vol. 45, no. 1, pp. 174–179, February 1996.

55. S. Agarwal and J. M. Holtzman, "Modeling and analysis of handoff algorithms in multi-cellular systems," *IEEE Vehicular Technology Conference (IEEE VTC 1997 Spring)*, vol. 1, pp. 300–304, May 1997.

56. N. D. Tripathi, J. H. Reed, and H. F. V. Landingham, "Handoff in cellular systems," *IEEE Personal Communications*, vol. 5, no. 6, pp. 26–37, December 1998.

57. M. N. Halgamuge, H. L. Vu, K. Ramamohanarao, and M. Zukerman, "Signal-based evaluation of handoff algorithms," *IEEE Communications Letters*, vol. 9, no. 9, pp. 790–792, September 2005.

58. A. E. Leu and B. L. Mark, "A discrete-time approach to analyze hard handoff performance in cellular networks," *IEEE Transactions on Wireless Communications*, vol. 3, no. 5, pp. 1721–1733, September 2004.

59. L. Dimopoulou, G. Leoleis, and I. S. Venieris, "Fast handover support in a WLAN environment: Challenges and perspectives," *IEEE Network*, vol. 19, no. 3, pp. 14–20, May 2005.

60. S. N. Moiseev, S. A. Filin, M. S. Kondakov, A. V. Garmonov, A. Y. Savinkov, Y. S. Park, D. H. Yim, J. H. Lee, S. H. Cheon, and K. T. Han, "Load-balancing QoS-guaranteed handover in the IEEE 802.16e OFDMA network," *IEEE Global Communications Conference (GLOBECOM 2006)*, November–December 2006.

61. S. Frattasi, H. Fathi, F. H. P. Fitzek, R. Prasad, and M. D. Katz, "Defining 4G technology from the user's perspective," *IEEE Network*, vol. 20, no. 1, pp. 35–41, January 2006.

62. N. Nasser, A. Hasswa, and H. Hassanein, "Handoffs in fourth generation heterogeneous networks," *IEEE Communications Magazine*, vol. 44, no. 10, pp. 96–103, October 2006.

63. F. Zhu and J. McNair, "Optimizations for vertical handoff decision algorithms," *IEEE Wireless Communications and Networking Conference (WCNC 2004)*, vol. 5, no. 1, pp. 867–872, March 2004.

64. S. Y. Hui and K. H. Yeung, "Challenges in the migration to 4G mobile systems," *IEEE Communications Magazine*, vol. 41, no. 12, pp. 54–59, December 2003.

65. S. N. Moiseev, S. A. Filin, M. S. Kondakov, A. V. Garmonov, A. Y. Savinkov, Y. S. Park, D. H. Yim, J. H. Lee, S. H. Cheon, and K. T. Han, "Load-balancing QoS-guaranteed handover in the IEEE 802.11 network," *IEEE International Symposium on Personal, Indoor and Mobile Radio Communications (PIMRC 2006)*, September 2006.

66. S. A. Filin, A. V. Garmonov, A. Y. Savinkov, V. B. Manelis, S. N. Moiseev, M. S. Kondakov, S. H. Cheon, D. H. Yim, K. T. Han, and Y. S. Park, "Joint fragment size and transmission rate optimization for basic access mechanism of IEEE 802.11b distributed coordination function," *European Conference on Wireless Technology (ECWT 2005)*, pp. 403–406, October 2005.

67. A. V. Garmonov, S. H. Cheon, D. H. Yim, K. T. Han, Y. S. Park, A. Y. Savinkov, S. A. Filin, V. B. Manelis, S. N. Moiseev, and M. S. Kondakov, "Joint fragment size and transmission rate optimization for request-to-send/clear-to-send mechanism of IEEE 802.11b distributed coordination function," *IEEE International Symposium on Personal, Indoor and Mobile Communications (PIMRC 2005)*, vol. 3, pp. 1453–1457, September 2005.

68. A. V. Garmonov, S. H. Cheon, D. H. Yim, K. T. Han, Y. S. Park, A. Y. Savinkov, S. A. Filin, V. B. Manelis, S. N. Moiseev, and M. S. Kondakov, "Joint fragment size, transmission rate, and request-to-send/clear-to-send threshold optimization for IEEE 802.11b distributed coordination function," *IEEE Vehicular Technology Conference (VTC 2005 Spring)*, vol. 3, pp. 2056–2060, May–June 2005.

Chapter 15

Multi-Service Opportunistic QoS-Enhanced Scheduler for the Downlink of IEEE 802.16 Point-to-Multipoint Systems

---

Yonghong Zhang and David G. Michelson

In IEEE 802.16-based broadband wireless metropolitan access networks (MANs), the packet scheduler that sits at the Medium Access Control (MAC) layer plays a key role in both maximizing system throughput and achieving a satisfactory QoS. Developing effective multi-service packet schedulers is challenging because both criteria have to be satisfied. Here, we introduce a downlink packet scheduling scheme that combines conventional multi-user diversity with an urgency function whose value increases rapidly as the time remaining to meet the quality of service (QoS) requirement associated with a particular service becomes shorter. Simulation results obtained using combinations of data flows subscribing to IEEE 802.16's UGS, rtPS, nrtPS, and BE services confirm that our multi-service opportunistic QoS-enhanced scheduler (M-OQES) algorithm outperforms existing multi-service scheduling schemes. Moreover, the computational complexity of the M-OQES algorithm compares favorably with that of the widely adopted proportional fair rule.

15.1 Introduction

Unlike wireline systems, IEEE 802.16 broadband wireless MANs must function effectively while contending with time-varying fading channels with limited bandwidth. Accordingly, effective management of scarce radio resources is required to maximize both system throughput and QoS. A key role in achieving these goals is played by the packet scheduler that sits at the MAC layer and which arbitrates among the multiple users that seek access to the shared medium.

The real-time scheduling problem has been intensively studied for many years in conjunction with applications as diverse as sharing of computer processor cycles or time slots on communication channels. For real-time data communications, the objective is to guarantee the delivery of each packet before it expires. In such cases, the earliest deadline first (EDF) or earlier due date policy has been considered optimal. For non-real-time data communications, such as file transfer, however, the objective is to fairly distribute the limited resource to each user, so that all users experience similar levels of service. The generalized processor sharing (GPS) algorithm is one of the several scheduling schemes that can guarantee such fairness in an error-free environment.

Although these schedulers function well when the channel is error-free, they are less effective when the channel experiences fading over time. Examples of past efforts to develop schedulers suitable for wireless applications are described in Reference [1] and the references therein. Instead of treating fading of the wireless channel as an impairment, multi-user diversity exploits this time-varying property by scheduling transmissions to a particular user only during instants of minimal fading [2]. The proportional fair (PF) rule [3] and the opportunistic scheduler [4] are examples of scheduler algorithms that focus on fairness for data services. The exponential rule (EXP) [5] and our own opportunistic QoS-enhanced scheduler (OQES) [6], on the other hand, are suitable for real-time traffic associated with time constraints.

When designing a scheduling scheme suitable for use with the IEEE 802.16 standard, another issue needs to be considered. The standard provides different levels of service to meet the needs of different classes of users. Some users require that each of their packets be delivered within a specific time interval, some require guaranteed minimum traffic rates, while others do not care about the individual behavior of their packets as long as they can receive reasonable service at low cost. These diverse and often contradictory requirements add another level of complexity to the design of scheduling schemes. A few studies (e.g., Ref. [7]) have focused on achieving QoS for the downlink, and some others (e.g., Ref. [8]) seek to optimize the uplink scheduling problem for IEEE 802.16-based systems. However, to the best of our knowledge, none that seek to maximize system throughput while ensuring QoS on the system downlink has been introduced. The existing multi-service scheduling schemes that are suitable for the downlink are mainly designed for wireless asynchronous transfer mode (ATM) systems (e.g., Refs. [9,10]) and are not readily portable to IEEE 802.16-based systems. Moreover, they are not able to exploit the multi-user diversity property of the wireless system. An EXP and PF rule combined algorithm [11] has recently been proposed. It seeks to maximize system throughput for both delay-sensitive traffic and Best Effort (BE) services. However, it does not include a mechanism to ensure minimum data rate.

In this chapter, we propose a multi-service opportunistic M-OQES that schedules multiple service flows from the Base Station (BS) to a multiplicity of Subscriber Stations (SS). It seeks to satisfy all classes of QoS requirements that are defined by the IEEE 802.16 standard while trying to maximize the overall system throughput. In Section 15.2, we review IEEE 802.16-supported services and their QoS requirements. In Section 15.3, we describe our M-OQES. In Section 15.4, we present simulation results that confirm that our M-OQES algorithm outperforms existing multi-service scheduling schemes. In Section 15.5, we summarize the main contributions of this work.

15.2 IEEE 802.16-Supported Services and Their QoS Requirements

Four services are defined in the IEEE 802.16 standard: the Unsolicited Grant Service (UGS), the Real-time Polling Service (rtPS), the Non-real-time Polling Service (nrtPS), and the BE service.

15.2.1 Unsolicited Grant Service

The UGS is designed to support real-time data streams that consist of fixed-length data packets issued at fixed intervals. Typical applications include T1/E1 data streams and Voice over IP (VoIP) without silence suppression. In such cases, the task of the downlink scheduler at the BS is to transmit the packet at fixed intervals with little or no variation so that the Maximum Latency (ML) and Tolerated Jitter (TJ) requirements can be satisfied.

15.2.2   Real-Time Polling Service

The rtPS is designed to support real-time data streams that consist of variable-length data packets that are issued at periodic intervals. Typical applications include video streaming and video conferencing. Such applications require that each packet arrives at the SS within a specified time interval. For example, for video conferencing, the tolerable delay from the BS to the SS is roughly between 40 and 90 ms. Because techniques for buffering at the receiver have improved in recent years, delay variance (jitter) is not as important as it is in the UGS. The main QoS service flow parameter that must be considered for rtPS in downlink scheduling is ML.

15.2.3   Non-Real-Time Polling Service

The nrtPS is designed to support delay-tolerant data streams such as File Transfer Protocol (FTP). Such streams usually consist of variable-length data packets for which a minimum data rate is required. The mandatory QoS service flow parameters that are considered by the BS downlink scheduler are: (1) the Minimum Reserved Traffic Rate (MRTR) and (2) the Traffic Priority (TP), which specify the priority among identical flows.

15.2.4   Best Effort

The BE service is designed to support data streams that do not have any minimum service level requirement and which can be handled on a space-available basis. As there is no minimum requirement, the only QoS service flow parameter that needs to be considered is TP.

15.2.5   Downlink QoS Requirements for
IEEE 802.16 Services

Table 15.1 summarizes the QoS requirements and the mandatory QoS service flow parameters that must be considered by the downlink scheduling scheme. The IEEE 802.16 standard does not specify how resources shall be allocated among same priority BE flows after the QoS requirements of the UGS, rtPS, and nrtPS flows have been satisfied. A reasonable rule is to ensure fairness, so that such flows can receive the same level of service. Another issue that the standard does not address is how nrtPS flows should be treated after their MRTRs have been met. A possible solution is to share the system resources with the BE flows with a certain fair share.

Table 15.1 Quality of Service (QoS) Requirements and Mandatory QoS Service Flow Parameter That Are Considered by the Downlink Scheduler

Service Type	QoS Requirements	Relevant Service Flow Parameters
Unsolicited Grant Service	Periodic transmission	Maximum Latency (ML), Tolerated Jitter (TJ)
Real-time Polling Service	Within certain delay	ML
Non-real-time Polling Service	Maintain minimum data rate; fair share of the remaining resource	Minimum Reserved Traffic Rate (MRTR)
Best Effort	Higher priority flow gets lower delay; fairness among same priority flows	Traffic Priority (TP)

15.3 Multi-Service Opportunistic QoS-Enhanced Scheduler

Our M-OQES scheduler is designed for use in conjunction with the time division multiple access (TDMA)/time division duplex (TDD) based MAC protocol described in the IEEE 802.16 standard. The system architecture is depicted in Figure 15.1. Throughout this chapter, we assume that an appropriate call admission control scheme has been implemented. The terms SS and user are used interchangeably.

In devising a scheme that will satisfy the QoS requirements (as summarized in Table 15.1) while maximizing system throughput, it is necessary to consider the characteristics of the wireless channel. The most important characteristic is the manner in which the signal strength received by wireless users randomly fades over time. Such fading is experienced by both mobile users and stationary or fixed users. In the latter case, fading is caused by movement of the scatterers in the vicinity of the users, including moving vehicles and wind-blown foliage. Scheduling packets to a user without considering the quality of the channel conditions that it is experiencing could lead to low channel utilization.

To maximize system throughput, it is beneficial to exploit multi-user diversity and aim to allocate resources to the user with the best channel quality at each given instant. However, only considering system performance would both sacrifice users with strict time constraints (e.g., real-time applications) and users at unfavorable locations (e.g., at the cell edge). To compensate for such users, we use an urgency factor along with the channel quality condition as the basis for our scheduling rule.

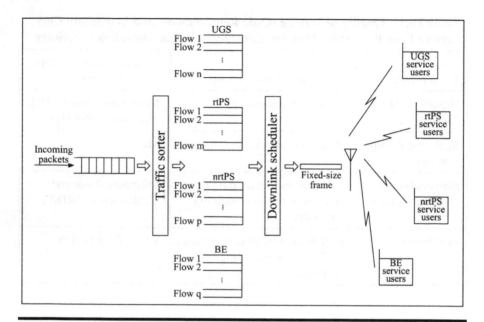

Figure 15.1 System architecture of the multi-service opportunistic QoS-enhanced (M-OQES) scheduling algorithm. UGS, Unsolicited Grant Service; rtPS, Real-time Polling Service; nrtPS, Non-real-time Polling Service; BE, Best Effort.

UGS flows have higher priority over other services, and are scheduled periodically regardless of the channel conditions.

Each time a scheduling decision is made, the flow with the maximum scheduling value $v = UF$ is given the opportunity to transmit its head-of-line (HOL) packet. Here, the factor U is determined by the data rate that the corresponding user can achieve relative to its historical channel conditions and the factor F specifies the urgency of the flow. Both factors are described in more detail next.

15.3.1 Relative Rate Function U

The relative speed of the user is defined as the sum of two terms. The first term is similar to the PF rule, that is, the ratio of the user's current rate to their average rate. The second term is the relative maximum rate. When a user reaches the maximum rate they ever reached in a past window, the user receives the maximum rate reward (MaxRR) λ. (The value of λ is chosen to exceed the maximum possible value of the PF term.) Otherwise, the user receives nothing. In this manner, we make the relative maximum rate the decision rule and thereby encourage users to transmit at their highest possible rate. When no user reaches its maximum rate or multiple users reach their maximum rate simultaneously, that is, the relative maximum rate term of multiple users are the same, the first term or PF rule becomes the decision rule.

15.3.2   Urgency Function F

As the name suggests, the urgency function F indicates how urgently the HOL packet of the flow must be transmitted to satisfy QoS requirements. Because each service has different QoS requirements, each defines urgency in a different way.

For rtPS traffic, every packet must be transmitted within the flow's maximum latency or ML. Accordingly, the urgency factor F^{rtPS} increases as the deadline for transmitting the HOL packet draws closer. If the deadline is far away, the urgency factor is set to its minimum value of 1 and the packet is said to be in its normal state. Once the waiting time passes a certain threshold, say T_{rt}, the flow goes into the urgent state. The closer the flow is to the deadline, the higher the value of the urgency parameter. The closeness to the HOL packet's deadline is expressed in terms of the frame time left, t, which is the difference between the flow's ML and its age. The HOL packet's age is the summation of the time the packet spent in the BS queue and the estimated transmission time. An example function of F^{rtPS} for the rtPS service is shown in Figure 15.2. If the remaining frame time, t, is

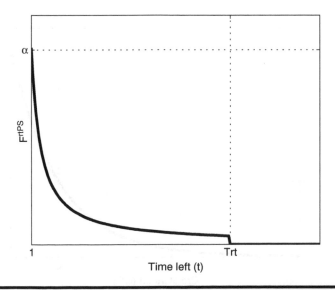

Figure 15.2   Example urgency function for Real-time Polling Service (rtPS) service (F^{rtPS} (t) = 1,t > T_{rt}; $\alpha e^{1/t}$, t < T_{rt}).

smaller than zero, there is a high chance that the packet will miss its deadline. To avoid wasting bandwidth by sending expired packets, we proactively drop such packets.

For BE traffic, the main issue is to ensure fairness among users. Accordingly, the urgency term F^{BE} indicates whether the flow has received its fair share of the channel. We use a credit-like approach to monitor the difference between a flow's fair share and the actual service that it gets. When a flow i is given the transmission opportunity for its HOL packet of length L bytes, all other flows that have packets waiting for transmission will receive credits equal to the fair share it would get if the L bytes had been shared fairly among all these flows. At the same time, flow i will get negative credits equal to the total credits the other flows receive. In such cases, we say that flow i has spent L credits.

For example, suppose there are three users in the system sharing the resources with a fair share of (2,1,1). If user 1 spends four credits, that is, transmits a packet of length four bytes, then −2 credits are given to user 1 because the system handled four bytes and user 1 is only entitled of two bytes. The remaining two bytes of channel usage were borrowed from the other users. The credits of users 2 and 3 are both 1 because that is the share they should have received. After user 3 spends eight credits, the credits for the users are 2, 3, and −5, respectively. We use the ratio of user's credit c to a constant C to evaluate the BE users' urgency. For the previous example, the credit factor for each flow becomes 0.02, 0.03, and −0.05 with $C = 100$. The higher the ratio, the more credits the flow has, and the more urgent the flow should be. An example function of F^{BE} for BE service is shown in Figure 15.3.

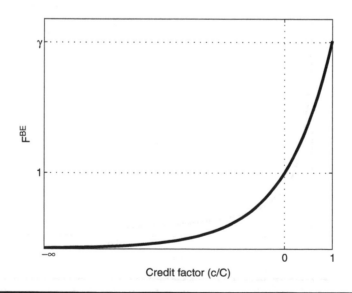

Figure 15.3 **Example urgency function for Best Effort (BE) service ($F^{BE}(c/C) = \gamma e^{c/C}$).**

When considering the priority of the flow, the credit factor becomes p^*c/C, where p is the TP of the flow.

For nrtPS, the urgency factor F^{hrtPS} shows how far the flow is away from its MRTR and its fair share. We use a token bucket to maintain the minimum reserved rate and the credit mechanism introduced before to control its fair share. The token bucket is a container of length S that holds the transferable token for a flow with each token representing one bit. As time goes by, the token is generated at the MRTR as the flow specified (in bits per second). When the HOL packet is transmitted, an amount equal to the value of the token is removed from the bucket. If the generated token reaches a certain percentage of the container, that is, T_{nrt}, we say that the flow is underserved, and we give it urgent status. This corresponds to a value of F^{hrtPS} greater than 1. We use the token factor to express the urgency in terms of the fullness of the token bucket, that is, s/S, where s is the value of the flow's token. The fuller the bucket, the more urgent the flow. An example function of the minimum reserved rate part of the urgency function $F^{hrtPS,MRTR}$ for nrtPS service is shown in Figure 15.4. When the token in the bucket is not enough for the current HOL packet, the packet is still allowed to be sent, the over-taken token is spent as credits, and the number of the token is set to zero. A flow with negative credit and nonurgent token factor is placed in the over-served state, and is temporarily degraded as BE traffic. The urgency value of a nrtPS flow takes into consideration both the rate reservation part and the fair share part, and the value is the greater one of the two.

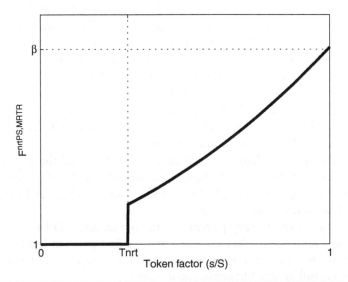

Figure 15.4 Example urgency function. The minimum reserved rate part of the urgency function for Non-real-time Polling Service (nrtPS) service ($F^{nrtPS, MRTR}$ (s/S) = 1, $s/S < T_{nrt}$; $\beta e^{s/S}$, $s/S > T_{nrt}$).

15.3.3 M-OQES Algorithm

The M-OQES algorithm attempts to schedule flows when the corresponding channel is capable of supporting its maximum rate but gives overriding priority to flows that are in danger of violating their QoS requirements. For rtPS flows, packets need to be sent within the ML. If poor channel quality causes rtPS packets to enter the urgent state, the rapid increase in the urgency factor will cause the scheduler to transmit them without waiting for channel conditions to improve. In such cases, channel utilization will be less efficient. nrtPS and BE flows allow some flexibility in their QoS requirements. However, when they are far off their normal token factor or fair share, the scheduler will hasten their transmission regardless of the channel conditions. We want to avoid such cases as much as possible to improve the overall system throughput. Ideally, if the corresponding user's channel changes quickly enough, that is, the Doppler rate is high enough, that the relative maximum rate can be reached before each packet goes into the urgent state and the system performance will be maximized. Opportunistic beam-forming using multiple antennas at the BS is one method for artificially increasing the Doppler rate of the channel [2].

Here, we describe the M-OQES scheduling algorithm in detail. During each frame preparation time:

1. If a UGS flow reaches its sending time (e.g., the associated timer times out), get the HOL packet of the flow and go to step 5. Otherwise, continue to step 2.
2. Calculate the scheduling value $v = UF$ for each rtPS flow, and select the flow with the largest v. If the flow's HOL packet is in the urgent state or the flow receives the MaxRR, get the HOL packet of the flow and go to step 5. Otherwise, mark the user as MAX_{rt} and continue to step 3.
3. Calculate the scheduling value v for each nrtPS flow whose credit is not negative or whose token factor is urgent ($s/S > T_{nrt}$) and select the flow with the largest v. If the flow's HOL packet is in the urgent state or the flow receives the MaxRR, get the flow's HOL packet and go to step 5. Otherwise, mark the user as MAX_{nrt} and continue to step 4.
4. Calculate the scheduling value v for each BE flow and the degraded nrtPS flow, select the flow with the biggest v and mark it as MAX_{BE}. Select the flow with the biggest v among MAX_{rt}, MAX_{nrt}, and MAX_{BE}, and get the flow's HOL packet. Continue to step 5.
5. If the packet can be fully packed into the frame, do so. Otherwise, break the packet into two parts in such a way that the first part can just fill the frame. Pack the first part into the frame and leave the remainder in the buffer. If the frame is full, finish. Otherwise, go to step 1.

The computational resource requirements of the M-OQES algorithm are not fixed and will vary depending on the circumstances. The computational load is least when the timer for one of the UGS flows expires. Because this indicates that it is time to send a UGS HOL packet, the scheduling scheme does not need to do any further calculations to make the scheduling decision. On the other hand, when none of the UGS flows reaches its sending time and neither rtPS nor nrtPS flows satisfy the selection criteria in steps 2 and 3, the scheduling algorithm must proceed until step 4 to make the decision. In such worst case scenarios, the computational complexity is $O(\log(n))$, where n is the number of flows. This is the same as the PF rule.

15.3.4 Implementation

Assigning appropriate values to the various parameters used in the algorithm is a very important step in successful implementation of the M-OQES algorithm. The relevant parameters, their meanings, their setting rules, and example values are summarized in Table 15.2.

15.4 Simulation Results

We evaluated the performance of our M-OQES algorithm by simulating the downlink of one sector of an IEEE 802.16 system. We conducted two types of MATLAB-based simulations. The first was used to measure the overall performance of M-OQES compared with existing schedulers. The second was used to determine how closely short-term fairness between the nrtPS and BE services is achieved by M-OQES.

15.4.1 Simulation Environment

Because UGS traffic has a higher priority than the other services and is always scheduled first, it has no effect on the scheduling schemes used for the other services. Accordingly, we did not include the UGS in our simulations and only evaluated the scheduling schemes designed for rtPS, nrtPS, and BE traffic.

The rtPS service was evaluated using simulated video traffic generated using the simple IPB composite model [12]. The movie *Star Wars* was simulated with an average data rate of 1 Mbps. There is one packet (frame) generated in every 40 ms, which is synthesized by combining three self-similar processes in a way similar to the Group of Pictures (GOP) structure. The self-similar processes adopted before are generated using the algorithm given in [13]. The delay requirement is 100 ms, and the maximum tolerable delay violation is 2 percent. In terms of nrtPS and BE services, to evaluate the minimum rate guarantee for nrtPS service and fairness among users, we generated continuous flows to simulate FTP for nrtPS service and World Wide Web (WWW) for BE service. For both cases, one packet is generated every

Table 15.2 Parameters of the Multi-Service Opportunistic QoS-Enhanced Scheduling (M-OQES) Algorithm

Parameter	Scope	Meaning	Setting Rules or Guides	Example Value
α, β, γ	System	The value of $F^{rtPS}(x)$, $F^{nrtPS, MRTR}(x)$, and $F^{BE}(x)$, when $x = 1$	$\alpha > \beta > \gamma > 1$	When using the urgent functions in Figures 15.2 through 15.4, we can set $\alpha = 6 * e$, $\beta = 4 * e$, $\gamma = 3 * e$
λ	System	Value of the maximum rate reward (MaxRR)	Bigger than the max/min rate ratio in the system	$rate_{max}/rate_{min}$
T_{rt}	Flow	The threshold of rtPS flows going to the urgent status	A function of the HOL packet length and the flow's maximum tolerable packet loss ratio	Estimated packet transmission time $* [-\log_{10}$ (maximum tolerable packet loss ratio)]
T_{nrt}	Flow	The threshold of nrtPS flows going to the urgent status	Set with S. $0 < T_{nrt} < 1$, the bigger the value, the more rate variance, and the bigger the system throughput	Typical value is between 0.2 and 0.7
C	Flow	The maximum tolerable credit loss for nrtPS and BE flows	Tolerable amount of lacking service	If a flow can stand 1 sec of no service, then $C = 1 * 100K$ with a rate of 100 kbs
S	Flow	Size of the token bucket for nrtPS flows	Set with T_{nrt}. The bigger the token bucket, the more variant of the receiving rate is allowed, and the bigger the system throughput	If a flow's rate is 1 Mbps, and the rate variation within 2 sec is tolerable, then $S = 2 * 1\ Mbps$

Table 15.3 Simulated Traffic and their Quality of Service (QoS) Requirements

Video	Frame dimension	240 * 252
	Max frame	26,955 bytes
	Mean rate	1 Mbps
	Delay requirement	100 ms
	Maximum delay violation	0.02
File Transfer Protocol (FTP)	Fair share among FTP	1
	Generating rate (Type 1)	300 kbps
	Packet length (Type 1)	150 bytes
	Minimum required traffic rate (Type 1)	250 kbps
	Generating rate (Type 2)	1 Mbps
	Packet length (Type 2)	500 bytes
	Minimum required traffic rate (Type 2)	500 kbps
World Wide Web (WWW)	Fair share among WWW	1
	Generating rate (Type 1)	300 kbps
	Packet length (Type 1)	150 bytes
	Generating rate (Type 2)	1 Mbps
	Packet length (Type 2)	500 bytes
Fair share between FTP and WWW (Type 1)		(1, 2)
Fair share between FTP and WWW (Type 2)		(2, 3)

four milliseconds. The ratio of rtPS flows, nrtPS flows, and BE flows is 1:2:2. Table 15.3 summarizes the traffic and their QoS requirements for both types of simulations.

The base station of the simulated system is located at the center of the cell with three 120° directional antennas. SSs use omnidirectional antennas, and are uniformly distributed in the cell. The increase in path loss with distance and the degree of shadow fading follows Erceg's suburban model [14] terrain type A. Fast fading follows the Rician distribution. The system uses multiple antennas at the BS [2] to generate faster fading in the channel to enhance the performance of multi-user diversity. The maximum Doppler frequency is uniformly distributed between 10 and 20 Hz.

The BS communicates with the users using the rates adapted to the user channel conditions so that the bit error rate QoS required by each service can be met. The transmission between the BS and the user is assumed to be error-free. When a user's received SNR ratio is lower than the lowest requirement, no transmission will be scheduled for the user.

All simulations last for two minutes, that is, 120,000 frames, which is considered to be long enough to avoid any specific scenarios.

15.4.2 Evaluation of M-OQES's Performance

We have evaluated the effectiveness of M-OQES by comparing its performance with that of some existing schedulers. We use EDF for the rtPS service, and RR for the nrtPS and BE services as baselines in view of these algorithms' popularity. To be fair, the EDF and RR algorithms are used in conjunction with adaptive modulation and coding. We also selected the EXP and PF [11] rules for comparison. The EXP rule is suitable for the rtPS service, and the PF rule is a good match for BE service. For the nrtPS service, however, there is no suitable rule, so we use two versions of the EXP/PF rule. EXP/PF-1 uses EXP for the nrtPS service, and EXP/PF-2 uses PF for the nrtPS service.

Figures 15.5 through 15.8 show the system throughput; mean, maximum, and minimum receiving ratio for video users, and mean, maximum, minimum throughput for FTP and WWW users. In each case, we ran the simulation with 200 users.

Figure 15.5 shows that the system throughput achieved using M-OQES is higher than the others. The improvement is roughly 10 percent over EXP/PF-1, EXP/PF-2, and 17 percent over AMC-EDF/RR. The difference between throughput and effective throughput, which is the throughput less any dropped bits, is not significant, although AMC-EDF/RR and M-OQES are both better than the two EXP/PF schedulers.

All four scheduling schemes are able to deliver most of the real-time video packets within the required time constraint while keeping the dropping ratio below 2 percent as suggested by Figure 15.6. AMC-EDF/RR and M-OQES are better than the two EXP/PF algorithms in that their packet receiving ratios are both 100 percent.

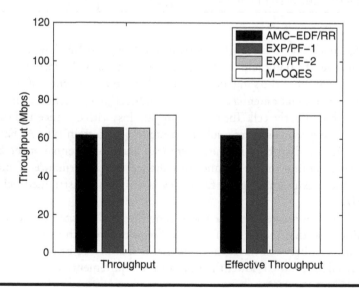

Figure 15.5 System throughput of the simulated system when using different scheduling schemes. M-OQES, multi-service opportunistic QoS-enhanced scheduling.

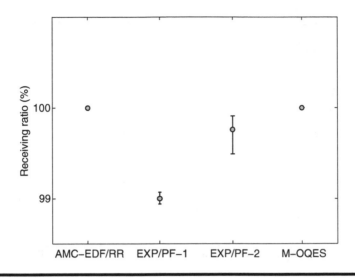

Figure 15.6 Maximum, mean, and minimum receiving ratios of video users of the simulated system when using different scheduling schemes. M-OQES, multi-service opportunistic QoS-enhanced scheduling.

In Figure 15.7, although AMC-EDF/RR achieves good long-term usage fairness for FTP users with little difference between the minimum and maximum user throughput, it fails to maintain the MRTR at 250 kbps. EXP/PF-2 is also not able to maintain FTP users' minimum rate requirements, and the users receive quite

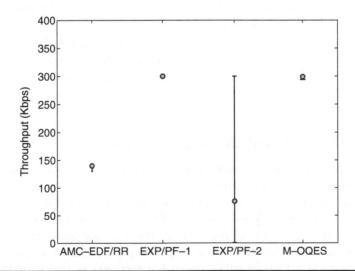

Figure 15.7 Maximum, mean, and minimum throughput of the Non-real-time Polling Service (nrtPS) users of the simulated system when using different scheduling schemes. M-OQES, multi-service opportunistic QoS-enhanced scheduling.

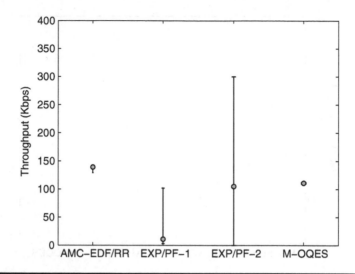

Figure 15.8 **Maximum, mean, and minimum throughput of Best Effort (BE) users of the simulated system when using different scheduling schemes. M-OQES, multi-service opportunistic QoS-enhanced scheduling.**

different data rates because of their different channel conditions, with the maximum rate at 300 kbps and the minimum at 0.26 kbps, which indicates a quite poor usage fairness. Both EXP/PF-1 and M-OQES are able to maintain the MRTR, however, the EXP/PF-1 rule has delivered every packet of the FTP flow no matter how starving the WWW users are.

In terms of WWW users as Figure 15.8 illustrated, both AMC-EDF/RR and M-OQES are able to achieve fairness among users. Both versions of the EXP/PF rule favor some users over others, which results in a best versus worst data ratio of 34 for EXP/PF-1 and 1154 for EXP/PF-2.

15.4.3 Evaluation of Short-Term Fairness

In this section we discuss the fairness issue in terms of usage, rather than proportional fairness or service time fairness.

For real-time traffic, fairness is important when the system is overloaded and cannot satisfy each user's QoS requirement. Here, each scheduler is able to keep the drop ratio within the QoS requirements of the real-time users, and specifically, M-OQES and AMC-EDF/RR are able to deliver every packet of the users on time. So, fairness is not an issue for rtPS services. For data services, that is, nrtPS and BE services, both long term fairness and short term fairness shall be considered. In Figures 15.7 and 15.8, we showed that M-OQES achieves better long-term fairness than the other two types of EXP/PF schedulers for nrtPS and BE services. In the following, we illustrate that M-OQES is also capable of achieving short-term fairness for data services.

To illustrate how short-term fairness is achieved among nrtPS and BE services, we did another type of simulation with seven users with mean SNRs of 13.8, 31.4, 18.0, 22.7, 12.2, 21.6, and 27.6 dB, respectively. User 1 is a real-time video user who requires rtPS service, and users 2 through 7 are FTP or WWW users that consume nrtPS and BE services. The simulated traffic and QoS requirement of each flow can be found in Table 15.3.

At the start, there were only five users in the system, including one video user, two FTP users, and two WWW users. After 40 sec, another FTP user joined the system, and later, at 80 sec, an extra WWW user was added. Figure 15.9 shows the time-series consumed bandwidth averaged over 10 sec for each user. Because the video user does not need to share any resources with the FTP and WWW users, we did not include the throughput of the video user in the figure. Throughout the simulated 120 sec, each of the video user's packets were delivered on time.

From Figure 15.9, we can see that throughout the simulation period, the rates received by FTP users are quite similar, and so are those for the WWW users. This shows that the bandwidth has been fairly shared among the same service consumers.

In the first 40 sec, after satisfying the QoS requirements of the video user and the MRTR (500 kbps) of the FTP users, the remaining resources is shared among the FTP and WWW users with a fair share of (2,3). Thus, apart from the

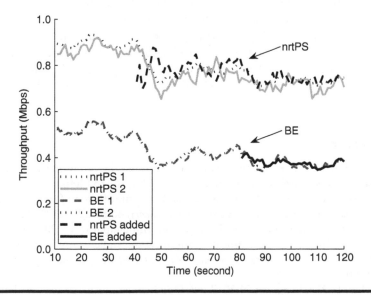

Figure 15.9 Short-term fairness of the multi-service opportunistic QoS-enhanced (M-OQES) scheduling algorithm. nrtPS, Non-real-time Polling Service; BE, Best Effort.

minimum rate, each FTP user received a total data rate of 860 kbps with a fair share contribution of 360 kbps, and each WWW user got the fair share of about 540 kbps. Apparently, the received fair share of FTP and WWW users do follow the specified fair share, which is (360,540) or (2,3). When the FTP user joined at 40 sec, the four existing users' data rate dropped accordingly. The data rates for the WWW users then became 390 kbps, and 760 kbps for FTP users, including the newly joined one. And then at 80 sec, another WWW user joined the system. Again, the existing FTP and WWW users' data rate dropped according to their MRTR and fair share. With seven users in the system, the fair share unit became 120 kbps, and the data rates for the FTP and WWW users became 740 kbps and 360 kbps, respectively.

The overall system throughput grew each time new users were added, from 3.8 Mbps at the beginning, to 4.06 Mbps after one FTP user joined, and finally to 4.3 Mbps (the video user has a data rate of roughly 1 Mbps). The observed increase in system throughput is mainly due to the contribution of multi-user diversity.

15.5 Conclusions

We have introduced a new downlink scheduler for IEEE 802.16 point-to-multi-point broadband wireless access systems. It aims to provide good performance for multiple services by achieving an effective compromise between maximizing throughput and satisfying each user's QoS requirements. It does so by combining conventional multi-user diversity with an urgency function whose value increases rapidly as the time remaining to meet a specified QoS requirement becomes short. Simulation results confirm that our M-OQES algorithm outperforms existing multi-service scheduling schemes. Moreover, the computational complexity of the M-OQES algorithm is comparable with that of the widely adopted PF rule.

References

1. H. Fattah and C. Leung, An overview of scheduling algorithms in wireless multime-dia networks. *IEEE Wireless Communications*, 9(5), 76–83 (2002).
2. P. Viswanath, D. N. C. Tse, and R. Laroia, Opportunistic beamforming using dumb antennas. *IEEE Transactions on Information Theory*, 48(6), 1277–1294 (2002).
3. A. Jalali, R. Padovani, and R. Pankaj, Data throughput of CDMA-HDR: a high efficiency-high data rate. Personal communication wireless system. In *Proc. of IEEE 51st Vehicular Technology Conference (VTC 2000-Spring)*, vol. 3, pp. 1854–1858 (2000).
4. X. Liu, E. K. Chong, and N. B. Shroff, Opportunistic transmission scheduling with resource-sharing constraints in wireless networks. *IEEE Journal on Selected Areas in Communications*, 19(10), 2053–2064 (2001).

5. S. Shakkottai and A. L. Stolyar, Scheduling for multiple flows sharing a time-varying channel: the exponential rule. *American Mathematical Society Translations,* 207 (2002).

6. Y. Zhang and D. G. Michelson, Opportunistic QoS enhanced scheduler for real-time traffic in wireless communication systems. In *Proc. of IEEE 64st Vehicular Technology Conference,* Montreal, Canada (September 2006).

7. C. Cicconetti, L. Lenzini, and E. Mingozzi, Quality of service support in IEEE 802.16 Networks. *IEEE Network,* 20(2), 50–55 (March 2006).

8. K. Wongthavarawat and A. Ganz, Packet scheduling for QoS support in IEEE 802.16 broadband wireless access systems. *International Journal of Communication Systems,* 16(1), 81–96 (2003).

9. J. F. Frigon, V. C. M. Leung, and H. Chan Bun Chan, Dynamic reservation TDMA protocol for wireless ATM networks. *IEEE Journal on Selected Areas in Communications,* 19(2), 370–383 (2001).

10. E. Hossain and V. K. Bhargava, Link-level traffic scheduling for providing predictive QoS in wireless multimedia networks. *IEEE Transaction on Multimedia,* 6(1), 199–217 (2004).

11. R. Jong-Hun, J. M. Holtzman, and K. Dong Ku, Performance analysis of the adaptive EXP/PF channel scheduler in an AMC/TDM system. *IEEE Communications Letters,* 8(8), 497–499 (2004).

12. N. Ansari, L. Hai, Y. Q. Shi, and Z. Hong, On modeling MPEG video traffics. *IEEE Transactions on Broadcasting,* 48(4), 337–347 (2002).

13. M. W. Garrett and W. Willinger, Analysis, modeling and generation of self-similar VBR video traffic. In *Proc. of ACM conference on Communications Architectures, Protocols and Applications,* pp. 269–280 (1994).

14. V. Erceg, L. J. Greenstein, S. Y. Tjandra, et al., An empirically based path loss model for wireless channels in suburban environments, *IEEE Journal on Selected Areas in Communications,* 7(7), 1205–1211 (1999).

Chapter 16

Effects of Rain on IEEE 802.16 Systems Deployment

Bernard Fong

Climatic phenomena such as rain, snow, and gas absorption cause attenuation and distortion to radio waves in an outdoor environment. The extent of signal degradation is primarily dependent on the carrier frequency and the actual rate of rainfall at a given time. In addition, the effect of rain is the main cause of attenuation at frequencies above several GHz commonly used in broadband wireless networking applications [1]. This chapter looks at how rain attenuation affects the operations of both fixed and mobile computing devices, and various ways of reducing their impacts on system performance.

16.1 Introduction

A considerable part of the work related to this chapter is based on extensive study of the effects of rain on broadband wireless access (BWA) systems and in building a statistical model on its impacts. The analysis of long-term results suggests that rainfall is the single most influential factor in the reliability and availability of radio links in the application of broadband access networks. In this chapter, a detailed analysis of the extent of signal degradation caused by rain in tropical regions, where its effects are most severe, is presented. The next section looks at the effects of rain. Being an uncontrollable factor, it is practically impossible to cover the entire propagation path of several kilometers path length to eliminate its effects. Section 16.3 presents a detailed report on a range of different systems, how rain affects wireless communication systems at different data rates, and the problems associated with deployment of alternatively polarized signals as a result of rainfall. In Section 16.4, we look at the effect of rain attenuation in two tropical areas and investigate its impact on different deployment scenarios for different applications. The importance of studying the deployment region's rainfall statistics and impacts on system operation is briefly discussed in Section 16.5. This is followed by Section 16.6, which concludes this chapter by investigating remedies to the effect of rain attenuation, namely by appropriate adjustment of the system fade margin.

16.2   Uncontrollable Factor

Over the past decades, fixed wire networks have evolved to provide a very high level of availability and reliability particularly with technological advancements in optical communications. Recent advancements in wireless communications technologies have made radio networks possible to match the reliability of fixed networks. A number of factors affecting wireless transmission of multimedia data discussed in [2,3] are considered. It becomes clear that more work still needs to be done to minimize the effects of rain. Unlike indoor environments where most parameters are controllable, signal degradation caused in an outdoor environment induces a much higher degree of uncertainty as many factors are virtually uncontrollable.

Rain causes attenuation, scattering, and cross-polarization to the propagating wave. While the former reduces signal strength significantly even over a short path, the latter makes the deployment of alternately polarized signals a challenging issue in many cases. Their combined effect makes it difficult to maintain an adequate level of system availability.

Alternately, polarized signals are often used to increase utilization of available bandwidth. The depolarization effect of rain makes utilization of orthogonally polarized signals a particularly challenging issue in heavy rainfall regions as horizontally polarized signals undergo a higher degree of attenuation than vertically polarized signals as they propagate through the wireless channel under the influence of rain. In general, higher frequencies are affected more severely by cross-polarization and signal attenuation. The effects of heavy rainfall on radio propagation path reduce the system availability because rain causes cross-polarization interference that subsequently reduces the polarization separation between signals of vertical and horizontal polarizations as the signals propagate through rain. The extent of radio link performance degradation is measured by cross-polarization diversity (XPD) that is determined by the degree of coupling between signals of orthogonal polarization. XPD typically results in a 10 percent reduction in coverage due to cell-to-cell interference.

16.3 Signal Degradation

This section looks at the impact of rain attenuation on various services. An investigation on the effect of rainfall on transmission data rate [4], effect on mobile computing in wireless local area networks (WLAN) [5], and on fixed BWA systems serving stationary subscribers [6], is discussed in this section.

16.3.1 Impact on Data Rate

The study of rain attenuation effect is best performed using wireless frame relay transports multimedia data due to its requirement of low BER transmission systems over microwave channels. Although it provides a reliable and economical means for point-to-multipoint distribution, the main challenge in operational aspects is guaranteeing quality of service (QoS) as wireless communication in an outdoor environment is often accomplished under harsh environments. Sudden change in weather may have a severe impact on microwave links.

Transmission of frame relay data requires careful consideration for burst errors as data frames are separated by flag symbols. Occurrence of such errors often leads to flag corruption that subsequently causes two adjacent frames being misinterpreted as a single larger frame. This section looks at the coverage range of wireless frame relay traffic transmission over a line-of-sight (LOS) link with a local multipoint distribution service (LMDS) system. Comparisons have been made between 10 and 26 GHz with the effects of rain attenuation as a function of data rate. These frequencies are studied because the lower frequency of 10 GHz is commonly used in Asia's tropical region due to its relatively good tolerance to heavy and persistent rainfall, whereas the 26-GHz band is widely used in Europe. Frame relay transport provides a low error rate means of data transfer. Outdoor radio links, however, often suffer from severe bit errors due to operating environments.

16.3.2 Case Study with an LMDS System

This example determines the effects of variation in rainfall on the range of radio links for wireless frame relay data transmission to maximize area coverage. As higher rainfall shortens link coverage giving rise to the need of increasing fade margin to maintain link availability of no less than 99.99 percent, it is inappropriate to determine a set of optimal parameters for LMDS as different systems operate at a different data rate with different carrier frequencies. A study on the effects of radio coverage when the rainfall varies could be developed to provide information for any compensation to rain attenuation, such as fade margin requirements for a guaranteed coverage.

16.3.2.1 System Description

An LMDS system provides point-to-point wireless connection between a transmitter–receiver pair and is used in evaluating the effects of rainfall on the link availability at different frequencies transmitting raw data at different bit rates. Comparison between transmitting at 20 and 40 Mbps is made at frequencies of 10 and 26 GHz. The maximum range is measured for a bit error rate (BER) of 10^{-6} with a system gain G of 110 dB, which is a ratio of the transmitted power to receiver sensitivity and is given by:

$$G = \frac{k}{R_B} \qquad (16.1)$$

where k is an arbitrary constant and R_B is the data rate in bits per second (bps). The equation shows that G is inversely proportional to the bit rate R_B. The practical link range is greatly affected by parameters such as antenna gains for both transmitter and receiver, and rainfall statistics. The theoretical range is computed as a function of increasing rain rate, from no rainfall (0 mm/h) up to ITU rain region P [7] (140 mm/h) where persistent heavy downpour can severely affect outdoor wireless links. For a basic IEEE 802.16 system deployed in Singapore, the fade margins for different carrier frequencies can be calculated based on the rain attenuation statistics as described in [8].

The system layout is illustrated in Figure 16.1. The transmitter is connected to the network backbone via an OC-3 link with the transmitting radio hub maintaining a LOS wireless link with the receiver. The range of radio link is varied and affected by rain attenuation, with statistical modeling similar to that used in [9]. The transmitter is intended to provide point-to-multipoint broadband service to multiple receivers located within the varying radio link range. In this system, only one receiver is used to measure the effects to the radio link due to attenuation. A pair of 0.6-m diameter dish antennas having narrow beam width of 4° with direct LOS are

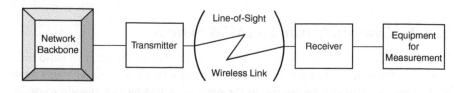

Figure 16.1 System block diagram for data rate comparison. The microwave link can be severely affected by changes in weather. There should be no physical obstacles between the respective antennas of transmitter and receiver.

used by the transmitter and the receiver. To utilize bandwidth efficiently, a multicast scheme with pointer forwarding [10] is used. The experiment has been carried out with two portable tipping-bucket rain gauges with the rate of rainfall recorded when measurements are taken with rate of rainfall exceeding 20 mm/h. No measurements are taken when the rain rate falls below 20 mm/h. Results are averaged over a four-month period to investigate the effect of rain on variation in data rates.

The system gain is dependent on its modulation scheme as a result of bandwidth limitation. Although 16-QAM offers a good compromise between bandwidth efficiency and receiver structure complexity, QPSK is chosen for performance comparison as it offers the best overall performance in LMDS [11].

16.3.2.2 Coverage Affected by Rain with Different Carrier Frequencies

The link coverage as a function of rainfall is used to determine the effects of rain on the system availability. The reduction in range is computed as a function of rain intensity and a comparison between 20 and 40 Mbps is made to investigate the effects of rain attenuation on the increase in data rate at each frequency.

The results for 10 and 26 GHz are shown in Figures 16.2 and 16.3, respectively. From these results, Figure 16.2 shows that 10 GHz provides a longer range than 26 GHz under the influence of rainfall and the difference becomes less as the rate of rainfall increases. It is noted that the frequency of around 26 GHz is subject to more severe absorption due to resonance.

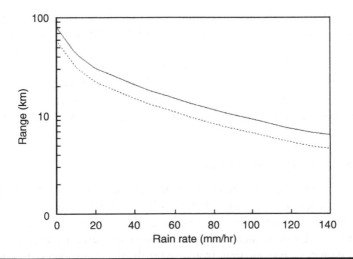

Figure 16.2   Maximum coverage of a 10-GHz carrier, supporting data rates of up to 40 Mbps. The range is severely reduced even with little rain when the transmitted power remains unchanged.

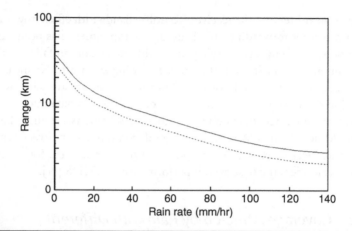

Figure 16.3 Maximum coverage of a 26-GHz carrier, system parameters are identical to those of 10 GHz illustrated in Figure 16.2, except that the antenna size is halved to 30-cm diameter for both transmitter and receiver.

A slight reduction in range is observed as the data rate increases from 20 to 40 Mbps in both cases. The range ratio d' is given by

$$d' = \sqrt{\frac{R_{max}}{R_{min}}} \qquad (16.2)$$

where R_{max} and R_{min} are the maximum and minimum data rates, respectively. Although a 10-GHz carrier offers radio range that is some three times longer than that of 26 GHz, it is generally not a good choice due to a number of issues described earlier like bandwidth constraints and multi-path fading.

It is also noted that both carriers exhibit a sharp drop in range when the rainfall is below 70 mm/h then become approximately linear at a reduction of some 1.5 km per every additional 20 mm/h rain for ITU rain regions M to P inclusive.

16.3.2.3 Summary of Results

These results compare the maximum range provided by an LMDS system at frequencies suitable for tropical regions and in Europe and results show that link distance reduces sharply from no rain to a rate of rainfall of about 75 mm. Although the link can be increased by adjusting parameters such as the fade margin and system gain, it is apparent from the results that the presence of rain significantly reduces radio link distance. Figure 16.2 shows that the range is reduced to only one-sixth that of free space loss under persistent rain in ITU rain region P and a reduction in frequency from 26 GHz down to 10 GHz provides a range of approximately threefold increase. In addition, results show that the effect of data rate increase is very little with some 15 percent difference by increasing from 20 to 40 Mbps. Frame relay network reliability of an LMDS is varied according to rainfall that gives rise to a difference in random bit errors.

16.3.3   Next-Generation Wireless Local Area Network Serving Mobile Receivers

Traditionally, WLAN in compliance with the IEEE 802.11 standards operates in the very congested portion of around 2.4–5 GHz of the spectrum. The 17 GHz one has been proposed for next-generation WLAN that offers a number of advantages such as better spectrum utilization with improved frequency reuse and operation in the less congested portion of the spectrum. Therefore, 17 GHz is particularly well suited for widespread use by wearable devices with a high degree of portability.

Rain-induced depolarization and attenuation is the main source of uncontrollable signal degradation for wireless links with carrier frequencies exceeding 10 GHz. Earlier work in [12] shows the significance in signal attenuation due to rain when the carrier frequency is increased from 10 GHz to over 20 GHz. In addition to the effect of attenuation per kilometer of distance covered by the signal, cross-polarization also increases with higher frequencies [13]. Utilizing orthogonal polarized signals increases frequency reuse while heavy rain can have a serious impact on the link availability due to reduction of cross-polarization isolation between signals of horizontal and vertical polarizations. Earlier work in [14] has presented some backgrounds on deployment considerations in the more controllable indoor environment while little is known about its performance when operating outdoor. The following example illustrates how such system can be set up, based on the experiments in the following example that analyze the impacts of rain-induced attenuation, phase rotation, and cross-polarization at 17 GHz using a telemedicine system.

16.3.3.1   Example by Using IEEE 802.16-Based System

An LMDS system described earlier was used to conduct the measurements, by providing two links at 17 GHz using the same LOS path for two stationary receivers set up each maintaining a 4 km distance from the Base Station constantly recording the received signal. Table 16.1 shows a summary of the parameters of this system. Large plastic sheets were used to prevent raindrops from reaching the antennas' surface on both sides so that the effect of wet antenna was minimized throughout the experiment while having no effect on the propagation paths. The rate of rainfall at the time measurement was taken was logged using a tipping-bucket rain gauge. A maximum difference of 0.5 dB between the two receivers has been set to verify the measured data and readings from the two receivers that differ by less than 0.5 dB are averaged

Table 16.1 System Parameters of the IEEE 802.16-Based System Operating at 17 GHz

System gain	110 dB
Transmitter antenna gain	18 dBi
Horn half power beam width	45° (azimuth)/2.5° (elevation)
Receiver antenna gain	28 dBi
Receiver sensitivity	−85 dBm (BER = 10^{-9})
Modulation scheme	QPSK
Date rate	45 Mbps
Carrier frequency	17.15 GHz

BER, bit error rate.

and recorded. The captured data is analyzed only when the rate of rainfall exceeds 20 mm/h because the accuracy of the rain gauge used is found to be around 10 percent with little rain. In addition, the extent of signal degradation caused by such small amount of rain is found to be insignificant.

A cross-polarization scheme using dual cross-coupling between orthogonally polarized signals proposed by [15] is used to study the cross-polarization effects due to variable rain rate as:

$$\text{XPD} = -20\log(l.\cos^2\varepsilon.k.\exp(-2\sigma^2)) \tag{16.3}$$

$$k = \sqrt{\alpha^2 + \theta^2} \tag{16.4}$$

where l is the path length in km, ε is the elevation angle between the transmitting Base Station and the receiving node, σ denotes the standard deviation of canting angle distribution, α is the attenuation in dB, and θ is the phase shift in degrees.

16.3.3.2 Results Analysis

The rate of rainfall is collected with the corresponding received signals measured. The data is used to evaluate the phase difference and the factor of cross-polarization experienced at 17 GHz under persistent rainfall in the tropical region of Singapore. Figure 16.4 shows the best-fit lines representing attenuation for the two signals of horizontal and vertical polarizations with all readings recorded within 0.5 dB of the plots. The attenuation difference between horizontal and vertical polarization from Figure 16.4 is plotted in Figure 16.5. With reference to the signal of vertical polarization, using the phase relative to the horizontally polarized signal measures the

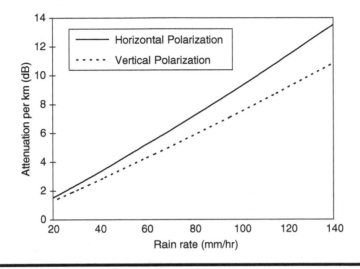

Figure 16.4 Signal attenuation due to rain; horizontally polarized signal undergoes higher attenuation at the same frequency.

phase difference and the phase rotation is derived using the method described in [16]. A plot of phase difference between the two polarizations is shown in Figure 16.6. Results of difference in both attenuation and phase show an approximate linear relationship with increasing rainfall. It is noted that the gradients change more steeply around the range of 70–100 mm/h rain rate.

The effects of heavy rainfall on radio propagation path reduce the system availability because rain causes cross-polarization interference that subsequently reduces the polarization separation between signals of vertical and horizontal polarizations as the

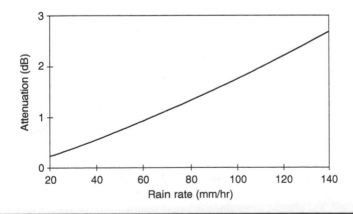

Figure 16.5 Attenuation difference between horizontal and vertical polarizations.

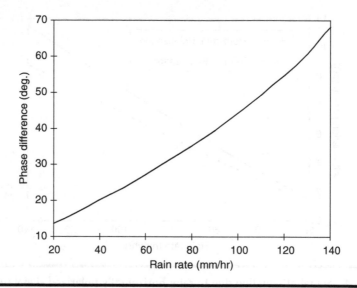

Figure 16.6   Relative phase.

signal propagates through rain. The extent of radio link performance degradation is measured by XPD that is determined by the degree of coupling between signals of orthogonal polarization. The amount of cross-polarization per kilometer path is shown in Figure 16.7. The cross-polarization factor is found to be insignificant at 17 GHz as

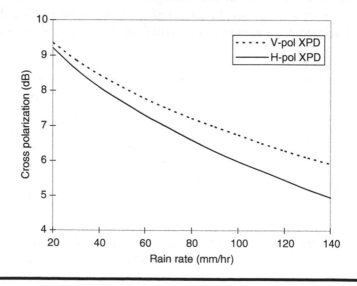

Figure 16.7   Cross-polarization diversity.

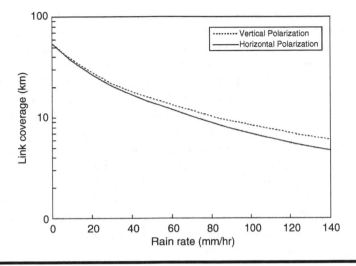

Figure 16.8   Variation of coverage at 17 GHz under the influence of rain, with a minimum link availability of 99.99 percent.

a difference of 1 dB between the horizontal and vertical polarization exists when the rain rate approaches 140 mm/h, which does not occur often in most areas.

The link coverage is obtained as a function of rainfall to determine the effects of rain on the system availability. It is assumed that the minimum link availability is 99.99 percent. The maximum range is measured for a BER > 10^{-6} as in Section 16.2.1. Figure 16.8 shows the link coverage at different rates of rainfall. The dotted curve with a smaller effect on range shows the vertically polarized signal as its attenuation is less. The link covered by horizontally polarized signal is shown in the solid curve of the figure. It is noted that the difference in link coverage increases much more significantly when the rainfall rate exceeds 40 mm/h, where the difference in attenuation between the two polarizations is only 0.58 dB; the difference in range is approximately 15 percent. From [8], the maximum coverage range d, as provided by vertically polarized signal under the influence of rain, is approximated by:

$$d = \sqrt{\frac{K\alpha^2}{10^{-0.1G}\lambda^2}} \qquad (16.5)$$

$$K = T \cdot R$$

where T and R measure the antennas' effective areas of transmitter and receiver, respectively. λ is fixed for 17 GHz at 17.6 mm. Figure 16.8 also shows that the range covered by vertically polarized signal is reduced to approximately one-fifth that of free space loss under persistent rain in ITU rain region P [7] and a reduction is even more significant with horizontally polarized signals. The range shown for the links provided by the LMDS system used in this experiment can be adapted for estimation of maximum available range given by a WLAN system by substitution of appropriate system parameters. In general, roughly 10 percent of the range can be offered by a typical WLAN system relative to that shown in Figure 16.4 for an LMDS system due to differences in attributes such as transmission power and antennas used.

Wideband point-to-multipoint measurements have been taken at 17 GHz using short links to study the effects of tropical rainfall on attenuation, phase rotation, and cross-polarization of orthogonally polarized signals [7]. It is found from the presented results that minimal signal degradation is caused when the rain rate is below around 70 mm/h as the rate of increase in attenuation difference and phase rotation is less than that at higher rain rates. The presented results may be used as system deployment guidelines to evaluate necessary fade margins for next-generation high-speed WLAN services using 17-GHz carriers. They can also be used for estimation of system performance for future systems.

16.3.4 Fixed Broadband Wireless Access Network Serving Stationary Receivers

IEEE 802.16 systems such as LMDS handle a large volume of traffic supporting a wide range of multimedia services for a large number of subscribers. The study of radio wave propagation in such environment is necessary for optimization of equipment design and system implementation. Further experimental studies have been carried out for a range of services in [17,18]. They provide further improvement on system optimization by measuring fixed point-to-multipoint links at 26 GHz under various rain rates to determine the extent of attenuation and phase rotation that provide some guidelines for future system deployment.

Although a large number of reports from CCIR and ITU give some indications on statistics related to rainfall-induced attenuation, very little is known about the effects of persistent heavy rainfall on the link available for delivery of high-speed multimedia traffic over 26-GHz channels. This 26-GHz band offers advantages such as relatively high bandwidth resulting in much higher bit rates available to individual users when compared with a carrier of around 10 GHz used typically in tropical regions. It also allows the reuse of spectrum with a higher density of smaller cells resulting in a higher network capacity in urban and rural areas [19]. These networks are appropriate for connection within homes and small offices to high-capacity public switching and backbone networks for wireless multimedia services.

16.3.4.1 Case Study

At a carrier frequency of 26 GHz, higher frequency reuse and capacity is offered as a result of higher spectral efficiency compared with 10-GHz systems usually preferred in tropical regions where the effects of rain are less significant at lower frequencies. We study the propagation of two alternately polarized signals in an LMDS system. The profiles of received signals at outdoor receivers are captured and the extent of signal degradation is calculated. The rate of rainfall at the time during each measurement was recorded with a portable tipping-bucket rain gauge. Measurements were conducted with two mobile receivers over two unobstructed LOS links at approximately 4 km from the transmitting hub throughout the experiment. A difference of no more than 0.5 dB is recorded between the two receivers and the average value is taken in each measurement. To minimize the effects of wet antenna without any noticeable effects on the outdoor propagation path, large plastic sheets were used above the transmitting hub and receivers to prevent raindrops from reaching the antennas' surface. Signals are transmitted in both horizontal and vertical polarizations and the impact on link availability due to rain attenuation and depolarization is studied.

16.3.4.2 Measurement Results

The rainfall rate was recorded with the received signals analyzed simultaneously. The data captured was used to evaluate the attenuation and phase difference at 25.9 GHz under persistent rainfall in the tropical region of Singapore. The attenuation A and phase ϕ is evaluated as described in [16] such that:

$$A = 8.69.\text{Im}(kl) \tag{16.6}$$

$$\phi = 57.3.\text{Re}(kl) \tag{16.7}$$

where l is the path length and k is the effective propagation constant as

$$k = k_0 + \frac{2\pi}{k_0} \int n(r)dr \tag{16.8}$$

with k_0 being the propagation constant for free space and $n(r)$ is a function of the raindrop size distribution.

The measured attenuation for both polarizations is shown in Figure 16.9 with the best-fit lines representing attenuation per kilometer range for the two vertical polarizations with most of the readings recorded found lying within 1 dB of the best-fit plots. The attenuation difference between horizontal and vertical polarization that results from Figure 16.9 is plotted in Figure 16.10. A fairly linear increase in attenuation difference is observed as the rate of rainfall increases. A 3 dB difference is found when the rain rate approaches 90 mm/h and almost 6 dB at 150 mm/h when the horizontally polarized signal can suffer from 25-dB attenuation in 1 km. The phase rotation is derived using the method described in [16]. Figure 16.11

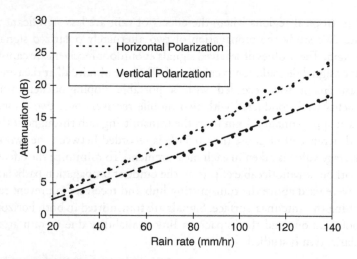

Figure 16.9   Attenuation measurement.

shows the phase difference between the two polarizations. A 45° difference occurs when the rain rate is approximately 90 mm/hr.

16.3.4.3   Recapitulation of Results

The effects of rain attenuation on an LMDS system operating at 26 GHz have been studied with alternately polarized carriers. Results show that the difference in signal

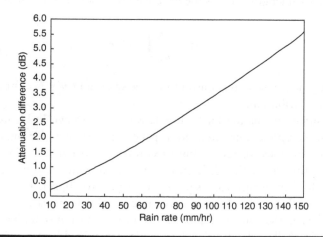

Figure 16.10   Attenuation difference.

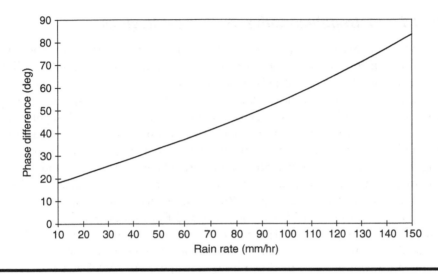

Figure 16.11   Phase difference.

attenuation between horizontal and vertical polarizations becomes more significant as the rate of rainfall increases. As much as 3-dB attenuation difference and 45° phase difference exist between the two polarizations when the rate of rainfall approaches 90 mm/h. The results offer some guidelines in the estimation of the necessary system fade margins that are greatly affected by the rain statistics. Increase in gain margin of both the system and antennas can improve link availability to compensate for the reduction in coverage caused by rainfall. It is apparent from the results that the presence of rain has a significant impact on both polarizations at 26 GHz.

16.4   Carrier Frequencies in Tropical Regions

In most areas, spectrum allocation for system deployment is governed by local authorities' regulations. Generally, higher frequencies are preferred for wider bandwidth availability due to operation in the less congested part of the spectrum. The single most important issue affecting link reliability of BWA systems is due to rain attenuation especially for tropical regions where persistent heavy rainfall is common.

We study the effects of rain attenuation on microwave channel in Singapore and Hong Kong, two tropical cites in Asia for wideband signals using a LOS LMDS link with orthogonal polarizations as illustrated next. These sites are selected primarily because they are both highly populated metropolitan cities having huge data densities and very often affected by persistent heavy tropical downpour. Measurements are performed by transmitting signals with horizontal and vertical polarizations with carrier frequencies commonly used in LMDS systems.

16.4.1 Metropolitan Cities in Tropical Regions

As described in Section 16.2, many LMDS systems utilize alternately polarized signals in conjunction with frequency result for improvement on bandwidth utilization. However, the effect of depolarization due to rainfall has a serious impact on frequency reuse that subsequently leads to a decrease in effective system capacity limiting both availability and scalability. Identical LMDS systems with slight variations in certain operating parameters were set up in two tropical metropolitan cities with independent climate patterns of Singapore and Hong Kong to analyze the long-term statistics of the difference in attenuation between horizontal and vertical polarized signals with the assumption that the accuracy of measurement is within 2 dB accuracy based on the finding in [20]. These two Asian cities, Hong Kong, and Singapore, are classified by ITU as regions N and P with mean rainfall rates of 95 and 145 mm/h, respectively [21]. These experiments determine the impact of rain attenuation at various carrier frequencies in vertical and horizontal polarizations by taking measurements under persistent heavy downpour. An LMDS system is set up to perform the measurements with system parameters shown in Table 16.2. Narrow beam width antennas of about 2° have been selected for this experiment. The system is measured with reference to a BER performance of 10^{-6}.

16.4.2 Measurement Results and Implications

All readings within an hour period were found to be very similar for each frequency with a typical spread of no more than 1 dB. We plot the average values for vertical and horizontal polarizations in Figures 16.12 and 16.13, respectively. In all these plots, a solid line represents results taken from Singapore and a dotted line for those from Hong Kong. Only about 1 dB difference is recorded between the attenuation

Table 16.2 System Parameters of Local Multipoint Distribution Service in Singapore and Hong Kong

System gain	110 dB
Transmitter antenna gain	18 dBi
Receiver antenna gain	28 dBi
Receiver sensitivity	−85 dBm (BER = 10^{-9})
Transmitter/receiver separation	5 km
Data rate	50 Mbps (point-to-point)
Modulation scheme	QPSK
Carrier frequencies (GHz)	10.5, 26.5, 28.5, and 39.5

BER, bit error rate.

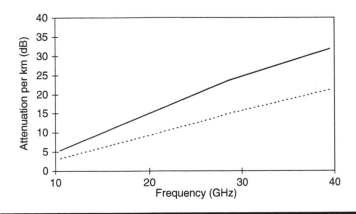

Figure 16.12   Vertical polarization.

of different polarizations when the carrier frequency is below 12 GHz indicating that negligible effect exists when the carrier frequency is relatively low. The difference increases at a substantial rate as the carrier frequency increases as shown in Figure 16.14. The dotted curve representing results taken from Hong Kong shows a noticeable yet insignificant attenuation difference between the two polarizations relative to the solid curve showing data taken from Singapore.

In Figures 16.12 through 16.14, best-fit lines for measured data is shown. The solid line above represents measurements taken from Singapore and the dotted line below represents measurements taken from Hong Kong for each polarization.

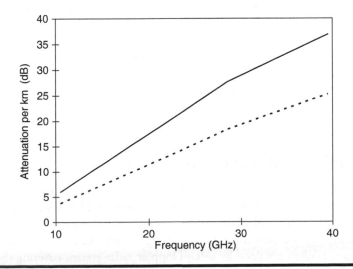

Figure 16.13   Horizontal polarization.

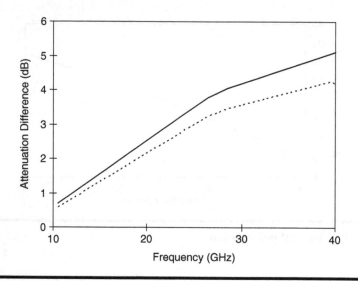

Figure 16.14   Attenuation difference.

The expected signal attenuation would be the same for a given carrier frequency in any location, but Singapore would statistically have a higher probability of heavy and more persistent downpour. The increase in rain rate has little effect on depolarization hence cross-polarization of orthogonally polarized signals. In both cases, the rain attenuation of a 10.5-GHz carrier is only about one-third that of a 26.5-GHz carrier. It is also noted that the difference in attenuation between the two polarizations is virtually identical at both locations.

All results indicate that signals undergo very severe signal attenuation at high frequencies. Interference between different polarizations becomes an important issue with higher frequencies. This effect is minimal at 10.5 GHz making this frequency suitable for LMDS deployment with orthogonal polarization in the tropical regions. Although effects of rain are reduced by the use of lower frequency, this results in more severe signal degradation due to multi-path with a longer path length.

16.5   Rainfall Statistics Considerations

The average rainfall statistics may vary considerably from month to month in many areas. With the significance of rain-induced degradation of BWA channels understood, a good understanding of the long-term rainfall pattern is necessary for network planning. Rainfall surveillance, giving a fairly accurate estimation of climate pattern, is often obtained by using Doppler radar system covering the area of concern. We are mostly interested in events of above 20 mm/h rainfall in the context of millimeter wave propagation as it is found that relatively low rain rate of below 20 mm/h has very little effect on wave propagation.

Reasonably accurate climate patterns can be easily obtainable from local weather observatories for many major cities, abnormal activities such as that caused by El Nin˜o may cause the rainfall statistics to differ significantly from that of previous years when averaged over a few months' period. The statistics is used in a region for which there is a likely impact of the predicted variations in rainfall patterns that may severely interrupt network performance. Adequate system fade margin (see next section) must therefore be dynamically adjusted based on accuracy and uncertainties assessment.

How sensitive is a certain system to weather changes depends very much depend on its supported applications. An Internet Service Provider (ISP) may easily meet its contractual agreement to its subscribers by ensuring a certain system up time. Temporary outage to other consumer networks may be acceptable during short bursts of very heavy downpour lasting several minutes. When system availability is not of great concern, system designers can rely on typical climate information for the region without worrying too much about any variations. However, life-saving systems such as wireless telemedicine network may require peak performance and reliability under persistent heavy rain, especially as many accidents are directly caused by rain. These systems must therefore be carefully designed to ensure that no part of the network fails as a result of any sudden change in weather.

Throughout the lifetime of a given system, climatic changes over long periods of time may require more than just adjustment of system fade margin. Therefore, we should not assume that once a system is completely set up it will operate as reliably as it has been over time. Besides precipitation statistics, wind sheer may pose a limitation on the type and size of antennas that can be mounted on exposed roof-top locations. The most useful type of climate information is the seasonally averaged rainfall and wind velocity in the system deployment region, not limited to its actual site. A good knowledge of the past record of rainfall and wind statistics is important for system planning. On the basis of such information, the most suitable types of antennas can be mounted in an optimal location for both signal reception and physically stability.

16.5.1 Past, Present, and Future Climate Information

Past statistical data, mostly available from appropriate government agencies, accurately describes the rainfall pattern, speed, and direction of wind that would have affected a system if one existed; so long as the existing data is long enough to have captured the full range of possible variations; usually for at least the past three to five years. Such data is particularly useful for initial system deployment so that system parameters are suitably determined to get started.

Climate changes over time due to factors such as global warming, changes in ocean currents, and cosmetic activities above the atmosphere. Certain system operating parameters may need to be fine-tuned for sudden changes in weather.

Certain unusual events may indicate a short-term deviation from any past statistic model that may affect the system for several months. For example, abnormally high sea surface temperature detected around the central Pacific Ocean may well indicate the effect of El Niño. From such observations, it is possible for us to carry out seasonal adjustment to systems that support critical missions to maintain a high level of availability.

Long-term climate forecasts may provide system designers' data for long-term planning. The range of possible events may be extracted from past data, confirmed by present information that determines whether the weather of a given period is within the statistical range. It is quite possible to predict future climate changes based on projection so that the system can be prepared for such anticipated changes. The underlying trend in the record of rainfall may provide indications about any pattern of increasing rainfall over the next few years. It is, however, important to understand that forecasts are by no means accurate; any projection about future climate can only be used for probabilistic analysis when setting system parameters by considering both known and predicted information.

16.6   Link Availability and Fade Margin

The topic of link availability measures the link outage time that leads to a certain percentage of time that a microwave link is available based on a statistical model. Generally, more critical applications such as aeronautics and biomedical systems require a much higher link available of at least 99.9999 percent. For consumer electronics application, where cost of providing services is a constraint, normally a link availability of 99.97 percent is deemed acceptable. The link availability is chiefly controlled by a number of system parameters such as gain, antennas used, and transmission power. A number of possible ways exist to improve link availability as well as transmission efficiency.

An appropriate fade margin must be set to maintain the radio link's availability. The fade margin, also known as system operating margin, measures the difference between the received signal strength and the signal minimum level to maintain link availability. The fade margin is usually maximized to support maximum reliability without excessive transmission power. It can be seen that the fade margin requirement increases when a higher-order modulation, which provides a higher spectrum efficiency in the expense of circuit complexity. Maintaining an adequate flat fade margin may result in more sophisticated transmission power control with 64-QAM or above which generally is not an issue with 16-QAM.

An example to illustrate the impact of rainfall on attenuation is plotted in Figures 16.15 and 16.16 to compare the effects at different carrier frequencies while other parameters remain unchanged. The attenuation is insignificant when the rainfall is low and minimal difference exists between these two graphs, which show the difference for vertically and horizontally polarized signals at 10 and 20 GHz, respectively.

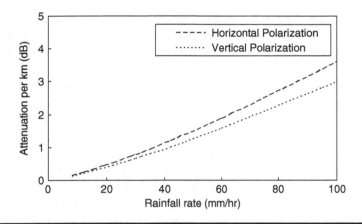

Figure 16.15   Rain attenuation at 10 GHz.

It is noted that vertical polarization is less severely affected by rain, and hence offers a longer range than horizontal polarization under identical operating environments. Figure 16.16 shows that horizontally polarized signals suffer from a high degree of attenuation under the same environment. The upper curve representing a 20-GHz carrier suffers more significant signal attenuation as the rainfall increases with a difference of 3 dB from 10 to 20 GHz experienced at a rate of 40 mm/h. Results show that the degree of attenuation caused by rain is largely dependent on polarization.

16.6.1   Experimental Set-Up

Each hub of an LMDS system often serves subscribers over several kilometers away. In practice, the operating range is limited to a very large extent by rain fading.

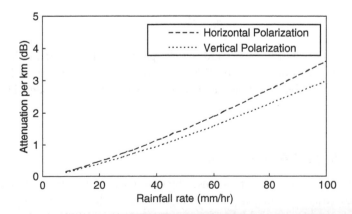

Figure 16.16   Rain attenuation at 20 GHz.

This section studies the link availability by extensive measurements conducted for polarization diversity reception in Singapore, classified as ITU rain region P [22]. Heavy rain severely affects the link availability as rain drops degrade the Signal-to-Noise Ratio (SNR) of the channel and transmission of data over these links may suffer from severe busty errors. Throughout the experiment, a target link availability of 99.99 percent is set as a reference.

An LMDS is set up with a fixed outdoor radio transmitter and a mobile receiver is used to measure the received signal that captures and stores the incoming data for analysis. The effects of channel mitigation are established by analyzing the received data under persistent heavy rainfall. The raw data is measured for the attenuation with respect to clear conditions with no rainfall (0 mm/h). The BRU with a 90° sector antenna converts the intermediate frequency signal from the Base Station into a variable carrier frequency for transmission over the wireless channel. The receiver maintains a direct LOS path with the hub and is placed closely to the subscriber system at 2 m apart to minimize any signal loss induced during the data transfer process from the receiving antenna to the capturing device.

An assumption that variations of readings taken on different days under different environmental conditions do not exceed 2 dB has again been made [20]. The hub that uses a transmitting antenna of 18 dBi gain is placed at the edge of a building at approximately 50 m above ground and a LOS is established between the BRU and SRU with the first fresnel zone free of any physical obstacles. The modulation scheme selected for this experiment is QPSK because of its properties that reduce the effects of cell-to-cell interference. Interference becomes most severe when two components of the same frequency and same polarization overlap. The operative path length d' as a result of rain attenuation is given by:

$$d' = \frac{d}{1 + d/d_0} \tag{16.9}$$

where

$$d_0 = 35.e^{-1.5 \times 10^{-2} R}$$

R is the rate of rainfall in mm/h and d is the path length measured in km that the signal can cover at a minimum of BER = 10^{-6} in free space. The transmitted signal is sent with a 90° Base Station antenna at a range of carrier frequencies between 10 and 40 GHz.

16.6.2 Signal Attenuation

The amount of signal attenuation caused by rain, measured in dB per km, as a function of the carrier frequency is shown in Figure 16.17. The amount of signal loss per km of path increases with higher frequencies as expected [23]. It also shows that attenuation increases at a more substantial rate in the 15–25-GHz range because

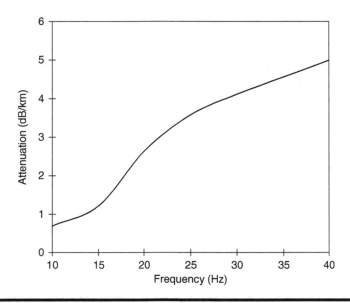

Figure 16.17   Attenuation due to rain.

water vapor emission is dominated by a molecular resonance in this frequency range with a local maximum at around 22 GHz. From 25 GHz, rain attenuation increases approximately linearly with increasing frequency. Further, we also found this relationship in signal attenuation to be independent of raindrop size distribution.

16.6.3 Link Margin

The effects of polarized transmission have been studied for cell planning [24]. Under a similar environment, we evaluate the fade margin of vertical and horizontal polarization in relation to the operating environment influenced by tropical rainfall. The fade margin measured is plotted in Figure 16.18 showing the necessary fade margin versus the corresponding range of coverage by each of the polarized component. The relationship between fade margin α and rain attenuation γ is governed by

$$\gamma = k \cdot R^{\alpha}. \tag{16.10}$$

The effects of depolarization between horizontal and vertical polarizations are compared by using a carrier frequency of 26 GHz. Figure 16.18 also shows that the fade margin with horizontal polarization is higher than with vertical polarization and the margin increases with the range, this is due to the fact that horizontally polarized signal is more severely affected by rain. While at lower frequency of around 10 GHz,

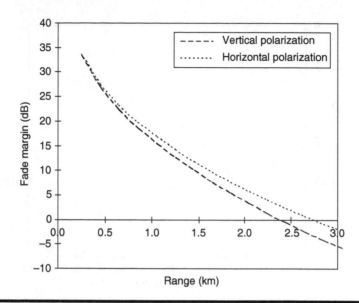

Figure 16.18   Fade margin with QPSK.

the link is much less affected by rain in this respect. However, it does come with a tradeoff of much more signal degradation due to multi-path fading.

The use of higher-order QAM modulation schemes have also been evaluated because they are often preferable due to higher SUE despite more significant

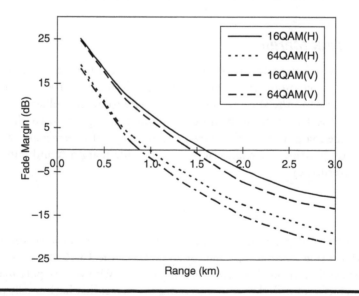

Figure 16.19   Fade margin with QAM.

cell-to-cell interference. Figure 16.19 shows a comparison between 16- and 64-QAM with reference to QPSK as plotted in Figure 16.18. The relative fade margins in dB of 16-QAM and 64-QAM are found to be –8.7 and –16.6 dB, respectively.

16.7 Summary

Signal attenuation due to rain is more severe as the carrier frequency increases where the rate of increase is greatest between 15 and 25 GHz due to water vapor resonance. While a lower frequency of around 10 GHz has much better performance with respect to rain attenuation, it is offset by multi-path due to the longer paths and the use of low gain antenna in the SRU. For this reason, 10 GHz is generally used in tropical regions where frequencies above 25 GHz are used in most parts of Europe and Americas.

The difference in horizontal and vertical polarizations becomes significant when the path length is over 1.5 km where the difference in fade margin reaches 2 dB. Finally, a reduction in range is resulted in changing the modulation scheme from QPSK to 16-QAM with the bandwidth efficiency doubled. An appropriate fade margin must be set based on statistical results to ensure that the system is not too severely affected by rain to cause an outage.

References

1. "Attenuation by Hydrometers," In Particular Precipitation and Other Atmospheric Particles, CCIR Report 721-3, ITU (1990).
2. L. Hanzo, "Bandwidth-Efficient Wireless Multimedia Communications," *Proceedings of the IEEE*, Vol. 86, No. 7, pp. 1342–1382 (July 1998).
3. E. P. Harris, W. E. Pence, and M. Sri-Jayantha, "Technology Directions for Portable Computers," *Proceedings of the IEEE*, Vol. 83, No. 4, pp. 636–657 (April 1995).
4. B. Fong, P. B. Rapajic, and G. Y. Hong, "Effects of Rain Attenuation on Wireless Transmission of Frame Relay Traffic," *IEEE International Conference on Communication Systems*, ICCS2002, pp. 1257–1259 (November 2002).
5. B. Fong, A. C. M. Fong, and G. Y. Hong, "On the Performance of Telemedicine System Using 17 GHz Orthogonally Polarized Microwave Links Under the Influence of Heavy Rainfall," *IEEE Transactions on Biomedicine and Information Technology*, Vol. 9, No. 3, pp. 424–429 (September 2005).
6. B. Fong, A. C. M. Fong, and G. Y. Hong, "Measurement of Attenuation and Phase on 26 GHz Wide-Band Point-to-Multipoint Signals Under the Influence of Tropical Rainfall," *IEEE Antennas and Wireless Propagation Letters*, Vol. 4, pp. 20–21 (2005).
7. "ITU-R recommendations," P-ser, Rec. ITU-R P. 618–628 (April 2003).
8. B. Fong, P. B. Rapajic, A. C. M. Fong, and G. Y. Hong, "Polarization of received signals for wideband wireless communications in a heavy rainfall region," *IEEE Communications Letters*, Vol. 7, No. 1, pp. 13–14 (January 2003).

9. S. A. Khan, A. N. Tawfik, E. Vilar, and C. J. Gibbins, "Analysis and Modelling of Rain Fade Durations at 97 GHz in 6.5 km Urban Link," *Electronics Letters*, Vol. 36, No. 5, pp. 459–461 (March 2, 2000).

10. G. C. Lee, T. P. Wang, and C. C. Tseng, "Resource Reservation with Pointer Forwarding Schemes for the Mobile RSVP," *IEEE Communications Letters*, Vol. 5, No. 7, pp. 298–300 (July 2001)

11. B. Fong and G. Y. Hong, "RF Net Scales Broadband to Local Area," *EE Times*, (June 17, 2002).

12. B. Fong, P. B. Rapajic, G. Y. Hong, and A. C. M. Fong, "Factors Causing Uncertainties in Outdoor Wearable Wireless Communications," *IEEE Pervasive Computing*, Vol. 2, No. 2, pp. 16–19 (April 2003).

13. M. O. Ajewole, L. B. Kolawole, and G. O. Ajayi, "Cross Polarization on Line-of-Sight Links in a Tropical Location: Effects of the Variation in Canting Angle and Rain Dropsize Distributions," *IEEE Transactions on Antennas and Propagation*, Vol. 47, No. 8, pp. 1254–1259 (August 1999).

14. M. Lobeira Rubio, A. Garcia-Armada, R. P. Torres, and J. L. Garcia, "Channel Modeling and Characterization at 17 GHz for indoor broadband WLAN," *IEEE Journal on Selected Areas in Communications*, Vol. 20, No. 3, pp. 593–601 (April 2002).

15. T. Oguchi, "Electromagnetic Wave Propagation and Scattering in Rain and Other Hydrometers," *Proceedings of the IEEE*, Vol. 71, pp. 1029–1078 (September 1983).

16. T. Oguchi, "Attenuation and Phase Rotation due to Rain: Calculations at 19.3 and 34.8 GHz," *Radio Science*, Vol. 8, No. 1, pp. 833–839 (February 1972).

17. H. Xu, T. S. Rappaport, R. J. Boyle, and J. H. Schaffner, "38-GHz Wide-Band Point-to-Multipoint Measurements under Difference Weather Conditions," *IEEE Communications Letters*, Vol. 4, No. 1, pp. 7–8 (January 2000).

18. V. Tralli, A. Verdone, and O. Andrisano, "Adaptive Time and Frequency Resource Assignment with COFDM for LMDS Systems," *IEEE Transactions on Communications*, Vol. 48, No. 2 (February 2001).

19. R. O. LaMaire, A. Krishna, P. Bhagwat, and J. Panian, "Wireless LANs and Mobile Networking: Standards and Future Directions," *IEEE Communications Magazine*, Vol. 34, No. 8, pp. 86–94 (August 1996).

20. K. I. Timothy, J. T. Ong, and E. B. L. Choo, "Performance of the Site Diversity Technique in Singapore: Preliminary Results," *IEEE Communications Letters*, Vol. 5 No. 2, pp. 49–51 (February 2001).

21. "Propagation Data and Prediction Methods Required for the Design of Terrestrial Line-of-Sight Systems," ITU-R 530-10 (November 2001).

22. "ITU-R recommendations," P-ser, Rec. ITU-R P. 618-5.

23. ITU-R P. 530-7.

24. D. Scherer, "Optimizing Frequency Re-use in Point-to-Multipoint Deployments," *Proceedings of the IEEE Radio and Wireless Conference*, pp. 83–86 2000.

Chapter 17

Resource Management in WiMAX Networks

Yi Qian, Shafaq B. Chaudhry, and Ratan K. Guha

In this chapter, we investigate resource management mechanisms in Worldwide Interoperability for Microwave Access (WiMAX) networks, including subcarrier assignment, adaptive power allocation, and call admission control schemes. Subcarrier assignment scheme deals with how to allocate subcarriers to form a subchannel. Adaptive power allocation allows the transmitter to grant different power levels to different subcarriers according to the channel state information. Call admission control highlights how to distribute a subscriber's access bandwidth over different types of applications such that the quality-of-service (QoS) requirement of each connection is guaranteed. We focus on orthogonal frequency division multiple access (OFDMA) with time division duplex (TDD) environment, which allows high-spectrum utility efficiency on the uplink and downlink channels in asymmetric "last-mile" Internet access. We conclude this chapter with an optimization strategy that finds a balance between the revenue of a WiMAX service provider and the satisfaction of WiMAX subscribers.

17.1   Introduction

To achieve broadband wireless access solution, the IEEE 802.16 standard subcommittee has released a series of standards for WiMAX, that supports high-throughput broadband connections over long distances (up to 30 miles), as detailed in [1]. There are still a large number of areas where wired infrastructures such as T1, Digital Subscriber Line (DSL), and cable are difficult to deploy, and hence expensive. To provide a broadband wireless access solution in these areas, proponents are advocating WiMAX as a feasible alternative to wired Internet access solutions such as cable modem and DSL. However, from a commercial viewpoint, whether the promise of WiMAX will materialize still depends on its deployment expense to telecommunication operators and its data transmission rate to subscribers.

Currently, WiMAX can provide single-channel data rate of up to 75 Mbps on both uplink and downlink. To achieve an even higher data rate, service providers can use multiple WiMAX channels for a single transmission to reach a bandwidth of up to 350 Mbps. The discussion in this book chapter is under the context of WiMAX orthogonal frequency division multiple access (OFDMA) with time division duplex (TDD), which allows on-demand bandwidth allocation between uplink and downlink channels to achieve high spectrum utility efficiency.

As one of its most anticipated features, WiMAX is expected to support a variety of services and traffic types, including data transfer, voice, video, and multimedia streaming. Correspondingly, four types of services have been defined by the WiMAX Forum*, namely, Unsolicited Grant Service (UGS), real-time Polling Service (rtPS), non-real-time Polling Service (nrtPS), and Best Effort (BE). Among them, UGS, rtPS, and nrtPS can be classified into the category of QoS guaranteed services. To handle a multi-service WiMAX access network of heterogeneous traffic load, a resource management scheme needs to be devised that can efficiently allocate radio and bandwidth resources to different subscribers and services. This chapter addresses several such resource management mechanisms, that is, subcarrier assignment (SCA) for subchannel formation, adaptive power allocation (APA) on WiMAX uplink and downlink, and call admission control (CAC) in a subscriber's local network. While the SCA schemes are standardized by the WiMAX Forum, the APA scheme can significantly affect the data transmission rate of each subscriber, and the CAC scheme can impact the data transmission rates allocated to difficult types of applications within the same subscriber. Thus, the APA or CAC algorithm is formulated as an optimization problem with a certain objective function. Here, the objective function is defined to maximize the revenue of a service provider or the satisfaction of the subscribers. Usually, there exists a contradiction between the expectations of the service provider and the subscribers. This chapter investigates the balancing strategy between WiMAX service providers' revenue and subscribers' satisfaction. Moreover, in a real WiMAX system, APA and CAC are often combined together to provide

---

* http: //www.wimaxforum.org

cross-layer resource management. This cooperation relationship between APA and CAC has also been discussed in this chapter.

17.2 WiMAX OFDMA-TDD System

WiMAX technology supports both mesh and point-to-multipoint (PMP) networks. WiMAX mesh network is usually used for creating wide-area wireless backhaul network for citywide wireless coverage and 3G Radio Network Controller to Base Station (BS) connection, whereas WiMAX PMP network aims at providing last-mile access to a broadband Internet Service Provider (ISP). Figure 17.1 demonstrates an example of a WiMAX PMP network involving one BS and N subscribers, as illustrated in [1]. The IEEE 802.16 standard specifies orthogonal frequency-division multiplexing (OFDM) as the modulation technique used in WiMAX to achieve its high data rates. Particularly, IEEE 802.16 has specified two flavors of OFDM systems: simple OFDM, and OFDMA. The first option is proposed for applications intended for use over short distances. It employs a Fast Fourier Transform (FFT) size of 256, which is a step further from 802.11a's 64 carriers. In OFDMA, however, a much higher FFT space of 2048 and 4096 carriers is available which is divided into subchannels. These subchannels are used to separate data into logical streams for downstream. Each of these streams may employ different modulation, coding, and amplitude to cater to subscribers with different channel characteristics.

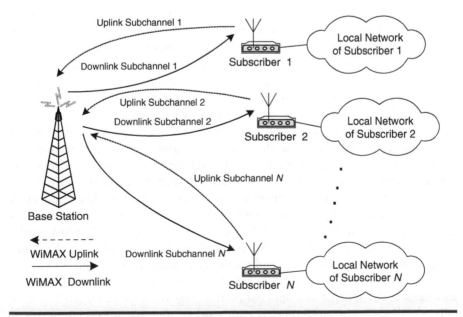

Figure 17.1 WiMAX point-to-multipoint (PMP) network.

Frequency division duplex (FDD) and TDD are the two most prevalent duplexing schemes used in broadband wireless networks. WiMAX can employ either of them to separate uplink and downlink communication signals. Both FDD and TDD have their own advantages depending on the application. In the FDD scheme, distinct frequency channels are assigned to uplink and downlink. In other words, at any particular instant, uplink and downlink traffic are transmitted simultaneously at different frequencies. Due to the symmetric nature, FDD uplink and downlink transmission channels are always of equal size. Hence, FDD is typically used in applications that require an equal uplink and downlink bandwidth. In contrast, TDD uses a single frequency to transmit signals in both downstream and upstream directions by arranging the uplink and downlink traffic to occupy the same frequency at different times. In implementation, TDD divides the data stream into frames and assigns different time slots to the forward and reverse transmissions within each frame. As devices use the same frequency channel to transmit or receive on-demand, TDD can flexibly control the amount of bandwidth allocated to uplink and downlink. For this reason, TDD has a much higher spectrum utility efficiency than FDD in asymmetric communication where uplink traffic is a fraction of downlink traffic. In this book chapter, WiMAX is supposed to be employed for last-mile Internet access, which is a typical case of asymmetric communication. Therefore, we only study the case of WiMAX OFDMA with TDD duplexing.

17.3 Subcarrier Assignment

Subchannels are used in upstream for multiple access. In practice, subscribers are assigned subchannels through media access control protocol (MAC) messages sent downstream. There are three types of OFDMA subcarriers, *data* subcarriers for data transmission, *pilot* subcarriers for various estimation and synchronization purposes, and *null* subcarriers for guard bands. A combination of pilot and data subcarriers makes up a subchannel. To accommodate different circumstances, the IEEE 802.16 OFDMA systems cater to two types of methods for allocation of subchannels: the distributed subcarrier permutation method, which is favored for high mobility environments, and the adjacent subcarrier permutation method.

In distributed allocation, data within a subchannel is assigned to multiple subcarriers which may or may not be adjacent to each other, so the carriers of a subchannel may be spread throughout the spectrum. The permutation formula that maps subchannels to physical subcarriers in the OFDMA symbol varies for uplink and downlink. For downlink, WiMAX OFDMA supports downlink-partial usage of subchannels (DL-PUSC), full usage of subchannels (FUSC), optional full usage of subchannels (OFUSC), and tile usage of subchannels 1 and 2 (TUSC1 and TUSC2). For uplink, two schemes are supported: uplink-partial usage of subchannels (UL-PUSC) and optional partial usage of subchannels (OPUSC). In adjacent allocation, a subchannel is composed of multiple adjacent subcarriers, whose positions are fixed in the frequency domain within an OFDMA symbol.

There is no variation for uplink and downlink. Band adaptive modulation and coding (AMC) is an example of an adjacent subcarrier allocation scheme. A brief outline of these schemes follows.

DL-PUSC: In DL-PUSC, as the name suggests, all logical subchannels available are not used, but only a subset is used. Subchannels are arranged into six groups that are assigned upto three segments (also called sectors) of a cell. This assignment is done such that probability of using the same subcarrier in adjacent segments or cells is minimized. If two groups are assigned to each segment, then the frequency reuse factor in the cell is three. If all groups are assigned to each segment, then the frequency reuse factor in the cell is one. Also, full channel diversity is employed by distributing subcarriers to subchannels, and this helps to counter the effects of fast fading channel response owing to user mobility. As specified by the IEEE 802.16 standard, every downlink subframe must start in the DL-PUSC mode. The scheme is cluster-based where the basic allocation unit has two symbols spread over 14 subcarriers.

FUSC: As opposed to PUSC, the allocated subcarriers are distributed among all available subchannels. Thus, frequency diversity can be employed to counter the effects of fading channels. Similar to PUSC, the subchannels are assigned to segments of a cell such that probability of using the same subcarrier in adjacent segments or cells is minimized. As an example, the subchannel formation in the case of FUSC is shown in Figure 17.2. The whole carrier space is divided into a number of N_G successive groups. Each group contains a number of N_E successive carriers, after excluding the initially assigned pilots. For example, as illustrated in [2], N_E is 32 in the OFDMA mode with 2048 carriers. A subchannel has one element from each group allocated through a pseudorandom process based on permutations, so the number of subchannel elements is N_G.

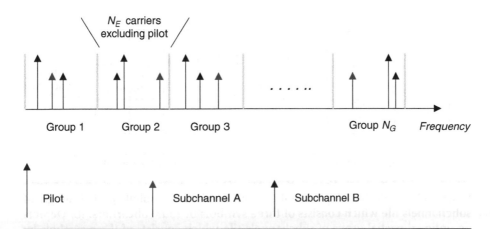

Figure 17.2   Principle of subchannel formation in full usage of subchannels (FUSC).

OFUSC: This scheme is similar to the FUSC, except it uses a bin structure and much larger number of subcarriers. The bin structure is explained under Adaptive Modulation and Coding (AMC).

TUSC1 and TUSC2: This permutation is similar to UL-PUSC. It is employed in adaptive antenna systems (AAS).

UL-PUSC: This uses a tile-based permutation, where the basic allocation unit has three symbols spread over four subcarriers. The four subcarriers at the corners are used as pilot subcarriers for channel estimation, whereas the rest are used as data subcarriers. The UL-PUSC OFDMA tile is shown in Figure 17.3a.

OPUSC: This scheme also uses a tile-based permutation, where a basic allocation unit has three symbols spread over three subcarriers. The subcarrier at the center is used as a pilot, whereas the rest are used as data subcarriers. Admittedly, it has less channel estimation capability than UL-PUSC, but it has much more data subcarriers owing to the smaller size of the tile. The OPUSC OFDMA tile is shown in Figure 17.3b.

Band AMC: In this permutation, a bin-structure is used, where a bin is a set of nine adjacent subcarriers within an OFDMA symbol. In 802.16e, an AMC slot is defined as two bins by three symbols and is shown in Figure 17.4. The permutation is the same for uplink or downlink. Because the subcarriers are adjacent, water-pouring type of algorithms can effectively be used. An example is shown in Figure 17.4.

Further details of the subchannel allocation can be found in [1,3,4].

The different schemes presented here provide a tradeoff to the designer between mobility and throughput. Distributed schemes employ frequency diversity to

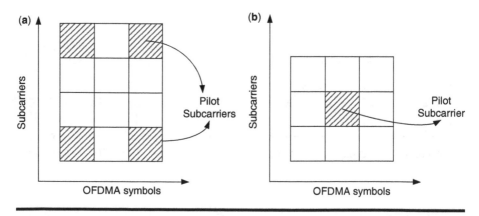

Figure 17.3 Tile-based permutation. (a) Depicts an uplink partial usage of subchannels tile which consists of three symbols by four subcarriers. (b) Depicts an optional partial usage of subchannels tile which consists of three symbols by three subcarriers. OFDMA, orthogonal frequency division multiple access.

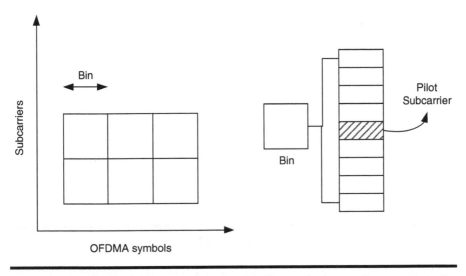

Figure 17.4 Band adaptive modulation and coding (AMC)—an adjacent subcarrier allocation scheme. An AMC slot is two bins by three symbols, where a bin is nine adjacent subcarriers. OFDMA, orthogonal frequency division multiple access.

counter slow fading in channels by means of selecting subcarriers pseudorandomly. Thus, it can cater for highly mobile users. On the other hand, due to the flat fading type of channel response, frequency selection cannot be used in adjacent schemes, and hence they make use of multi-user diversity to maximize throughput and are much suited for AAS. As distributed subcarrier permutation has good performance in both fixed and mobile environments, it becomes the dominant subcarrier permutation strategy for WiMAX.

17.4 WiMAX APA Optimization

17.4.1 Uplink APA Optimization

Once the subcarrier assignment is determined by a certain subchannel allocation scheme, the WiMAX uplink and downlink transmission can be optimized by APA, which allows the transmitter to allocate different power levels to different subcarriers according to the channel state information (CSI) from the physical layer, as discussed in [5]. To make the analysis simple, in this chapter we only examine the most common scenario where each WiMAX subscriber occupies exactly one subchannel. On the uplink, the WiMAX subscriber's transmitter only needs to allocate power resources to the different carriers inside its own subchannel. Therefore, the APA optimization on WiMAX uplink has exactly one goal, that is, achieving the best data transmission rate. This problem has been comprehensively studied by previous work, mentioned in [6,7], as single-user water-filling.

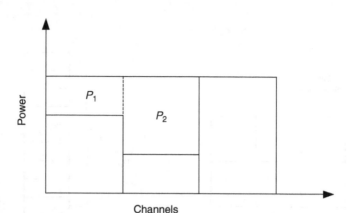

Figure 17.5 Water-filling for multiple channels.

The water-filling concept is depicted in Figure 17.5, where the vertical lines represent a channel's noise level. Power is allotted to the channel with the lowest noise. As the available signal power increases, power is spilled over to the next channel with the lowest noise. Let $R_{u,v}$ be the channel-gain-to-noise ratio of a subchannel v experienced by user u, then by the water-filling method, it can be shown that the power allocation for user u on subchannel v, $P_{u,v}$ is $[l_w - R_{u,v}^{-1}]^+$, where $[x]^+$ is max$\{0,x\}$, l_w is the water level selected such that $P_{tot} = \sum P_{u,v}$ [8,9]. Efficient algorithms for solving the multi-user water-filling problem have been presented in [10,11]. For multiple users, the authors in [12], present a frequency-power allocation for uplink which follows from multi-user water-filling, but this is applicable only when the channel-gain-to-noise ratios for the users is same. In contrast, the algorithm in [9] is suited for mobile environment, where power constraints are distributed to each user. It presents an iterative power allocation algorithm based on water-filling, which maximizes the rate-sum capacity in uplink OFDMA systems.

The aforementioned work assumes that knowledge of multiple users' CSI is readily available at the BS at every instant. This assumption may no longer be viable when the channel fades very fast and thus, other channel characteristics have been explored. For example, a power allocation scheme based on tap correlation has been proposed in [13]. Tap correlation is attributed to the fact that multi-path signals bounce from scatterers located in a narrow angular range, at a considerable distance from the transmitting source and hence, can be correlated temporally in the delay domain. The scheme shows that using tap correlation, multi-user diversity can be efficiently exploited, resulting in increase in the rate-sum capacity of an OFDMA uplink.

17.4.2 *Downlink APA Optimization*

In contrast to the uplink APA optimization, the WiMAX downlink APA optimization has to take care of the carriers of all subscribers, and different strategies may be employed in different circumstances. In this section, the criteria of APA optimization on WiMAX downlink is discussed, which has been described in detail in [14,15]. As already mentioned in the introduction, the IEEE 802.16 standard provides a wide range of services aimed at supporting various applications. These services are detailed in [16] and are summarized next.

UGS: UGS is designed to support real-time data streams consisting of fixed-sized data packets issued at periodic intervals, such as T1/E1 and Voice-over-IP (VoIP).

rtPS: rtPS is designed to support real-time data streams consisting of variable-sized data packets that are issued at periodic intervals, such as Motion Picture Experts Group (MPEG) video.

nrtPS: nrtPS is designed to support delay-tolerant data streams consisting of variable-sized data packets for which a minimum data rate is required, such as File Transfer Protocol (FTP).

BE: BE service is designed to support data streams for which no minimum service level is required and which can be handled on a space-available basis.

Let $H_k(f)$ and $N_k(f)$ respectively denote the channel frequency response function and noise power density function of the kth subscriber. Then, the N-subscriber WiMAX downlink channel model can be represented as in Figure 17.6.

The quality of each subscriber's subchannel is indicated by the Signal-to-Noise ratio (SNR) function $\rho_k(f)$, which is defined as:

$$\rho_k(f) = \frac{|H_k(f)|^2}{N_k(f)}. \tag{17.1}$$

Here, $\rho_k(f)(1 \leq k \leq N)$ is the CSI that downlink APA needs.

To make the analysis simple, the most common scenario where each WiMAX subscriber occupies exactly one subchannel is investigated. The WiMAX downlink

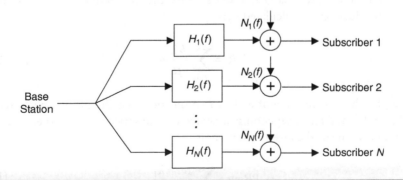

Figure 17.6 WiMAX downlink channel model.

APA optimization has to take care of the subcarriers of all subscribers, and different strategies may be employed to achieve different objectives.

To solve the downlink APA optimization problem, a fairness-constrained greedy revenue algorithm is developed in [15]. The main idea of the algorithm is to maximize the revenue of service provider and provide fairness among all subscribers. The rest of this section is organized as follows. The design criteria of APA optimization are first discussed, in which revenue and fairness are defined quantitatively. Then, the constrained greedy algorithm is described, which can efficiently solve the APA optimization problem.

17.4.2.1 Optimal Revenue Criterion

Optimal revenue criterion is based on the fact that service providers prefer a downlink power allocation scheme that can maximize their revenue. Let APA revenue be defined as the revenue that a certain APA scheme can potentially produce per time unit. APA revenue is not the final revenue achieved by a service provider, because the final revenue can also be affected by other resource allocation steps such as CAC. Nevertheless, we only consider the APA revenue in this section. To investigate the APA revenue of a certain scheme, revenue rate is defined for each type of service as the revenue generated by a bandwidth unit during a time unit. Let rer^{UGS}, rer^{rtPS}, rer^{nrtPS}, and rer^{BE} be the revenue rates of UGS, rtPS, nrtPS, and BE service, respectively. Moreover, to maximize the APA revenue, a potential revenue of each subscriber is defined as the revenue generated by all potential traffic load in a time unit. As different WiMAX subscribers can have different amounts of UGS, rtPS, nrtPS, and BE traffic in their local networks, they hold distinct downlink traffic rates and potential revenue. To quantitatively describe all the concepts, the downlink traffic and revenue model is defined next. For a given subscriber, suppose M classes of traffic share B units of downlink bandwidth resource in the local network. For traffic class i, assume that requests arrive from a random process with average rate λ_i; each connection request demands b_i bandwidth resources; and the average connection holding time is $1/\mu_i$ seconds. Thus, the total traffic rate, TR, from the given subscriber can be calculated by:

$$\text{TR} = \sum_{i=1}^{M} \frac{b_i \lambda_i}{\mu_i}. \tag{17.2}$$

Let rer_i be the revenue rate of a class i connection, which can be one of rer^{UGS}, rer^{rtPS}, rer^{nrtPS}, and rer^{BE}, depending on which type of service is used. Then, the total potential revenue of the given subscriber is:

$$\text{PRev} = \sum_{i=1}^{M} \frac{rer_i b_i \lambda_i}{\mu_i}. \tag{17.3}$$

Suppose the kth WiMAX subscriber has a downlink traffic rate TR_k^D and a potential revenue PRev_k^D. If tr_k^D is the downlink data transmission rate of the subscriber k, then the downlink bandwidth capacity is $\mathrm{DB}_k = x\% \mathrm{tr}_k^D$. Accordingly, the downlink revenue-to-bandwidth ratio for subscriber k is defined as:

$$\mathrm{RBR}_k^D = \begin{cases} \dfrac{\mathrm{PRev}_k^D}{\mathrm{TR}_k^D} & \text{if } \mathrm{DB}_k \leq \mathrm{TR}_k^D \\ 0 & \text{otherwise} \end{cases} \tag{17.4}$$

and the downlink APA revenue for subscriber k is defined as:

$$\mathrm{AP\,AR}_k^D = \mathrm{RBR}_k^D\,\mathrm{DB}_k. \tag{17.5}$$

Furthermore, the overall downlink APA revenue of the WiMAX network is given by:

$$\mathrm{AP\,AR}^D = \sum_{k=1}^N \mathrm{AP\,AR}_k^D. \tag{17.6}$$

17.4.2.2 Fairness-Constrained Optimal Revenue Criterion

In practice, the power allocation scheme also needs to guarantee fairness among different WiMAX subscribers. Here, consider the fairness criterion as a measure to assure that no subscriber's data transmission rate becomes too low to be acceptable. Thus, the fairness-constrained optimal revenue criterion requires that either $x\%\mathrm{tr}_k^D > \mathrm{TR}_k^D$ or $\mathrm{tr}_k^D > F_{\mathrm{th}}\mathrm{tr}_{\mathrm{avg}}^D$, for any subscriber $k(1 \leq k \leq N)$, where $\mathrm{tr}_{\mathrm{avg}}^D$ is the average downlink data transmission rate of all subscribers, and F_{th} is the fairness threshold factor. The condition $x\%\mathrm{tr}_k^D > \mathrm{TR}_k^D$ means that the downlink traffic load of the kth subscriber could be totally accepted by the network; the condition $\mathrm{tr}_k^D > F_{\mathrm{th}}\mathrm{tr}_{\mathrm{avg}}^D$ means that the downlink data transmission rate of subscriber k is higher than a certain fairness threshold.

17.4.2.3 Fairness-Constrained Greedy Revenue Algorithm

To implement the fairness-constrained optimal revenue criterion, a fairness-constrained greedy revenue algorithm is proposed. The key idea of this algorithm is to keep the fairness constraint while allocating bits and the corresponding power in each iteration such that the revenue argument per power unit is maximized. To explain the algorithm clearly, a set of notations is defined as shown in Table 17.1.

According to [6,7,17], the achievable efficiency of the kth subscriber on subcarrier j can be expressed as:

$$c_k[j] = f(\log_2(1 + \beta p[j]\rho_k[j])) \tag{17.7}$$

Table 17.1 Notations for Fairness-Constrained Greedy Revenue Algorithm

N	Number of subscribers
S_N	Set of all subscribers, defined as 1,2,..., N
J	Number of subcarriers
S_J	Set of all subcarriers, defined as 1,2,...,J
$N(j)$	The subscriber that subcarrier j belongs to
F	Total physical bandwidth
Δf	Physical bandwidth of each subcarrier, which is equal to F/J
P_C	Total transmit power constraint
D_k	Subcarrier set assigned to subscriber k
$H_k[j]$	Channel gain of subscriber k on subcarrier j
$N_k[j]$	Noise power of subscriber k on subcarrier j
$\rho_k[j]$	Channel condition on subcarrier j of subscriber k, defined as $H_k[j]^2/N_k[j]$
$P[j]$	Transmit power on subcarrier j
$C_k[j]$	Achievable transmission efficiency (data rate per Hertz) of subscriber k on subcarrier j

where β is a constant determined by:

$$\beta = \frac{1.5}{-\ln(5\mathrm{BER})} \tag{17.8}$$

and $f(.)$ depends on the adaptation scheme. For example, if continuous rate adaptation is used, $f(x) = x$; and if variable M-ary quadrature amplitude modulation (MQAM) with modulation levels 0,2,4,6,... is employed, then $f(x) = 2\left\lfloor\frac{1}{2}x\right\rfloor$. To assign the subcarrier set D_k to subscriber k, assume a nonoverlapped partition:

$$D_i \cap D_j = \phi, i \neq j; \quad \bigcup_{k=1}^{N} D_k \subseteq S_J. \tag{17.9}$$

Consequently, the downlink data transmission rate of the kth subscriber is:

$$\mathrm{tr}_k^D = \sum_{j \in D_k} c_k[j]\Delta f. \tag{17.10}$$

Let $f_{\text{bit}}(j, b)$ be the power required to transmit b b/s/Hz on subcarrier j. From Equation 17.7, if the continuous rate adaptation is employed, we can derive that:

$$f_{\text{bit}}(j,b) = \frac{2^b - 1}{\beta \rho_{N(j)}[j]}.$$ (17.11)

With these definitions, the constrained greedy algorithm is illustrated in Algorithm 17.1. In the initialization phase, this algorithm assigns zero bits to all subcarriers. During each bit loading iteration, power is allocated on some subcarrier such that the increase of revenue per power unit is maximized. The algorithm converges when the total transmission power constraint is reached. Besides the consideration of the service provider's revenue, the algorithm takes into account the fairness constraint while choosing the subcarrier to allocate power. Note that, if the fairness constraint is removed, that is, if $F_{\text{th}} = 0$, then the fairness-constrained algorithm reduces to a pure greedy revenue algorithm with optimal revenue criterion.

17.5 Call Admission Control

17.5.1 Architecture of CAC Deployment

To handle a multi-service WiMAX network, it is very important to employ the CAC mechanism. First, CAC is a critical step for the provision of a QoS-guaranteed service, because it can prevent the system capacity from being over-used. Second, CAC can help a WiMAX network provide different classes of traffic load with different priorities by manipulating their blocking probabilities. Figure 17.7 illustrates the architecture of CAC deployment in a WiMAX network.

In this architecture, the CAC manager carries out the admission control for each subscriber's local network. As the CAC manager is located at the BS, it can easily know the uplink/downlink bandwidth capacity of subscriber k from other modules of the BS. When an application in subscriber k's local network initiates a connection to access the Internet, it sends connection request to the CAC manager with an upstream bandwidth requirement bU and a downstream bandwidth requirement bD. The CAC manager then performs admission control check on both uplink and downlink of subscriber k. In this respect, the CAC in WiMAX network is a two-dimensional CAC problem.

One way to tackle this problem is to decompose it into two independent one-dimensional CAC problems. In other words, the CAC manager employs an uplink CAC policy and a downlink CAC policy to run an admission test on the incoming request and only the connection request that passes both the uplink and downlink admission tests can be accepted eventually. For last-mile access, downlink admission control plays a much more important role than uplink admission control. Correspondingly, the downlink CAC optimization is discussed here.

Input: P_C, F_{th}/* P_C: total power constraint, F_{th}: fairness constraint */
for all j **do**
  $b_j \leftarrow 0$
end for
for all k **do**
  $tr_k^D \leftarrow 0$
end for
$p_{\text{assigned}} \leftarrow 0$
$Set_F \leftarrow S_j$ /*Set_F: set of subcarriers that can meet the fairness constraint */
for $j = 1$ to J **do**
  $\Delta p_j \leftarrow f_{\text{bit}}(j, b_j + \Delta b_j) - f_{\text{bit}}(j, b_j)$; /* assuming MQAM is employed, b_j is the current modulation level of subcarrier j, Δb_j is the difference between the next modulation level and the current modulation level, Δp_j is the power increment */
  if $p_{\text{assigned}} + \Delta p_j > P_C$ **then**
    $\Delta p_j \leftarrow \infty$
  end if
  $\Delta R_j \leftarrow x\% \Delta b_j \, \Delta f \, RBR_{N(j)}^D$ /* ΔR_j is the APA revenue */
end for
while $Set_F \neq \phi$ **do**
  $\min_{tr} \leftarrow \min_{j \in S_j} \{ tr_{N(j)}^D \mid (\Delta p_j \neq \infty) \ \& \ (x\% tr_{N(j)}^D < TR_{N(j)}^D) \}$
  $Set_F \leftarrow \{$subcarrier $j \mid (\Delta p_j \neq \infty) \ \& \ (x\% tr_{N(j)}^D < TR_{N(j)}^D) \ \& \ (\min_{tr} \geq F_{th} \, (tr_{avg}^D + \Delta b_j \, \Delta f / N)$ or $tr_{N(j)}^D = \min_{tr}) \}$
  if $Set_F \neq \phi$ **then**
    $j^* \leftarrow \arg \max_{j \in Set_F} \{ \Delta R_j / \Delta P_j \}$
    /* Assigning the power resource to subcarrier j^* */
    $p_{\text{assigned}} \leftarrow p_{\text{assigned}} + \Delta P_{j^*}$
    $tr_{N(j^*)}^D = tr_{N(j^*)}^D + \Delta b_{j^*} \, \Delta f$
    $b_{j^*} \leftarrow b_{j^*} + \Delta b_{j^*}$
    $\Delta P_{j^*} \leftarrow f_{\text{bit}}(j^*, b_{j^*} + \Delta b_{j^*}) - f_{\text{bit}}(j^*, b_{j^*})$
    $\Delta R_{j^*} \leftarrow x\% \Delta b_{j^*} \, \Delta f \, RBR_{N(j^*)}^D$
    for $j = 1$ to J **do**
      if $p_{\text{assigned}} + \Delta p_j > P_C$ **then**
        $\Delta p_j \leftarrow \infty$
      end if
    end for
  end if
end while

Algorithm 17.1   Fairness-constrained greedy revenue algorithm.

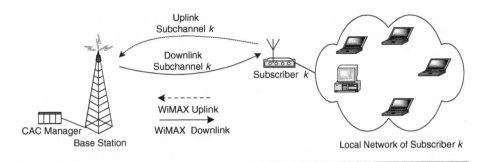

Figure 17.7   Call admission control (CAC) WiMAX network.

The objective of the downlink CAC optimization can be chosen to maximize the revenue of service providers or the satisfaction of subscribers. Usually there exists a contradiction between expectations of the service providers and the subscribers. Thus, balancing strategies for CAC optimization need to be devised and one such strategy is delineated next. As for the uplink admission control, suppose that the CAC manager employs an easy-to-implement policy, such as complete sharing (CS) which accepts a connection request if and only if there is sufficient resource.

In a WiMAX system, CAC is used to accept or reject connection requests based on the CSI and the QoS requirements of these connections. To formulate the downlink CAC optimization problem, a CAC policy is defined as follows. First, let B be the overall bandwidth resources (for subscriber k, $B = DB_k$) and let M be the number of traffic classes. Then, define the system state vector as $\vec{n} = (n_1, n_2, \ldots, n_M)$, where n_i is the number of class i connections in the system. Assuming that the bandwidth requirement of a class i connection is set to b_i, the bandwidth requirement vector can be represented by $\vec{b} = (b_1, b_2, \ldots, b_M)$. Based on these parameters, Ω_{CS} is defined as the set of all possible system states, that is, $\Omega_{CS} = \{\vec{n} | \vec{n} \cdot \vec{b} \leq B\}$. In this definition, the subscript CS stands for complete sharing, which means that an incoming connection will be accepted if sufficient bandwidth resources are available on the downlink of a subscriber. Now define a CAC policy, denoted by Ω, as an arbitrary subset of Ω_{CS}. Given Ω, a connection request will be accepted if and only if the system state vector remains in Ω after the connection has been accepted.

In the rest of this section, first, the design criteria of the CAC optimization will be discussed. Then, constrained greedy algorithms will be presented to solve the problem of CAC.

17.5.2   CAC Design Criteria

In general, service providers expect a CAC policy that can produce the maximal revenue. Let Ω^* denote the optimal revenue policy. Continuing in the same vein as the previous section, Ω^* prefers the traffic load with a high revenue output. Details can be found in [18–24]. Here, the long-run average revenue of a policy can be calculated by:

$$R(\Omega) = \sum_{\vec{n} \in \Omega} (\vec{n} \cdot \vec{r}) P_\Omega(\vec{n}) \tag{17.12}$$

where $P_\Omega(n)$ is the steady-state probability that the system is in state $\vec{n}$, $\vec{r} = (r_1, r_2, ..., r_M)$ is the reward vector, and r_i is the revenue generated by accepting a type i connection calculated as $r_i = \mathrm{rer}_i b_i$.

17.5.2.1 Optimal Utility Criterion

On the other hand, subscribers prefer the CAC policy that can provide the maximal utility, or equivalently, the maximum access bandwidth. To satisfy the demands of subscribers, consider the optimal utility CAC policy, denoted as Ω^+. Compared with Ω^*, Ω^+ will allocate more bandwidth resources to the traffic load that can yield high utility. Resembling Equation 17.12, the statistical bandwidth (SB) that policy Ω can achieve can be derived as:

$$SB(\Omega) = \sum_{\vec{n} \in \Omega} (\vec{n} \cdot \vec{b}) P_\Omega(\vec{n}) \tag{17.13}$$

where the reward vector $\vec{r}$ in Equation 17.12 is replaced with the bandwidth requirement vector $\vec{b}$. Based on Equation 17.13, the utility of policy Ω is given by:

$$U(\Omega) = \frac{1}{B} SB(\Omega) = \frac{1}{B} \sum_{\vec{n} \in \Omega} (\vec{n} \cdot \vec{b}) P_\Omega(\vec{n}). \tag{17.14}$$

17.5.2.2 Utility-Constrained Optimal Revenue Criterion

In a commercial WiMAX network, notice that the CAC policy must provide a trade-off between demands of the service provider and the subscriber. In other words, a CAC policy is needed that can balance the optimal revenue criterion and the optimal utility criterion. This leads to the concept of a utility-constrained optimal revenue policy. Here the utility constraint is defined such that the utility of the designated CAC policy must be higher than the utility lower bound LB^U, where $LB^U = lb^U U(\Omega^+)$; lb^U is the utility lower bound factor such that $(0 < lb^U < 1)$, and $U(\Omega^+)$ is the utility of Ω^+. Here, Ω^{U*} is the utility-constrained optimal revenue policy.

17.5.3 CAC Optimization Algorithms

Based on the discussion before, brute force searching is a straightforward method to achieve the optimal solution. Specifically, to locate Ω^*, one needs to calculate the long-run average revenue of each possible policy through Equation 17.12. In parallel, the utility of each possible policy shall be calculated by Equation 17.14 to locate Ω^+. Both long-run average revenue and utility of each possible policy has to be calculated to locate Ω^{U*}. Nevertheless, previous studies in [18,20] pointed out that the brute force searching has an unbearable complexity, and it could only be used in off-line scenarios. Consequently, many researchers endeavored to develop simple structured approximate solutions. For example, a finite-stage dynamic programming algorithm has been proposed in [18], which has a complexity of $O(B^2 M)$, to find the optimal revenue policy of the complete partition (CP) structure. Here, the complexity is measured by the size of the search space that a CAC optimization algorithm has to explore. In the following subsections, a series of CP-structured heuristic algorithms, which have a complexity of $O(BM)$, is described.

17.5.3.1 CP-Structured Admission Control Policy

CP policy allocates each class of traffic a certain amount of nonoverlapped bandwidth resources. In this manner, the blocking rate of one traffic class will not influence that of others. Because of this partitioning characteristic, a CP policy can be decomposed into M independent subpolicies, and the ith sub-policy takes care of class i traffic. From a mathematical perspective, the CP policy separates the overall bandwidth resource B into M nonoverlapped parts, denoted by $B_{CP}^1, B_{CP}^2, \dots, B_{CP}^M$, where B_{CP}^i belongs to class i traffic. Notice that, if a policy Ω satisfies the coordinate convex condition and the arrival and service processes are both memoryless, then $P_\Omega(\vec{n})$ can be calculated by the following formula, as presented in [25]:

$$P_\Omega(\vec{n}) = \frac{1}{G(\Omega)} \prod_{i=1}^{M} \rho_i^{n_i} / n_i!, \vec{n} \in \Omega \qquad (17.15)$$

where,

$$G(\Omega) = \sum_{\vec{n} \in \Omega} \prod_{i=1}^{M} \rho_i^{n_i} / n_i! \quad \text{and} \quad \rho_i = \lambda_i / \mu_i \qquad (17.16)$$

and the blocking probability of class i traffic is:

$$Pb_i(\Omega) = \frac{G(\Omega_i^b)}{G(\Omega)} \qquad (17.17)$$

where $\Omega_i^b = \{\vec{n} | \vec{n} \in \Omega \ \& \ \vec{n} + \vec{e}_i \notin \Omega\}$ and $\vec{e}_i$ is the M dimensional vector of all zeros except for a one in the ith component. For a CP policy, the ith subpolicy can be viewed as an $M/M/N/N$ queueing system, in which the number of servers is

$s_i = B^i_{CP}/b_i$. Therefore, according to Equations 17.12 and 17.15, the long-run average revenue obtained from class i traffic is given by:

$$R_i(CP) = \sum_{j=0}^{s_i} r_i j \frac{\rho_i^j / j!}{\sum_{k=0}^{s_i} \rho_i^k / k!} \tag{17.18}$$

and the overall long-run average revenue of CP policy is:

$$R(CP) = \sum_{i=1}^{M} R_i(CP) = \sum_{i=1}^{M} \sum_{j=0}^{s_i} r_i j \frac{\rho_i^j / j!}{\sum_{k=0}^{s_i} \rho_i^k / k!}. \tag{17.19}$$

The SB of class i traffic, the overall statistical bandwidth of CP policy, and the utility of the CP policy, are given by:

$$SB_i(CP) = \sum_{j=0}^{s_i} b_i j \frac{\rho_i^j / j!}{\sum_{k=0}^{s_i} \rho_i^k / k!} \tag{17.20}$$

$$SB(CP) = \sum_{i=1}^{M} SB_i(CP) = \sum_{i=1}^{M} \sum_{j=0}^{s_i} b_i j \frac{\rho_i^j / j!}{\sum_{k=0}^{s_i} \rho_i^k / k!} \tag{17.21}$$

$$U(CP) = \frac{1}{B} \sum_{i=1}^{M} SB_i(CP) = \frac{1}{B} \sum_{i=1}^{M} \sum_{j=0}^{s_i} b_i j \frac{\rho_i^j / j!}{\sum_{k=0}^{s_i} \rho_i^k / k!}. \tag{17.22}$$

According to the Erlang B Formula, the blocking probability of class i traffic, as detailed in [26], is given as:

$$Pb_i(CP) = B(s_i, \rho_i) = \frac{\rho_i^{s_i} / s_i!}{\sum_{k=0}^{s_i} \rho_i^k / k!}. \tag{17.23}$$

Notice that, as given in [27], the Erlang B Formula can be calculated by the following recursion as:

$$B(s_i + 1, \rho_i) = \frac{\rho_i B(s_i, \rho_i)}{s_i + 1 + \rho_i B(s_i, \rho_i)} \tag{17.24}$$

where $B(0, \rho_i) = 1$.

From this discussion, the optimal CP problem deals with finding the best bandwidth partitioning scheme. Similar to the definition of Ω^* and Ω^+, let CP* be the policy of maximum revenue, which is a simple structured approximation of Ω^*; CP$^+$ be the CP policy of maximum utility; and CP$^{U^*}$ be the CP policy of maximum revenue under utility constraint.

17.5.3.2 Greedy Approximation Algorithm for CP*

To locate CP* exactly in the space of all CP policies, a finite-stage dynamic programming algorithm with complexity of $O(B^2M)$ is presented in [18]. To reduce the complexity, a greedy approximation algorithm is developed here. According to the theory of marginal economic analysis, as detailed in [26], for the class i traffic, the load carried by the jth server in the $M/M/N/N$ queueing system is formulated as:

$$F_S^i(j,\rho_i) = \rho_i[B(j-1,\rho_i) - B(j,\rho_i)], \quad 1 \le j \le s_i. \tag{17.25}$$

Then, the revenue brought by the jth server is given by:

$$R_S^i(j,\rho_i) = r_i F_S^i(j,\rho_i) = r_i \rho_i[B(j-1,\rho_i) - B(j,\rho_i)]. \tag{17.26}$$

Thus, the revenue rate of the jth server is:

$$r_S^i(j,\rho_i) = \frac{R_S^i(j,\rho_i)}{b_i} = \frac{r_i \rho_i[B(j-1,\rho_i) - B(j,\rho_i)]}{b_i}. \tag{17.27}$$

Also, from Equation 17.26 we have:

$$R_i(CP) = \sum_{j=1}^{s_i} R_S^i(j,\rho_i). \tag{17.28}$$

From [28], we know that $F_S^i(j,\rho_i)$ is decreasing in j, therefore, $R_S^i(j,\rho_i)$ and $r_S^i(j,\rho_i)$ are also decreasing in j.

Based on these equations, a greedy algorithm is presented to approximate CP* in Algorithm 17.2 which can exactly locate CP* if the number of possible error iterations E is equal to 0. An iteration step may incur an error, when and only when the capacity boundary condition $B_{\text{free}} < \max\{b_i\}$ becomes true, where $1 \le i \le M$, B_{free} is the amount of free bandwidth resources. If the capacity boundary condition holds, the greedy method cannot be implemented thoroughly because, in this case, the capacity limit can interfere with the process of locating the traffic class of maximal revenue rate. Regarding the discussion before, compared with CP*, the revenue

/* Initializations */
$B_{\text{free}} \leftarrow B$; /* B_{free}: the amount of free bandwidth resources */
$T \leftarrow M + 1$; /* T: the selected traffic class */
$E \leftarrow 0$; /* E: the number of iteration steps with possible approximation error */
for $i = 1$ to M **do**

$B_{\text{CP}}^i \leftarrow 0$ /* B_{CP}^i: the bandwidth assigned to class i traffic */
$s_i \leftarrow 0$ /* s_i: number of servers assigned to class i traffic */
$B(s_i, \rho_i) \leftarrow 1$ /* Initialization of Elang B formula */

$$B(s_i + 1, \rho_i) \leftarrow \frac{\rho_i B(s_i, \rho_i)}{s_i + 1 + \rho_i B(s_i, \rho_i)}$$ /* Elang B recursion */

$$r_S^i \leftarrow \frac{r_i \rho_i \left[B(s_i, \rho_i) - B(s_i + 1, \rho_i) \right]}{b_i}$$ /* r_S^i: revenue rate of adding a new server to class i traffic */

end for
/* Allocating bandwidth resources by greedy method */
while $T > 0$ **do**

$T \leftarrow \arg\max_{1 \le i \le M} \{r_S^i\}$
if $b_i|_{i=T} \le B_{\text{free}}$ **then**

$B_{\text{free}} \leftarrow B_{\text{free}} - b_i|_{i=T}$
$B_{\text{CP}}^i|_{i=T} \leftarrow B_{\text{CP}}^i|_{i=T} + b_i|_{i=T}$
$s_i|_{i=T} \leftarrow s_i|_{i=T} + 1$
$$B(s_i + 1, \rho_i)|_{i=T} \leftarrow \frac{\rho_i B(s_i, \rho_i)}{s_i + 1 + \rho_i B(s_i, \rho_i)} \bigg|_{i=T}$$
$$r_S^i|_{i=T} \leftarrow \frac{r_i \rho_i [B(s_i, \rho_i) - B(s_i + 1, \rho_i)]}{b_i} \bigg|_{i=T}$$

else

$r_S^i|_{i=T} \leftarrow 0$
if $B_{\text{free}} > 0$ **then**
$E \leftarrow E + 1$ /* an iteration step with possible approximation error */
end if
if $\sum_{i=1}^{M} r_S^i = 0$ **then**
$T \leftarrow -1$ /* the algorithm is completed */
end if
end if
end while
Return $\{B_{\text{CP}}^i, 1 \le i \le M\}$ as the final bandwidth allocation decision

Algorithm 17.2   Greedy approximation algorithm for CP*.

error generated by the greedy method can be strictly bounded by the following inequality:

$$R_{\text{err}} < R(\text{CP}^*, B) - R(\text{CP}^*, B - \max_{1 \le i \le M}\{b_i\}). \qquad (17.29)$$

Here, R_{err} stands for the revenue error; the function $R(\Omega, C)$ stands for the revenue obtained from policy Ω with bandwidth capacity C. Clearly, there are $O(B)$ iteration steps in the greedy approximation algorithm for CP^*. As this algorithm searches through M possible system states, during each step, to locate the traffic class that yields the maximal revenue rate, the size of the whole searching space or the complexity of the greedy approximation algorithm is $O(BM)$.

17.5.3.3 Utility-Constrained Greedy Approximation Algorithm for CP^{U*}

The utility constraint requires that the utility of an admission control policy be higher than the lower bound $LB^U = lb^U U(\Omega^+)$. Hence, before designing a heuristic algorithm to approximate CP^{U*}, it must find Ω^+ to calculate $U(\Omega^+)$ and LB^U. Note that, when $\text{rer}_1 = \text{rer}_2 = \cdots = \text{rer}_M$, that is, all classes of traffic have the same revenue rate, $U(\Omega^*)$ turns into $U(\Omega^+)$. As a result, the greedy approximation algorithm for CP^* can also be used to locate CP^+ and then achieve the approximate value of LB^U. Similar to the previous algorithm, for class i traffic, the statistical bandwidth brought by the jth server can be defined as:

$$\text{SB}_S^i(j, \rho_i) = b_i F_S^i(j, \rho_i) = b_i \rho_i [B(j-1, \rho_i) - B(j, \rho_i)]. \qquad (17.30)$$

Then, the utility of the jth server is given by:

$$U_S^i(j, \rho_i) = \frac{\text{SB}_S^i(j, \rho_i)}{b_i} = F_S^i(j, \rho_i) = \rho_i[B(j-1, \rho_i) - B(j, \rho_i)]. \qquad (17.31)$$

From Equation 17.30, we have:

$$\text{SB}_i(\text{CP}) = \sum_{j=1}^{s_i} \text{SB}_S^i(j, \rho_i). \qquad (17.32)$$

Based on these discussions, a utility-constrained greedy approximation algorithm for CP^{U*} is presented in Algorithm 17.3.

To strictly guarantee the utility constraint, the previous algorithm searches CP^+ and CP^{U*} simultaneously. The searching process of CP^+ is always one step ahead of

Input: lb^U /* lb^U: utility lower bound factor */
/* The notations with "+" are for CP⁺ approximation */
/* The notations without "+" are for CP^{U^*} approximation */
/* Initializations */
$B_{\text{free}} \leftarrow B;\ B^+_{\text{free}} \leftarrow B;\ T \leftarrow M+1;\ T^+ \leftarrow M+1;$
for $i = 1$ **to** M **do**

$\quad B^i_{\text{CP}} \leftarrow 0;\ B^i_{\text{CP+}} \leftarrow 0;\ s_i \leftarrow 0;\ s^+_i \leftarrow 0;$

$\quad B(s_i, \rho_i) \leftarrow 1;\ B(s^+_i, \rho_i) \leftarrow 1;$

$\quad B(s_i + 1, \rho_i) \leftarrow \dfrac{\rho_i\, B(s_i, \rho_i)}{s_i + 1 + \rho_i\, B(s_i, \rho_i)};$

$\quad B(s^+_i + 1, \rho_i) \leftarrow \dfrac{\rho_i\, B(s^+_i, \rho_i)}{s^+_i + 1 + \rho_i\, B(s^+_i, \rho_i)};$

$\quad r^i_S \leftarrow \dfrac{r_i\rho_i[B(s_i, \rho_i) - B(s_i + 1, \rho_i)]}{b_i};$

$\quad U^i_S \leftarrow \rho_i\, [B(s_i, \rho_i) - B(s_i + 1, \rho_i)];$

$\quad U^i_{S+} \leftarrow \rho_i\, [B(s^+_i, \rho_i) - B(s^+_i + 1, \rho_i)];$

$\quad Set_U \leftarrow \Phi;$ /* Set_U: the set of traffic classes qualified for utility constraint */
end for
$U_{ref} \leftarrow \max_{1 \le i \le M}\{U^i_{S+}\};$ /* U_{ref}: the reference for utility constraint */
/* Allocating bandwidth resources */
while $T > 0$ **do**

$\quad Set_U \leftarrow \{i\,|\,i\ \text{satisfies}\ U^i_S > lb^U U_{ref}\};$
$\quad T \leftarrow \arg\max_{i \in Set_U}\{r^i_S\};$
$\quad$**if** $b_i|_{i=T} \le B_{\text{free}}$ **then**

$\quad\quad B_{\text{free}} \leftarrow B_{\text{free}} - b_i|_{i=T};$
$\quad\quad B^i_{\text{CP}}|_{i=T} \leftarrow B^i_{\text{CP}}|_{i=T} + b_i|_{i=T};$
$\quad\quad s_i|_{i=T} \leftarrow s_i|_{i=T} + 1;$
$\quad\quad B(s_i + 1, \rho_i)|_{i=T} \leftarrow \dfrac{\rho_i B(s_i, \rho_i)}{s_i + 1 + \rho_i B(s_i, \rho_i)}\bigg|_{i=T};$
$\quad\quad r^i_S\big|_{i=T} \leftarrow \dfrac{r_i\rho_i[B(s_i, \rho_i) - B(s_i + 1, \rho_i)]}{b_i}\bigg|_{i=T};$
$\quad\quad U^i_S\big|_{i=T} \leftarrow \rho_i\, [B(s_i, \rho_i) - B(s_i + 1, \rho_i)]|_{i=T};$

$\quad\quad$**while** $B_{\text{free}} \le B^+_{\text{free}}$ **do**

$\quad\quad\quad T^+ \leftarrow \arg\max_{1 \le i \le M}\{U^i_{S+}\};$
$\quad\quad\quad U_{ref} \leftarrow U^i_{S+}|_{i=T^+};$
$\quad\quad\quad B^+_{\text{free}} \leftarrow B^+_{\text{free}} - b_i|_{i=T^+};$
$\quad\quad\quad B^i_{\text{CP}}|_{i=T} \leftarrow B^i_{\text{CP}}|_{i=T} + b_i|_{i=T^+};$
$\quad\quad\quad s^+_i|_{i=T^+} \leftarrow s^+_i|_{i=T^+} + 1;$

$$B(s_i^+ + 1, \rho_i)|_{i=T^+} \leftarrow \frac{\rho_i B(s_i^+, \rho_i)}{s_i^+ + 1 + \rho_i B(s_i^+, \rho_i)}\Big|_{i=T^+};$$

$$U_{S^+}^i\big|_{i=T^+} \leftarrow \rho_i \,[B(s_i^+, \rho_i) - B(s_i^+ + 1, \rho_i)]|_{i=T^+};$$

   end while

else

   /* Entering the capacity restricted region */

   $r_S^i|_{i=T} \leftarrow 0;\ U_S^i|_{i=T} \leftarrow 0;$

   $lb^U \leftarrow 0;$ /* change to use pure greedy algorithm */

   if $\sum_{i=1}^{M} r_S^i = 0$ **then**

      $T \leftarrow -1;$ /* the algorithm is completed */

   end if

 end if

end while

Return $\{B_{\mathrm{CP}}^i, 1 \le i \le M\}$ as the final bandwidth allocation decision

---

Algorithm 17.3   Utility-constrained greedy approximation algorithm for CP^{U^*}.

the searching process of CP^{U^*}, as CP^+ is used as the reference to compute the utility constraint. Moreover, this algorithm can also achieve high revenue because it still employs the greedy method in terms of revenue. Clearly, Algorithm 17.3 has the same complexity as the greedy approximation algorithm for CP^*, that is, $O(BM)$.

17.6   Performance Results

In this section, simulation results are presented to illustrate the performance of downlink APA and CAC optimization algorithms developed in the previous sections. In the WiMAX system, APA and CAC should work cooperatively to provide an integrated APA–CAC downlink resource management system. Hence, the numerical results of combined APA–CAC optimization are also demonstrated in this section.

At first, the downlink APA optimization is investigated in the OFDMA-FUSC mode of 32 subscribers. The channel is assumed to have a bad-urban (BU) delay profile, as presented in [29], and it suffers from shadowing with 8 dB standard deviation. Let the amount of available subcarriers be 1024 where each subcarrier occupies a 10-kHz physical bandwidth. Distances between subscribers and the BS are randomly chosen from 2 to 10 km, the acceptable BER is set to be 10^{-6}, and TDD downlink proportion factor x percent is configured as 80 percent. As for the traffic pattern, 32 subscribers are programmed to have different traffic loads, which are

Table 17.2   Proportion of BE, UGS, rtPS, and nrtPS Traffic

PP_{BE}	Uniformly distributed in [10 percent, 30 percent]
PP_{UGS}	Uniformly distributed in [10 percent $(1-PP_{BE})$, 30 percent $(1-PP_{BE})$]
PP_{rtPS}	Uniformly distributed in [20 percent $(1-PP_{BE})$, 60 percent $(1-PP_{BE})$]
PP_{nrtPS}	$(1-PP_{BE}-PP_{UGS}-PP_{rtPS})$

Abbreviations: BE, Best Effort; UGS, Unsolicited Grant Service; rtPS, Real-Time Polling Service, nrtPS, Non-Real-Time Polling Service.

uniformly distributed in [50 Mbps, 120 Mbps]. Among the downlink traffic load of a certain subscriber, the proportions of BE, UGS, rtPS, and nrtPS traffic, denoted by PP_{BE}, PP_{UGS}, PP_{rtPS}, and PP_{nrtPS}, respectively are random variables defined in Table 17.2.

The revenue rate of UGS, rtPS, nrtPS, and BE traffic is assumed to be $rer^{UGS} = 5$, $rer^{rtPS} = 2$, $rer^{nrtPS} = 1$, and $rer^{BE} = 0.5$, respectively. Moreover, the fairness constraint is set to 80 percent, that is, $F_{th} = 80$ percent.

Using the algorithms developed in the previous sections, the performance of optimal revenue criterion and fairness-constrained optimal revenue criterion is illustrated in Figure 17.8. To facilitate the analysis, in Figure 17.8a, the achieved APA revenue of a certain APA criterion is normalized by the overall potential revenue generated by the downlink traffic load of all 32 subscribers. Although using the criterion of optimal revenue can achieve maximum revenue, it makes about 50 percent subscribers out of fairness constraint. In contrast, the fairness-constrained optimal revenue criterion can guarantee that 100 percent of the subscribers satisfy the fairness constraint.

Next, with respect to the downlink CAC optimization, numerical results are presented in Figure 17.9 to demonstrate the performance of Ω^{U^*} and CP^{U^*}. Here, Ω^{U^*} is obtained by the method of brute force searching and CP^{U^*} is obtained by the dynamic programming algorithm. In this simulation scenario, the overall downlink bandwidth capacity B is set to be 70 Mbps, the revenue rate is priced as $rer^{UGS} = 5$, $rer^{rtPS} = 2$, $rer^{nrtPS} = 1$, and the downlink traffic load is configured as in Table 17.3. Moreover, in Figure 17.9, revenue and utility are normalized by $R(\Omega^*)$ and $U(\Omega^+)$, respectively. As illustrated in Figure 17.9, when $lb^U = 0$, Ω^{U^*} turns out to be Ω^*, yielding a solution of high revenue but low utility. Similarly, when $lb^U = 1$, Ω^{U^*} turns out to be Ω^+, yielding a solution of high utility but low revenue. Thus, an appropriate value should be chosen for lb^U (90 percent in this case) to give Ω^{U^*} a balanced revenue and utility. In addition, Figure 17.9 also shows that CP^{U^*} has almost the same performance as Ω^{U^*}.

To locate CP^{U^*} with less complexity, a utility-constrained greedy approximation algorithm has been developed in the previous section. Numerical results are shown

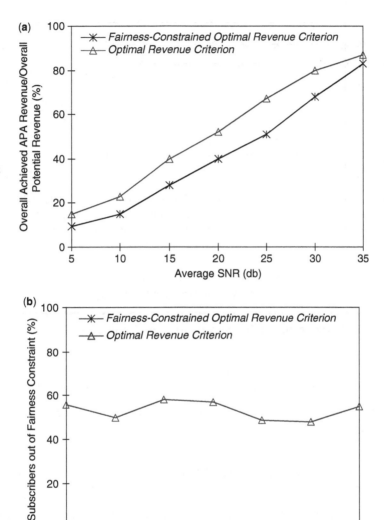

Figure 17.8 Comparison between optimal revenue criterion and fairness-constrained optimal revenue criterion. SNR, Signal-to-Noise Ratio; APA, adaptive power allocation.

in Figure 17.10 to illustrate the approximation error of this algorithm. Here, the traffic load configuration in Table 17.3 is reused. Furthermore, the revenue and utility approximation errors are normalized by $R(CP^{U^*})$ and $U(CP^{U^*})$, respectively. From Figure 17.10, we can conclude that the utility-constrained greedy approximation algorithm yields a close approximation of CP^{U^*}.

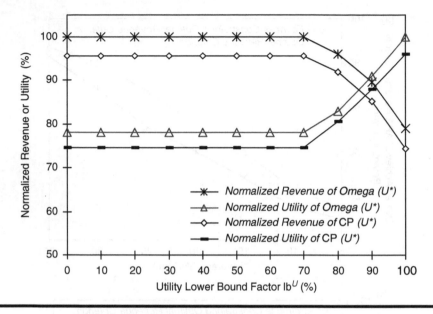

Figure 17.9   Revenue and utility performance of *U** and CP^*U**.

Finally, the performance of the combined APA–CAC optimization is evaluated. To simplify the expression, the following notations are employed.

APA1: equal power allocation criterion, which means each subcarrier acquires the same power;

Table 17.3   Downlink Traffic Load Configuration

Traffic Type	Qos Level	Application Type	Bandwidth Requested	Arrival Rate (Calls/h)	Service Time (min/call)
1	UGS	PSTN Call	64 kbps	400	25
2	UGS	MPEG-2	4 Mbps	3	60
3	rtPS	Windows Player	500 kbps	96	25
4	rtPS	MPEG-2	6 Mbps	3	90
5	nrtPS	Slow-ftp	200 kbps	120	60
6	nrtPS	Fast-ftp	1 Mbps	48	25

Abbreviations: QoS, quality of service; UGS, Unsolicited Grant Service; rtPS, Real-Time Polling Service; nrtPS, Non-Real-Time Polling Service.

APA2: optimal revenue criterion, which is implemented by the pure greedy revenue algorithm;

APA3: fairness-constrained optimal revenue criterion, which is implemented by the fairness-constrained greedy revenue algorithm.

CAC1: CS policy;

CAC2: optimal revenue policy, which is implemented by the greedy approximation algorithm for CP*;

CAC3: utility-constrained optimal revenue policy, which is implemented by the utility-constrained greedy approximation algorithm for CP^{U*}.

Let the combined APA–CAC optimization strategy be (APA$_m$, CAC$_n$), where APA$_m$ (m = 1,2,3) indicates the selected APA optimization criterion and CAC$_n$ (n = 1,2,3) indicates the selected CAC optimization policy. To measure the performance of a combined APA–CAC optimization strategy, two metrics are used: Final revenue—revenue of the downlink traffic load that is accepted by the combined APA–CAC strategy; and average subscriber satisfaction—the satisfaction of the kth subscriber is given by:

$$SAT_K = \frac{\text{ATR}_k^D}{\text{TR}_k^D} \tag{17.33}$$

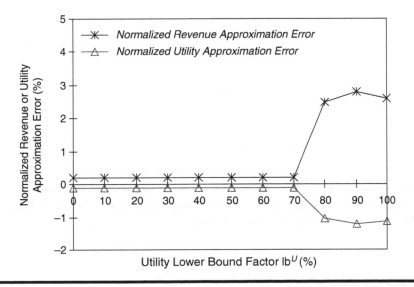

Figure 17.10 **Approximation error of utility-constrained greedy approximation algorithm.**

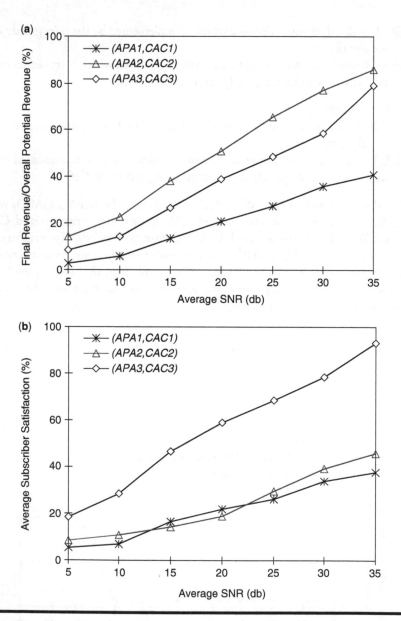

Figure 17.11 Performance comparison among (APA1, CAC1), (APA2, CAC2), and (APA3, CAC3). SNR, Signal-to-Noise Ratio.

where ATR_k^D is the traffic rate that is accepted by the combined APA–CAC strategy in subscriber k's local network. Correspondingly, the average subscriber satisfaction is given by:

$$\text{SAT}_{\text{avg}} = \frac{1}{N} \sum_{k=1}^{N} \text{SAT}_k. \tag{17.34}$$

Simulation results are presented in Figure 17.11 to demonstrate the performance of (APA1, CAC1), (APA2, CAC2), and (APA3, CAC3). In this simulation, the same wireless channel model and traffic load pattern is employed as in the simulation scenario of Figure 17.8. Moreover, for APA optimization, the fairness constraint F_{th} is set to 80 percent; for CAC optimization, the utility constraint lb^U is configured as 90 percent. Figure 17.11 shows that (APA1, CAC1) has low performance in terms of both final revenue and average subscriber satisfaction; (APA2, CAC2) has high revenue but low subscriber satisfaction; (APA3, CAC3) has both high revenue and high subscriber satisfaction.

17.7 Conclusion

In this chapter, we investigated subcarrier assignment, APA, and CAC resource management schemes for OFDMA–TDD-based multi-service WiMAX networks. The subcarrier assignment is currently being standardized by the activities of IEEE 802.16. To successfully deploy a commercial WiMAX system, design criteria for APA or CAC have been studied to strike a balance between the revenue of the service provider and the satisfaction of the subscribers. Although much effort is required to fully understand the theories behind these advanced adaptive resource management techniques, their implementation is quite simple and effective.

References

1. IEEE standard for local and metropolitan area networks—Part 16: Air interface for fixed broadband wireless access systems. IEEE Std. 802.16 (October 2004).
2. I. Koffman and V. Roman, Broadband wireless access solutions based on OFDM access in IEEE 802.16. *IEEE Communications Magazine*, pp. 96–103 (April 2002).
3. IEEE standard for local and metropolitan area networks—Part 16: Air interface for fixed broadband wireless access systems—Amendment 2: Physical and medium access control layers for combined fixed and mobile operation in licensed bands. IEEE Std. 802.16e (February 2006).
4. J. Yun and M. Kavehrad, PHY/MAC cross-layer issues in mobile WiMAX. *Bechtel Telecommunications Technical Journal* (1) (January 2006).
5. S. Shakkottai, T. S. Rappaport, and P. C. Karlsson, Cross-layer design for wireless networks. *IEEE Communications Magazine*, 41(10), 74–80 (October 2003).

6. G. Song and Y. G. Li, Cross-layer optimization for OFDM wireless networks—Part I: Theoretical framework. *IEEE Transactions on Wireless Communications*, 4(2), 614–624 (March 2005).

7. G. Song and Y. G. Li, Cross-layer optimization for OFDM wireless networks—Part II: Algorithm development. *IEEE Transactions on Wireless Communications*, 4(2), 625–634 (March 2005).

8. T. M. Cover and J. A. Thomas, *Elements of Information Theory* (John Wiley & Sons, Inc., New York, NY, USA, 1991). ISBN 0471062596.

9. K. Kim, Y. Han, and S.-L. Kim, Joint subcarrier and power allocation in uplink OFDMA system. *IEEE Communications Letters*, 9(6), 526–528 (June 2005).

10. C. Zeng, L. Hoo, and J. Cioffi. Efficient water-filling algorithms for a Gaussian multi-access channel with ISI. In *IEEE Vehicular Technology Conference*, vol. 3, pp. 1072–1077 (Boston, MA, USA, 2000).

11. G. Münz, S. Pfletschinger, and J. Speidel. An efficient water-filling algorithm for multiple access OFDM, In *IEEE GLOBECOM*, vol. 1, pp. 681–685 (November 2002).

12. W. Yu and J. M. Cioffi, FDMA capacity of Gaussian multiple-access channels with ISI. *IEEE Transactions on Communications*, 50, 102–111 (January 2002).

13. E. Yoon, D. Tujkovic, A. Paulraj, Subcarrier and power allocation for an OFDMA uplink based on tap correlation information. In *IEEE International Conference on Communications*, vol. 4, pp. 2744–2748 (May 2005).

14. B. Rong, Y. Qian, and H.-H. Chen. Adaptive power allocation and call admission control in multiservice WiMAX access network. *IEEE Wireless Communications*, 14(1), 14–19 (February 2007). To be published.

15. B. Rong, Y. Qian, and K. Lu, Integrated downlink resource management for multiservice WiMAX network. *IEEE Transactions on Mobile Computing*, 6(6), 621–632 (June 2007).

16. R. Marks, C. Eklund, K. Stanwood, and S. Wang. The 802.16 WirelessMAN MAC: It's Done, but What Is It? (November 2001). URL http://www.ieee802.org/16/docs/01/80216-01_58rl.pdf.

17. X. Qiu and K. Chawla, On the performance of adaptive modulation in cellular system. *IEEE Transactions on Communication*, 47(6), 884–895 (June 1999).

18. K. W. Ross and D. H. K. Tsang. The stochastic knapsack problem. In *Proceedings of the 27th IEEE Conference on Decision and Control*, vol. 1, pp. 632–633 (December 1988).

19. K. W. Ross and D. D. Yao, Monotonicity properties for the stochastic knapsack. *IEEE Transactions on Information Theory*, 36(5), 1173–1179 (September 1990).

20. A. Gavious and Z. Rosberg, A restricted complete sharing policy for a stochastic knapsack problem in B-ISDN. *IEEE Transactions on Communications*, 42(7), 2375–2379 (July 1994).

21. E. Altman, T. Jimnez, and G. Koole, On optimal call admission control in resource sharing system. *IEEE Transactions on Communications*, 49(9), 1659–1668 (September 2001).

22. E. L. Ormeci, Dynamic admission control in a call center with one shared and two dedicated service facilities. *IEEE Transaction on Automatic Control*, 49(7), 1157–1161 (July 2004).

23. B. C. Dean, M. X. Goemans, and J. Vondrdk. Approximating the stochastic knapsack problem: the benefit of adaptivity. In *The 45th Annual IEEE System on Foundations of Computer Science*, pp. 208–217 (October 2004).

24. V. Sarangan, D. Ghosh, N. Gautam, and R. Acharya, Steady state distribution for stochastic knapsack with bursty arrivals. *IEEE Communications Letters*, (2) 187–189 (February 2005).

25. J. S. Kaufman, Blocking in a shared resource environment. *IEEE Transactions on Communications*, pp. 1474–1481 (October 1981).

26. A. Jensen, *Moe's Principle*. The Copenhagen Telephone Company, Copenhagen (K.T.A.S) (1950).

27. D. L. Jagerman, Some properties of the Erlang loss function. *The Bell Systems Technical Journal*, 53(5), 525–551 (1974).

28. R. W. Wolff and C. L. Wang, On the convexity of loss probabilities. *Journal of Applied Probability*, 39(2), 402–406 (2002).

29. G. L. Stüber, *Principles of Mobile Communication, 2nd edn.* (Kluwer Academic Publishers, Norwell, MA, USA, 2001). ISBN 0-7923-7998-5.

Index

409

X